Extraction and Purification of Bioactive Compounds

Extraction and Purification of Bioactive Compounds

Editors

Zongbi Bao
Qianqian Xu

MDPI

Basel • Beijing • Wuhan • Barcelona • Belgrade • Novi Sad • Cluj • Manchester

Editors
Zongbi Bao
College of Chemical and Biological Engineering, Zhejiang University
Hangzhou, China

Qianqian Xu
School of Pharmaceutical and Materials Engineering, Taizhou University
Taizhou, China

Editorial Office
MDPI
St. Alban-Anlage 66
4052 Basel, Switzerland

This is a reprint of articles from the Special Issue published online in the open access journal *Processes* (ISSN 2227-9717) (available at: https://www.mdpi.com/journal/processes/special_issues/extraction_bioactive).

For citation purposes, cite each article independently as indicated on the article page online and as indicated below:

Lastname, A.A.; Lastname, B.B. Article Title. *Journal Name* **Year**, *Volume Number*, Page Range.

ISBN 978-3-0365-8998-5 (Hbk)
ISBN 978-3-0365-8999-2 (PDF)
doi.org/10.3390/books978-3-0365-8999-2

Contents

processes MDPI

Editorial

Special Issue on "Extraction and Purification of Bioactive Compounds"

Qianqian Xu [1] and Zongbi Bao [2,3,*]

[1] School of Pharmaceutical and Materials Engineering, Taizhou University, Taizhou 318000, China; xuqq@tzc.edu.cn
[2] Key Laboratory of Biomass Chemical Engineering of Ministry of Education, Zhejiang University, Hangzhou 310058, China
[3] Institute of Zhejiang University-Quzhou, 99 Zheda Road, Quzhou 324000, China
* Correspondence: baozb@zju.edu.cn

Citation: Xu, Q.; Bao, Z. Special Issue on "Extraction and Purification of Bioactive Compounds". *Processes* **2023**, *11*, 2034. https://doi.org/10.3390/pr11072034

Received: 4 July 2023
Accepted: 6 July 2023
Published: 7 July 2023

Bioactive compounds, which are abundant in flora, fauna, and microorganisms, can potentially be used by various industries, including pharmaceuticals, functional foods, dietary supplements, and cosmetics. Their extraction and purification, however, pose significant challenges due to the presence of impurities, interference from other compounds, and the need to maintain their stability and activity. This Special Issue, "Extraction and Purification of Bioactive Compounds" (https://www.mdpi.com/journal/processes/special_issues/extraction_bioactive accessed on 4 July 2023), delves into the complexities of these processes, presenting one review paper and eleven original research articles that explore the enrichment, identification, and various aspects of bioactive compounds.

The issue begins with the extraction of bioactive compounds from living organisms. Corrêa et al. [1] provide a comprehensive review of the current methodologies used for obtaining bioactive products from microalgae, discussing technologies used for cell disruption, selective extraction, recovery, and purification. Ghafoor et al. [2] compare conventional and green extraction techniques for natural antioxidants from Saudi date fruit, optimizing the supercritical extraction process using a response surface methodology and regression analysis techniques. A team from Serbia [3] explores the carotenoid extraction protocol using yeast fermentation technology, testing various methods for cell lysis, extraction, and solvents.

Additionally, the utilization of resins for the adsorption of bioactive compounds is addressed. Wu et al. [4] discuss the ion exchange strategy with Dowex® HCR-S resin for the isolation and purification of ectoine, a high-value bioactive compound, from a fermentation broth. Zain et al. [5] examine three macroporous resins for the enrichment of four bioactive compounds from an acid-hydrolyzed oil palm leaves extract, selecting the best-performing resin based on its sorption capacities.

The identification of bioactive compounds is another key focus of this issue. Chu et al. [6] extract phenolic compounds from ten common commercial mushroom species, identifying different phenolic compounds using LC-ESI-QTOF-MS/MS. Cao et al. [7] analyze the constituents of acidic compounds in the particulate and gaseous phases of mainstream cigarette smoke. Other researchers investigate the composition of bioactive compounds in organically produced cereals [8], dried sea cucumber viscera [9], and the medicinal plant, Clinacanthus nutans [10].

The issue concludes with an overview of the other aspects of bioactive compounds. Ramakrishnan et al. [11] optimize the operation parameters of enzymatic transesterification for biodiesel production from salmon oil. Xu et al. [12] measure the solubility of the high-value biocompound, 2,5-furandicarboxylic acid (FDCA), and its synthetic intermediate in binary solvent mixtures of water and 1,4-dioxane.

This Special Issue comprehensively explores the extraction and purification of bioactive compounds, shedding light on the complexities of these processes and the potential these compounds hold for various industries. We hope that the insights will inspire future research and innovation in this exciting field.

Author Contributions: Conceptualization, writing—review and editing, writing—original draft preparation, methodology, formal analysis, resources, investigation, data curation, Q.X.; validation, supervision, project administration, funding acquisition, Z.B. All authors have read and agreed to the published version of the manuscript.

Conflicts of Interest: The authors declare no conflict of interest.

References

1. Corrêa, P.S.; Morais Júnior, W.G.; Martins, A.A.; Caetano, N.S.; Mata, T.M. Microalgae Biomolecules: Extraction, Separation and Purification Methods. *Processes* **2021**, *9*, 10. [CrossRef]
2. Ghafoor, K.; Sarker, M.Z.I.; Al-Juhaimi, F.Y.; Babiker, E.E.; Alkaltham, M.S.; Almubarak, A.K.; Ahmed, I.A.M. Innovative and Green Extraction Techniques for the Optimal Recovery of Phytochemicals from Saudi Date Fruit Flesh. *Processes* **2022**, *10*, 2224. [CrossRef]
3. Šovljanski, O.; Saveljić, A.; Tomić, A.; Šeregelj, V.; Lončar, B.; Cvetković, D.; Ranitović, A.; Pezo, L.; Ćetković, G.; Markov, S.; et al. Carotenoid-Producing Yeasts: Selection of the Best-Performing Strain and the Total Carotenoid Extraction Procedure. *Processes* **2022**, *10*, 1699. [CrossRef]
4. Wu, Y.-C.; Wei, Y.-H.; Wu, H.-S. Adsorption and Desorption Behavior of Ectoine Using Dowex® HCR-S Ion-Exchange Resin. *Processes* **2021**, *9*, 2068. [CrossRef]
5. Che Zain, M.S.; Lee, S.Y.; Teo, C.Y.; Shaari, K. Adsorption/Desorption Characteristics and Simultaneous Enrichment of Orientin, Isoorientin, Vitexin and Isovitexin from Hydrolyzed Oil Palm Leaf Extract Using Macroporous Resins. *Processes* **2021**, *9*, 659. [CrossRef]
6. Chu, M.; Khan, R.D.; Zhou, Y.; Agar, O.T.; Barrow, C.J.; Dunshea, F.R.; Suleria, H.A.R. LC-ESI-QTOF-MS/MS Characterization of Phenolic Compounds in Common Commercial Mushrooms and Their Potential Antioxidant Activities. *Processes* **2023**, *11*, 1711. [CrossRef]
7. Lu, X.; Zhang, H.; Cao, Y.; Pang, Y.; Zhou, G.; Huang, H.; Li, J.; Jiang, J.; Yang, Q. A Comprehensive Study on the Acidic Compounds in Gas and Particle Phases of Mainstream Cigarette Smoke. *Processes* **2023**, *11*, 1694. [CrossRef]
8. Hussain, A.; Larsson, H.; Johansson, E. Carotenoid Extraction from Locally and Organically Produced Cereals Using Saponification Method. *Processes* **2021**, *9*, 783. [CrossRef]
9. Liu, Y.; Dave, D.; Trenholm, S.; Ramakrishnan, V.V.; Murphy, W. Effect of Drying on Nutritional Composition of Atlantic Sea Cucumber (*Cucumaria frondosa*) Viscera Derived from Newfoundland Fisheries. *Processes* **2021**, *9*, 703. [CrossRef]
10. Cheng, A.Y.F.; Teoh, P.L.; Jayasinghe, L.; Cheong, B.E. Volatile Profiling Aided in the Isolation of Anti-Proliferative Lupeol from the Roots of *Clinacanthus nutans* (Burm. f.) Lindau. *Processes* **2021**, *9*, 1383. [CrossRef]
11. Ramakrishnan, V.V.; Dave, D.; Liu, Y.; Routray, W.; Murphy, W. Statistical Optimization of Biodiesel Production from Salmon Oil via Enzymatic Transesterification: Investigation of the Effects of Various Operational Parameters. *Processes* **2021**, *9*, 700. [CrossRef]
12. Xu, L.; Fu, J.; Du, C.; Xu, Q.; Liu, B.; Bao, Z. Solubility of Biocompounds 2,5-Furandicarboxylic Acid and 5-Formylfuran-2-Carboxylic Acid in Binary Solvent Mixtures of Water and 1,4-Dioxane. *Processes* **2022**, *10*, 2480. [CrossRef]

processes

MDPI

Review

Microalgae Biomolecules: Extraction, Separation and Purification Methods

Priscila S. Corrêa [1,2], Wilson G. Morais Júnior [2], António A. Martins [1], Nídia S. Caetano [1,2,*] and Teresa M. Mata [1]

[1] LEPABE—Laboratory for Process Engineering, Environment, Biotechnology and Energy, Faculty of Engineering, University of Porto (FEUP), R. Dr. Roberto Frias s/n, 4200-465 Porto, Portugal; eng.pscorrea@gmail.com (P.S.C.); amartins@fe.up.pt (A.A.M.); tmata@fe.up.pt (T.M.M.)

[2] CIETI, School of Engineering (ISEP), Polytechnic Institute of Porto (P.Porto), R. Dr. António Bernardino de Almeida 431, 4249-015 Porto, Portugal; wjunioreq@gmail.com

* Correspondence: nsc@isep.ipp.pt or wgdmj@isep.ipp.pt; Tel.: +351-22-834-0500; Fax: +351-22-832-1159

Abstract: Several microalgae species have been exploited due to their great biotechnological potential for the production of a range of biomolecules that can be applied in a large variety of industrial sectors. However, the major challenge of biotechnological processes is to make them economically viable, through the production of commercially valuable compounds. Most of these compounds are accumulated inside the cells, requiring efficient technologies for their extraction, recovery and purification. Recent improvements approaching physicochemical treatments (e.g., supercritical fluid extraction, ultrasound-assisted extraction, pulsed electric fields, among others) and processes without solvents are seeking to establish sustainable and scalable technologies to obtain target products from microalgae with high efficiency and purity. This article reviews the currently available approaches reported in literature, highlighting some examples covering recent granted patents for the microalgae's components extraction, recovery and purification, at small and large scales, in accordance with the worldwide trend of transition to bio-based products.

Keywords: biomolecules; chromatography; cell disruption; microalgae; purification; supercritical fluids

Citation: Corrêa, P.S.; Morais Júnior, W.G.; Martins, A.A.; Caetano, N.S.; Mata, T.M. Microalgae Biomolecules: Extraction, Separation and Purification Methods. *Processes* **2021**, 9, 10. https://dx.doi.org/10.3390/pr9010010

Received: 1 December 2020
Accepted: 19 December 2020
Published: 22 December 2020

1. Introduction

Microalgae are unicellular or simple multicellular photosynthetic microorganisms, which can normally be found in aquatic environments such as freshwater, seawater, or hypersaline lakes. These organisms can be eukaryotic or prokaryotic, the latter being the cyanobacteria, which are commonly referred to as microalgae. There are numerous microalga species, some of which can grow in high salinity waters, having the advantage of reducing the contamination incidence and expanding its applications [1].

Microalgae can be cultivated photoautotrophically in closed photobioreactors or open ponds, using sunlight, CO_2 and inorganic nutrients to grow, producing biomass and O_2. However, conventional large-scale photoautotrophic production suffers from low biomass density due to light restrictions that hinder the cells growth, which dramatically increases production cost. Alternatively, microalgae can be grown under mixotrophic or heterotrophic conditions, respectively in the presence and absence of light, adding organic carbon as a nutrient, which has become a common practice for commercial production, increasing the productivity of algal biomass [2].

When compared to higher plants, microalgae have some advantages, such as higher productivity, lack of seasonality and, in the case of biofuel production, do not compete with human food [3,4]. Their remarkable biotechnological potential to produce a range of biocomponents such as pigments [5], lipids [6], polysaccharides [7], biopolymers [8], proteins [9], and vitamins [10], among others, has been widely addressed in the literature. Most of these bioproducts are stored intracellularly, which requires rupture of the cell wall for their recovery (Figure 1).

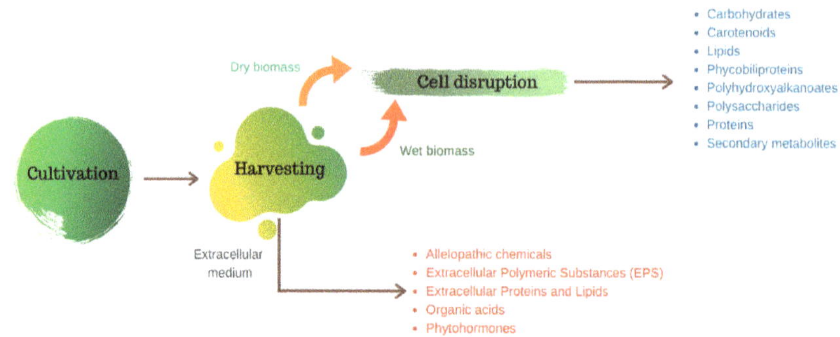

Figure 1. First steps in microalgae production, from cultivation to harvesting and cell disruption, for extraction or recovery of intracellular and extracellular components [11,12].

There are several techniques for cell disruption, which can be applied with greater or lesser degrees of success, depending on the characteristics of the cell wall of a given microalgae species. The rigidity of the cell wall can be provided, for example, by high levels of polysaccharides in the cell wall structure, such as glucose and mannose, present in *Chlorella zofingiensis*, or by complex sugars composition such as arabinose, galactose, rhaminose, mannose and xylose, as found in *Tetraselmis suecia* and *T. striata*. Another extremely resistant component is a non-hydrolyzable biopolymer, called algaenan or sporopollein. This biopolymer is composed of long ω-hydroxy fatty acids chains linked by several types of chemical bond, which confer its rigid properties. Algaenan can be found in some species such as *Chlorella* spp., *Nannochloropsis galditana* and *Scenedesmus* spp. On the other hand, cell walls composed of peptidoglycan, as found in *Arthorspira* spp., are less rigid and, consequently, more susceptible to degradation [13,14].

In this sense, several authors have studied the effect of different cell disruption techniques in order to enhance the recovery of the intracellular target product. Larrosa et al. [15] tested three different cellular rupturing techniques (i.e., milling, microwave oven and autoclaving) to enhance the recovery of phycocyanin and phenolic compounds from *Spirulina* sp. (strain LEB-18). In addition, Gim and Kim [16] tested several experimental conditions to optimize six cell disruption methods applied to *Botryococcus braunii* LB572 biomass, including mechanical (sonication, bead-beater, autoclave, French-press and microwave) and non-mechanical (osmotic shock) methods. More recently, Martínez et al. [17] proposed the use of pulsed electric fields (PEF) for astaxanthin extraction from *Haematococcus pluvialis*, comparing the extraction efficiency of PEF with classical methods, such as bead-beating, ultrasound, freezing–thawing, thermal and chemical treatments.

Biomolecule recovery and purification may be limiting factors for the establishment of biotechnological processes, by making them economically less attractive than chemical ones [18]. However, the development of new, more efficient and sustainable technologies and processes can be the key to promote the transition to bio-based products [19]. In the following sections, the main technologies for cell disruption, selective extraction, recovery and purification of microalgae biomolecules are discussed. Hence, this paper reviews the most common methods for extracting and purifying different components from microalgal biomass. It presents examples of various microalgae species and the methods used to extract the desired components with each one.

2. Cell Disruption Methods

Methods for microalgae cell disruption comprise mechanical, physical or non-mechanical techniques, usually employed to disrupt or disintegrate the cellular membrane, this way increasing the recovery yield of the desired component (e.g., lipids, pigments, proteins) that can be isolated from biomass. Figure 2 shows the general differences between mechanical and non-mechanical cell-disruption methods.

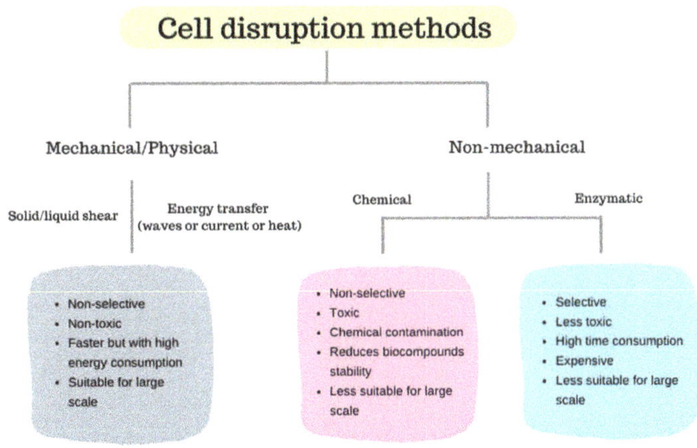

Figure 2. Comparison of different cell-disruption methods.

2.1. Mechanical and Physical Methods

Mechanical and physical methods may promote cell lysis through solid or liquid-shear forces (e.g., bead milling, high-speed or high-pressure homogenization) or energy transfer through waves (e.g., microwave irradiation, ultrasonication or laser), current (e.g., pulsed electric fields) or heat (e.g., autoclaving, freeze/thaw cycles, thermolysis).

2.1.1. Bead Milling

The cell disruption principle in bead mill machines is to promote mechanical cell damage by forcing the collision between the cells and the beads. This collision is promoted by a rotating shaft present in the grinding chamber. The diameter and load of the beads are important parameters with a direct influence on the cell disruption effectiveness [20]. The most common materials used in the beads are zirconium (high-density beads) and glass (low-density beads). Zirconium is preferred to process high viscosity media, while glass beads are more suitable for media with low viscosity [21].

In large-scale processes, techniques based on solid/liquid-shear forces are commonly employed due to their high efficiency and scale-up easiness. Furthermore, these methods avoid the contamination of chemical methods and preserve most of the biomolecules' functionality when compared to chemical and thermal treatments. An optimized mechanical method for *Chlorella*'s cell wall disruption at industrial scale (i.e., milling chamber volume $\geq$500 dm^3 and/or flow rate >1 m^3·h^{-1} and/or batch from 1 to 200 m^3) has been described in the US patent n° 10465159B2 [22]. The inventors evaluated the effect of specific parameters, such as bead material (glass, zirconium silicate and zirconium oxide), diameter of milling beads (0.3 to 1.7 mm), chamber filling rate (80% to 90%), operational scheme (single or multiple mills in series), peripheral speed of milling disks (8 to 12 m·s^{-1}, limited to avoid abrasion issues) and cell density (20%, 25.2% and 31.9%) on the specific energy demand and productivity. The configuration recommended by the applicant company in order to combine lower energy consumption with higher productivity is performed by lower diameter-zirconium silicate beads (0.3 to 0.6 mm) at 85–90% chamber filling rate, peripheral speed of milling disks between 11 and 12 m·s^{-1} and moderate biomass concentration (25.2%) with several mills in series. It was demonstrated that moderate conditions are preferable to reduce energy consumption to achieve a target degree of milling. Despite high-density beads (based on zirconium) presenting high specific energy, the glass beads (low density) were not efficient, requiring more passes to achieve the same degree of milling, which also increases the specific energy. Therefore, in order to overcome this issue, the inventors combined the use of zirconium silicate, which is less dense than zirconium oxide, with lower diameter to reduce the number of passes required. The same criteria were applied to biomass concentration; higher concentrations lead to higher productivities, but also increase the energy consumption. Additionally, moderate peripheral speed was recommended to avoid excessive abrasion and the filling rate, however, did not present a significant impact on energy consumption among the tested conditions. It is also possible to reduce the specific energy consumption and the process cost by using higher dry cell weight concentrations (0.5–8% w/w) and higher biomass flow rates, but it will also have a negative impact on the cell disruption efficiency [21]. Thus, despite the many advantages of using bead milling, including this being a suitable technique for large-scale production, and the aforementioned optimizations, this process still has high energy consumption [23].

2.1.2. High-Speed Homogenization

High-speed homogenization (HSH) is a simple and effective method, in which the cells' disruption is based on hydrodynamic cavitation caused by stirring at high speed (10,000–20,000 rpm) and shear forces at the solid–liquid interphase [21]. Despite presenting some drawbacks such as high energy consumption and protein denaturation, this technique is suitable for industrial scale and requires short processing time. It has been reported operational times of 30 or 60 s at 10,000 or 14,000 rpm for lipid and antioxidant extraction in *Nannochloropsis* sp., *Phaeodactylum tricornutum* and *Pavlova lutheri* [24,25].

González-Delgado and Kafarov [26] compared solvent assisted extraction with HSH and other solvent-based extraction methods for a microalgae biorefinery. The authors tested five microalgae species (*Nannochloropsis* sp., *Guinardia* sp., *Closterium* sp., *Amphiprora* sp. and *Navicula* sp.), concluding that despite the higher extraction yields achieved by combining polar and non-polar solvents, the solvent method presents high toxicity and lowest solvent recovery, increasing the process costs.

2.1.3. High-Pressure Homogenization

The high-pressure homogenization (HPH) method for cell disruption employs high pressure (≈20–120 MPa) to promote turbulence, liquid-shear stress and friction. According to the cell wall properties, parameters such as operating pressure and number of homogenization passes can be optimized to enhance the process efficiency [27]. Additionally, other variables such as dry cell weight concentration, microalgae species and growth conditions, impacts on the specific energy consumption [21].

Bernaerts et al. [28] studied the impact of (ultra) high-pressure homogenization (U)HPH on the rigid cells of *Nannochloropsis* sp., achieving similar lipid extraction efficiency, using 250 MPa in half of the homogenization passes compared to 100 MPa. However, despite the effective reduction of homogenization passes, the high pressure also heated the sample, resulting in aggregation of the intracellular components released. Besides the reduction of specific energy in (U)HPH by using biomass concentrations up to 25% (w/w), the energy consumption of this technique is still high. However, Elain et al. [29] demonstrated a satisfactory specific energy consumption (0.41 $kWh{\cdot}kg^{-1}$ biomass dry weight) of HPH in a study comparing the performance of this technique in mild conditions (e.g., of room temperature, neutral pH, shorter time, etc.) with conventional thermal treatment (hot water) in cell disruption of *Arthrospira platensis*, also increasing 2.5-fold the yield of polysaccharides extraction.

Thus, the major drawbacks of HPH comprise the non-selectivity, the formation of undesirable cell debris and the limitation to break harder cell walls. However, despite these disadvantages, HPH is, together with bead milling and HSH, the preferred method for the industrial scale.

2.1.4. Microwave Irradiation

Microwave irradiation is a simple and scalable method for cell disruption. This method has a well-established optimal operational value for heating (2450 MHz) and the cell walls are disrupted by the electromagnetic effect induced by the microwave irradiance that interacts with polar (e.g., water) and dielectric molecules, also promoting local heating [30].

This method is not suitable when the target component is volatile, but it has been successfully reported as an effective cell disruption technique for lipid extraction. Also, the combination of microwave with solvents, called microwave-assisted extraction (MAE), has been reported as the technique with lower operational costs and extraction time than the conventional techniques, and higher lipid extraction than other non-conventional methods (e.g., ultrasound-assisted extraction) [31].

By comparing different cell disruption methods (autoclaving, bead-beating, microwaves, sonication and osmotic shock), followed by chemical lipid extraction, Lee et al. [32] proved that microwave was the most effective method for cell disruption of three microalgae species (*Botryococcus* sp., *Chlorella vulgaris* and *Scenedesmus* sp.). In addition, by using a microwave method, Dai et al. [33] achieved the highest lipids extraction yield (18 wt%) from commercial microalgae dry powder, in comparison with 14 wt% by heating and 5 wt% by ultrasound. Also, Viner et al. [34] compared microwave with several cell disruption methods (freeze-drying, ultrasonication, cooling, liquid nitrogen grinding, osmotic shock and switchable osmotic shock) prior to lipid extraction in *Scenedesmus* sp. using liquid CO_2 and methanol. The highest total lipids extraction yield (9.6 wt% of dry algae) was achieved using microwave in the presence of water. Recently, the use of ionic liquids in MAE has been studied as a greener technology to overcome the intrinsic toxicity of the conventional solvents (e.g., chloroform, methanol) [35,36].

2.1.5. Ultrasonication

The ultrasonication method for cell disruption is based on liquid-shear forces caused by emission of high frequency wave sounds (up to 15–20 kHz). In liquid, these sound waves create gas bubbles or cavities that, after a certain number of cycles, achieve a critical size, collapsing and releasing large amounts of energy. Additionally, acoustic cavitation occurs by increasing local temperature and forming hydroxyl radicals that damage the cell wall [20]. Besides being a scalable technique with low operational cost, it is possible to optimize some parameters (e.g., temperature, cell concentration, acoustic intensity and time) to partially disrupt the cells, resulting in selective release of proteins [27]. Moreover, the promising use of ultrasonication for large-scale treatment of microalgal biomass has been previously pointed out by Adam et al. [37] who suggested that, the large-scale ultrasound extraction reactors used in food and chemical industries, can be easily modified to perform an ultrasound-assisted extraction of microalgae biomolecules in amounts up to 200 $kg \cdot h^{-1}$ of biomass dry weight. However, this technique is not very effective for some microalgae species and it is commonly combined with chemical treatments for efficiency improvement and to reduce energy demand [23,27].

2.1.6. Pulsed Electric Field

A pulsed electric field (PEF) has been described as an alternative method to overcome high energy consumption of classical mechanical methods based on solid/liquid-shear forces. Besides being energetically efficient and scalable, PEF also presents selectivity and fast processing time. However, despite the low operational costs, equipment is expensive and the technique depends on medium conductivity, limiting its use [38]. The disruption mechanism induced by PEF is based on electroporation as a result of transient membrane-permeabilization and electrophoretic movements into the cell caused by charged species [39]. The electroporation can be reversible (0.5–1.5 $kV \cdot cm^{-1}$, 0.5–5 $kJ \cdot kg^{-1}$), mostly used for genetic engineering or chemotherapy, or irreversible (10–20 $kV \cdot cm^{-1}$, 50–200 $kJ \cdot kg^{-1}$), being applicable for cell disruption and food processing [40].

Several parameters can influence PEF efficiency such as the electric field strength, pulse (shape, width), frequency, physicochemical parameters (temperature, pH and conductivity), operational time and cell wall properties [41]. Lam et al. [42] tested the use of PEF for protein release from *Chlorella vulgaris* and *Neochloris oleoabundans* achieving the maximum of 13% even through use of 10–100 times higher energy than bead milling, which released 45–50% of proteins. On the other hand, Käferböck et al. [43] reported a 90% increase in phycocyanin extraction efficiency from *Arthrospira maxima* by combining freeze-thawing and PEF, in comparison to bead milling method. However, a recent study comparing PEF, high voltage electrical discharges (HVED) and ultrasonication on aqueous extraction of *Nannochloropsis* sp., *Phaeodactylum tricornutum* and *Parachlorella kessleri* components have demonstrated HVED is most effective for carbohydrates and ultrasonication for proteins and chlorophyll a extraction from these species [44]. These results demonstrate that despite being a promising technology, there are still challenges to overcome for its establishment as a suitable technology for mild or large-scale microalgae biorefinery. However, PEF is a widely employed technology in the food industry that counts on specialized companies that are also involved in projects, approaching the use of PEF to stimulate growth and improve extraction of high-value compounds from microalgae. The German company ELEA Technology, for example, has been running the project iAlgaePro (https://elea-technology.de/project/ialgaepro/) since 2014. In this project the effectiveness of low-intensity PEF treatments to stimulate growth and also enhance the extraction of several compounds (e.g., phycocyanin, vitamins, polyphenols, lipids, among others) were demonstrated. The group reported significant difference between the phycocyanin extraction yield in PEF-treated (66.4 mg mL^{-1}) and non-treated ($\leq$0.2 mg·mL^{-1}) *Spirulina* biomass. Thus, this kind of initiative may accelerate the implementation of PEF in microalgae biorefineries.

2.1.7. Thermal Treatments

Thermal treatments are physical methods that use heat to promote cell disruption, such as thermolysis [45], autoclaving [15] and steam explosion [46,47]. Despite being simple technologies with low maintenance cost, the physical disruption is frequently associated with low efficiency, high energy consumption, generation of large amounts of undesirable cell debris and applicability limited by thermal resistance of the target product to be extracted. However, as shown in Table 1, steam explosion has many advantages compared to conventional thermal treatments. In this technique the biomass is exposed to high temperatures (160–290 °C), however, to pretreated microalgae biomass it is recommended to use lower temperatures to avoid degradation of the bioproducts, at vapor pressure between 1.03 and 3.45 MPa. The cell disruption occurs when the system is depressurized to return to room conditions [48,49]. Lorente et al. [46] tested four pretreatments (steam explosion, autoclaving, microwave and ultrasound) in three microalgae species (*Nannochloropsis gaditana*, *Phaeodactylum tricornutum* and *Chlorella sorokiniana*) to enhance lipid extraction using the Bligh and Dyer method. In this study, steam explosion as pretreatment showed the highest lipid yield for all species, especially for *N. gaditana* and *C. sorokiniana*. Furthermore, this technique promotes carbohydrates hydrolysis, forming aqueous phase rich in monomeric sugars suitable for subsequent fermentation.

Table 1. Main mechanical and physical methods for cell disruption: mechanism of disruption, advantages, disadvantages and remarks.

Cell Disruption Method	Principle of Cell Disruption	Advantages	Disadvantages	Remarks	Equipment Available	Specifications
Bead milling [21,49,50]	Mechanical deformation by compaction and shear	• High disruption efficiency • Processes high loads of biomass • Good temperature control • Easily scalable • Equipment commercially available	• High energy demand • Non-selective procedure • Formation of very fine cell debris	Suitable for large-scale	Bead mill for cell disruption—model EDW (ELE® Company) [a]	Chamber volume: 5–400 L Power: 11–500 kW Speed: 0–480 to 0–1500 rpm; Dimension: various. Flow: 30–200 to >3000 L h^{-1}Weight: 400–11,700 kg
High-speed homogenization [21]	Cavitation and shear	• Simple • High disruption efficiency • Short contact times	• High energy demand • Protein denaturation	Preferable for large scale (not indicated for mild scale)	High-speed homogenizer and disperser (Intertech®) [b]	Volume: 750–1150 L Power: 5–75 hp Speed: 1000–2880 rpm Flow rate: 650–5200 L min^{-1} Dimension: various

Table 1. *Cont.*

Cell Disruption Method	Principle of Cell Disruption	Advantages	Disadvantages	Remarks	Equipment Available	Specifications
High-pressure homogenization [17,20,50]	Cavitation and shear	• Easily scalable • Does not require cell drying • Suitable for processing large volumes	• Non-selective procedure • Use low dry cell weight concentrations, increasing energy demand and water footprint • Formation of very fine cell debris • Not effective to break hard cell walls • Reduces protein disgestibility • No indicate for fragile compounds isolation	• Suitable for emulsification processes • Preferable for large scale (not indicated for mild scale)	Ariete Series Homogenizers(©GEA Group) [c] DeBEE 2000 series (©BEE International) [d]	Pressure: 100–1500 bar Flow: 35–80,000 L h^{-1} Power: 10 hp Pressure: 1333–45,000 bar Flow rate: 0.5–2 L min^{-1}
Microwave irradiation [17,21,50]	Increases temperature and molecular energy	• Easily scalable • Simple • Can be combined with selective extraction (microwave assisted extraction)	• No indicate for volatile compounds isolation • Limited to polar solvents	Not recommended for mild microalgae biorefinery	MARS™6 Extraction (©CEM Corporation) [e]	Capacity: 55 L, up to 40 vessels Wattage: 2000 W Power density: 36 W L^{-1}

Table 1. *Cont.*

Cell Disruption Method	Principle of Cell Disruption	Advantages	Disadvantages	Remarks	Equipment Available	Specifications
Ultrasonication [17,20,27]	Cavitation and free radical formation	• Easily scalable • Low operational costs	• Low cell disruption efficiency for some microalgae species • Heat production	• Combination of ultrasound with different solvents can improve the effectiveness of cell disruption and reduce the energy demand • Equipment already available for industrial scale	Industrial ultrasonic devices UIP series (Hielscher Ultrasonics) [f]	Power: 0.5–16 kW Frequency: 18 or 20 kHz Flow rate: 0.25–10 m^3 h^{-1}
Pulsed electric fields [17,21]	Irreversible pore formation in cell membrane caused by short electrical pulses (electroporation)	• High disruption efficiency • Low operational costs • Scalable • Selective • Fast process time	• Can promote radical formation and undesired reactions, reducing the quality of the product • Depends on the media conductivity • Expensive equipment	Needs improvements for cell disruption in large-scale	ELEA PEFPilot™ dual trial system (ELEA Technology) [g]	Power: 400 V, 50 Hz; Water and air cooled Dimensions: 1,45 × 1,79 × 1,13 (W × D × H) Capacity: 10 kg per batch, up to 250 L h^{-1}

Table 1. *Cont.*

Cell Disruption Method	Principle of Cell Disruption	Advantages	Disadvantages	Remarks	Equipment Available	Specifications
Autoclaving [49,50]	Exterior heat diffusion through cell membrane to intracellular environment	• Simple • Low maintenance costs	• High energy consumption • Formation of large amount of cell debris • Not indicated for thermal-sensible compounds	Not indicated for large-scale	Systec V series (Systec GmbH) [h]	Chamber volume: 45/40–166/150 $L_{total/nominal}$
Steam explosion [49,50]	High temperature, vapor pressure and depressurization	• Low maintenance costs • Relatively low energy consumption • Low corrosion potential • Residual steam can be used to reduce process costs	• Variable efficiency according to microalgae species	Very suitable for commercial applications	Steam generator various models (Garioni Naval) [i]	Lab-scale:15–180 kW, 9–170 kg_{steam} h^{-1}, 7–8 barg Up scales: 300–6000 kg_{steam} h^{-1}, 3 passes, up to 18 barg; 3000–25,000 kg_{steam} h^{-1}, 2 passes, up to 21 barg

Websites: [a] www.ele-mix.com, [b] www.intertechglobal.com, [c] www.gea.com/en/index.jsp, [d] www.beei.com, [e] www.cem.com, [f] https://www.hielscher.com, [g] www.elea-technology.de, [h] www.systec-lab.com, [i] www.garioninaval.com. (W × D × H): Width × Depth × Height.

In Table 1 are summarized the main mechanical and physical methods for cell disruption, highlighting the principle of cell disruption and the main advantages and disadvantages of each of them.

2.2. Non-Mechanical Methods

Non-mechanical methods comprise chemical methods that may use acid or alkaline treatments [51–53] and detergents [54], osmotic shock [16] or enzymatic treatments [55,56].

2.2.1. Chemical Methods

Chemicals such as solvents, acids, alkali, hypochlorites, antibiotics, detergents, among others, can interact with components of the microalgal cell wall causing deformations and promoting cell disruption. Despite being a simple and well-known technique, the use of chemicals raises several environmental and economic concerns, especially for industrial scale. Further, the chemical contamination of the target product limits its application, once the active compound is generally classified as non-food grade [20]. However, the use of surfactants, which can both help harvesting biomass and promote cell disruption, is an interesting option in large scale for species whose harvest is a limiting factor. Surfactants interact with the cell membrane's phospholipids, causing distortions and consequently, cell disruption, improving release of intracellular components and bioproducts recovery [54]. The most commonly used surfactants are long-chain alkyl groups (C12 to C16) containing quaternary-ammonium cation. These compounds have hydrophobic ends capable of adsorbing or attaching to cell membranes, and once this happens, the quaternary cation makes the cell charge to become less negative, favoring cell aggregation [57]. Lai et al. [58] evaluated this synergistic effect by using cationic surfactants for flocculation and lipid extraction from *C. vulgaris*. The authors tested three cationic surfactants: dodecyltrimethylammonium bromide (DTAB), myristyltrimethylammonium bromide (MTAB) and hexadecyltrimethylammonium bromide (CTAB), showing that the pretreatment with the surfactant CTAB resulted in the most effective cell disruption, with the highest lipid recovery (nearly 90%) without changing the fatty acid methyl esters (FAME) profile. Moreover, small amounts of CTAB (0.45 mM) were required to improve flocculation and harvesting. Recently, Alhattab et al. [59] tripled the amount of total FAME extracted by 24 hours' pretreatment with the surfactant CTAB, followed by SFE with supercritical CO_2 (sc-CO_2) of *Chlorella saccharophila* biomass. However, they also observed that although the extraction was higher, the FAME composition changed significantly. This possibility of modulating FAME composition may be interesting depending on the desired application, but for biodiesel production, they found that the most suitable composition was obtained with pure sc-CO_2. Additionally, in the US patent n° 9994791B1, Zhang et al. [56] described a cell disruption method for microalgae *Nannochloropis salina* by using sodium dodecylbenzenesulfonate as anionic surfactant associated to pH adjustment and low pressurization, extending this application to other anionic and non-ionic surfactants.

2.2.2. Osmotic Shock

The presence of a high concentration of solute (salt, dextran or polyethylene glycol (PEG)) leads to a decrease of osmotic pressure, causing cell wall damage, increasing its permeability and, consequently, allowing the release of intracellular compounds. In this respect, Rakesh et al. [60] compared the use of autoclaving, microwave, osmotic shock, and pasteurization to *Chlorococcum* sp. MCC30, *Botryococcus* sp. MCC31, *Botryococcus* sp. MCC32, and *Chlorella sorokiniana* MIC-G5 to facilitate lipid extraction. They found that by applying osmotic shock improved lipid extraction could be achieved for *Botryococcus* sp. MCC32 (at 15% NaCl) and for *C. sorokiniana* MIC-G5 (at 5% NaCl). Furthermore, the composition of the extracts varied with the treatment used to facilitate the extraction. Rakesh et al. [60] also found different palmitate (C16:0) contents (25.64% and 34.20%) using osmotic shock (15% NaCl) treatment for *Botryococcus* sp. MCC32 and microwave (6 min) for *Botryococcus* sp. MCC31, respectively, while the use of *Botryococcus* sp. MCC32 as source of oil blends or nutraceuticals was proposed after osmotic shock of 15% NaCl treatment due to its oleic acid and unsaturated fatty acid content (19.95% and 38.17%, respectively). González-González et al. [61] also applied osmotic shock to *Chaetoceros muelleri* and *Dunaliella salina*, having achieved a lipid recovery efficiency of 72% and 21% respectively. They also found that the lipid-spent biomass of *C. muelleri* add one of the highest methane yields reported for microalgae of 484 mL CH_4 g VS^{-1}, showing that osmotic shock adds a double positive effect on lipid extraction and biomass quality for anaerobic digestion. López and Morales [62] extracted astaxanthin from *Haematococcus pluvialis* applying osmotic shock by highly concentrated sacarose solution, or syrup, at high temperatures. They concluded that astaxanthin extracted using osmotic shock remained available for consumption in the syrup. Koyande et al. [63] studied the recovery of whole proteins from *Chlorella vulgaris* FSP-E using osmotic shock through a liquid biphasic flotation (LBF) system, having concluded that protein recovery of 92.98% with a separation efficiency of 64.91%, partition coefficient of 1.47 and a volume ratio of 9 could be achieved using osmotic shock, whereas without osmotic shock the corresponding values were of only 84.84%, 69.68%, 1.89 and 2.96.

Although simple, the major drawbacks of this technique are that it takes longer than other processes such as autoclaving and microwave irradiation [32,64], being economically unfeasible on a large scale [65].

2.2.3. Enzymatic Methods

Enzymatic cell lysis is a high-selective method for cell disruption that requires low energy and operates at mild conditions [66]. The commercial enzymes such as cellulases, proteases, lysozyme and glucanases are vastly employed and commonly used in the immobilized form to increase their lifetime and stability, preventing reduction in catalytic activity [27]. The main drawbacks of using enzymes compared to mechanical or chemical methods are the long process time, low production capacity and the possible product inhibition. In addition, the high cost of the enzymes limits their applications in a microalgae biorefinery [21]. Liang et al. [55] tested the combination of ultrasound with enzymatic lysis (snailase and trypsin) for lipid extraction in three microalgae species, achieving the maximum lipid yield of 49.82% in *Chlorella vulgaris*, 46.81% in *Scenedesmus dimorphus* and 11.73% in *Nannochloropsis* sp. Zhang et al. [56] achieved 86.4% of lipid recovery in *Scenedesmus* sp. using a mixture of enzymes (cellulase, xylanase and pectinase), also increasing the fatty acid methyl esters (FAME) yield compared to the untreated biomass. However, despite improving the lipid yield from *Scenedesmus* sp., in the study by Zhang et al., the enzymes were used just as pretreatment followed by an organic solvent extraction, while Liang et al. used a more sustainable method based on enzyme-assisted aqueous extraction. Regarding to recent patents approaching solventless extraction of microalgae biomolecules, Bai et al. [67], US patent n° 10196600B2, described a method to induce self-lysis in microalgae cells (e.g., *Chlorella* sp., *Micractinium* sp., *Tetraselmis* sp., *Isochrysis galbana* and *Dunaliella* sp.). The active substance for self-lysis induction was extracted from *Bacillus thuringienses* ITRI-G1 suspension by vacuum distillation and isolated by high-performance liquid chromatography (HPLC). Once mixed with microalgal cells, the active substance triggers biochemical responses that induce self-lysis. The cell disruption effectiveness was estimated in terms of released-protein content by A_{280nm} (absorbance at 280 nm) measurements. After one hour of use of the active substance the absorbance (A_{280nm}) was four-fold greater (approximately 1.6) when compared to control (approximately 0.4). The results also demonstrated an increase in protein concentration (of almost two-fold) by mixing the active substance without stirring, which represents a desirable economic aspect.

Table 2 shows several cell disruption methods (mechanical, physical and non-mechanical) employed for processing a range of microalgae species and obtaining target bioproducts.

Table 2. Methods employed for microalgal cell disruption and components extraction.

Species	Description [a]	Cell Disruption Method	Target Bioproduct	Main Results
Arthrospira (*Spirulina*) sp. [15,68]	Filamentous cyanobacteria, no heterocystes or alkinetes, helical shape. Cell wall composed by four layer (L-I and III: fibrillar material, L-II: peptidoglycan and L-IV: lipopolysaccharides) [69]	Milling in a ball jar with porcelain balls at 60 rpm, for 120 min	Phycocyanin and phenolic compounds	74.98 mg C-PC g^{-1}/41.60 mg GAE g^{-1}
		Microwave oven 2450 MHz and 1400 W, 2 min		85.43 mg C-PC g^{-1}/41.90 mg GAE g^{-1}
		Autoclaving 121 °C and 200 kPa, for 30 min		1.17 mg C-PC g^{-1}/41.55 mg GAE g^{-1}
		Sonication 20% power at 35 kHz, 50% duty cycle for 7 min	Phycocyanin	94.89% (P_f: 6.17)
		Homogenisation, speed 3 for 3 min		89.51% (P_f: 5.59)
		Freeze-thawing, 8 h		77.10% (P_f: 4.15)
Botryococcus braunii [16]	Non-filamentous, pyriform shape (7 × 14 µm), colonies can vary from 30 µm to >2 mm), cell wall composed by polysaccharide with hydrocarbons between [70]	Ultrasonication 5–60 kHz, for 3–15 min	Lipids	28–30%
		Bead-beating at 2000–3500 psi, for 15 min		35–38%
		Autoclave 121 °C and 0.15 MPa, 5–90 min		38–40%
		French-press 500–3000 psi		29–43%
		Microwave oven 0–1250 W at 20–200 °C, under 2450 MHz, for 0–25 min		25–50%
		Osmotic Shock 0–2 MNaCl, stirred for 1 min and maintained 48 h		18–22%
Chlorella vulgaris [55]	Non-filamentous, spherical format (3–4 µm) and cell wall composed by extracellular polysaccharides, rhamnose, galactose, xylose [59]	Ultrasound at 600 W for 15 min and enzymatic lysis with snailase and trypsin (37 °C, pH 4.0)	Lipids	49.82%

Table 2. *Cont.*

Species	Description [a]	Cell Disruption Method	Target Bioproduct	Main Results
Haematococcus pluvialis [17]	Non-filamentous. Cell wall mostly composed by cellulose. Under favorable growth conditions can present flagella and a gelatinous thick extracellular matrix. In motile cells, the lost of the flagella result in changes on the extracellular matrix that become amorphous. Under stress conditions the cells can transform into cysts or aplanospores and a secondary wall is formed [71]	Freezing-thawing in liquid nitrogen, during 5 cycles	Astaxanthin	38–95%
		Dimethyl sulfoxide and glass beads. Cycles until pellet became colorless (5 or 10 cycles)		
		PEF (1 kV cm^{-1}, 50 ms, 50 kJ kg^{-1}) + 6 h incubation		
		Ultrasound at 80% of amplitude in a 450 W ultrasound, 10 times during 10 s (biomass diluted in ethanol)		
		Thermal treatment at 70 °C for 1 h		
Nannochloropsis sp. [55]	Non-filamentous, round shape (2–4 μm) and cell wall composed by glucose, cellulose, mannans, rhamnose, fucose, galactose and galacturonic acid [59]	Ultrasound at 600 W for 15 min and enzymatic lysis with snailase and trypsin (37 °C, pH 4.0)	Lipids	11.73%
Scenedesmus dimorphus [55] *Scenedesmus* sp. [56,72]	Non-filamentous, bean shape (10–12 μm) and cell wall composed by crystalline glycoprotein and algaenan (non-hydrolyzable structure) [59]			46.81%
		Cellulase (20 mg g^{-1}), xylanase (14 mg g^{-1}) and pectinase (10 mg g^{-1}) at 45 °C and pH 4.4 and chemical treatment with chloroform:methanol (1:1 *v/v*)		13.8 g 100 g^{-1} (86.4% recovery)
		Hydrothermal treatment with water 1:13 (*w/v*) at 147 °C for 40 min	Glucose	14.22 g L^{-1} (89.32% recovery)

Table 2. *Cont.*

Species	Description [a]	Cell Disruption Method	Target Bioproduct	Main Results
Synechocystis sp. [73]	Non-filamentous cyanobacteria. Although uncommon, it may be surrounded by a thin, colorless, diffluent mucilaginous envelope. Cell wall composed by a peptidoglycan layer and an outer membrane (mostly proteins and lipopolysaccharide) [74]	Ultrasound at 20–25 kHz for 30 min (cycles of 5 s on/5 s off)	Proteins	94.4% cell disruption efficiency 1.88 mg mL^{-1} protein
		Bead milling in glass beads for 10 min with cycles of 30 s vortexing and 30 s cooling on ice		54.4% cell disruption efficiency 1.09 mg mL^{-1} protein
		Silicon carbide (200–450 mesh) grinding, 3 cycles of 1 min grinding/1 min cooling on ice		93.3% cell disruption efficiency 1.89 mg mL^{-1} protein
		3 cycles of freezing at −80 °C for 10 min and thawing at 37 °C for 5 min		43.3% cell disruption efficiency 0.19 mg mL^{-1} protein
Phaeodactylum tricornutum [75,76]	Pleiomorphic diatom with poorly silicified cell walls (up to 10 silica bands), can present different shapes (fusiform, triradiate and cruciform) and the size range from 8–25 μm [59,77]	5 cycles·min^{-1} of sonication at 20 kHz for 15 min	Carotenoids	81.7 μg g^{-1} β-carotene; 679.2 μg g^{-1} zeaxanthin; 5163.4 μg g^{-1} fucoxanthin
		Soaking in ethanol at room temperature for 24 h, Cryogrinding in a ceramic mortar with liquid nitrogen and deionized water, Planetary micro mill, 2 cycles of 4 min at 400 rpm with 1 min of relaxation time, Potter homogenizer with ethanol for 1–5 min, Homogeniser at 18,000 rpm, 10–180 s, 2–4 cycles and 30 s of relaxation time, Sonication, 2–4 pulsed cycles (10 s on/5 s off), 30% power (500 W) and 30 s relation time, Mixer mill stainless steel grinding jars or propylene grinding tubes, bead-beating with ethanol for 1–4 min and 2–4 cycles.	Metabolites	Positive effect in cell disruption: bead-beating, planetary micro mill, sonication and mixer mill (both) Negative effect in cell disruption: soaking, cryiogrinding and Potter homogeniser

Table 2. *Cont.*

Species	Description [a]	Cell Disruption Method	Target Bioproduct	Main Results
Mixed microalgae feedstock (*Ankistrosdesmus* sp., *Chlamydomonas* sp., *Chlorella* sp., *Micromonas* sp. and *Scenedesmus* sp.) [52]	*Ankistrosdesmus*: Non-filamentous with mucilaginous envelopes present or absent, commonly find as colonies, fusiform cells (curved, straight or sigmoid), smooth cell wall [78]	Acid hydrolysis H_2SO_4 1.5 M at 80–90 °C for 80 min	Carbohydrates	10.2 g maltose, 103.1 g glucose and 68.8 g xylose/galactose per Kg of dry biomass
	Chlamydomonas, *Chlorella* and *Scenedesmus*: Non-filamentous. *Chlamydomonas* present a complex multilayer cell wall composed by 20–25 proteins and glycoproteins (rich in hydroxyproline) [79]			
	Micromonas: Flagellate with absent cell wall			
Freeze-dried mixed biomass (95% *Scenedesmus obliquus*, 4% *Scenedesmus quadricauda* and 1% *Nitzschia* sp.) [53]	*Nitzschia* sp.: diatom that can occur in three cell types: normal (fusiform, straight or curved, no longer than 35 μ), oval (8 μ long and 3–4 μ broad) or triradiate (arms varying from 6 to10 μ) [80]	Enzymatic lysis with cellulase (from *Tricoderma reesei*), β-glucosidase (from *Aspergillus niger*), pH 4.9, 50 °C and 300 rpm for 48 h	Carbohydrates (sugars) and byproducts (alcohols and organic acids)	Total sugars: 9.84 g per 100 g of dry biomass Total byproducts: 1.09 g L^{-1}
Freeze-dried mixed biomass (61% *Aphanothece* sp. and 39% *Scenedesmus obliquus*) [53]	*Aphanothece* sp.: cells can occur in many shapes (oval, ellipsoidal, straight or slightly curved) with absent of mucilaginous envelope			Total sugars: 0.02 g per 100 g of dry biomass Total byproducts: 7.38 g L^{-1}

[a] Complementary data available on: http://algaebase.org. PEF: pulsed electric field. C-PC: C-phycocyanin. GAE: gallic acid equivalents. PUFA: polyunsaturated fatty acids. Pf: purity factor.

Larrosa et al. [15] improved phycocyanin extraction in *Arthrospira platensis* using microwave irradiation from 85.43 mg·g^{-1} to 74.98 mg·g^{-1} by milling and 1.17 mg g^{-1} by autoclaving. The expressive low content of phycocyanin obtained by autoclaving is probably due to protein denaturation caused by the operational conditions. In addition, Chia et al. [68] showed a variation in phycocyanin recovery ranging from 77.10% to 94.89% and purity from 4.15 to 6.17 in *A. platensis* by changing the cell-disruption method (e.g., freeze-thawing, microwave, homogenization and sonication). Zhou et al. [73] also tested different cell disruption methods in *Synechocyst* is sp. for protein release, achieving the highest cell disruption with ultrasound (94.4%) followed by silicon carbide grinding (93.3%), bead-milling (54.4%) and freeze-thawing (43.3%). Thus, as shown in Table 2, the efficiency of the cell-disruption method can significantly change according to the microalgae species and the properties of the target product. In this sense, the use of two cell-disruption methods is a way to enhance cell disruption efficiency and, consequently, the recovery of the target biocompound [81]. Moreover, it is important to emphasize that besides the extraction efficiency and quality of bioproducts, the chosen cell-disruption method can also directly influence the subsequent purification steps.

3. Extraction Methods

As stated before, microalgal cells may contain many compounds of interest and value that must be separated from other less valuable components. Several different extraction methods can be used combined with cellular disruption or directly applied over the whole cell. These include solvent extraction, using organic solvents, ionic liquids, deep eutectic solvents, supercritical fluids, among others.

3.1. Organic Solvent Extraction

The use of organic solvents is a well-known technique for the extraction of microalgae biomolecules. Frequently, this kind of extraction is associated with a previous cell-disruption step to facilitate the access of solvent to inner cell compounds, thus enhancing the extraction yield. Several filled patents available on the US Patent database have also reported the use of organic solvents for microalgae biocomponents extraction (Table 3).

Table 3. Recent US-granted patents approaching cell disruption, extraction and recovery of microalgae biomolecules using organic solvents (FAME: fatty acid methyl esters).

Patent Purpose	Pretreament	Solvents	Main Results
Production and extraction of squalene from microalgae of the *Thraustochytriales* sp. Family [82]	Alkaline lysis (KOH 45%)	Hexane/ ethanol	6.7 g L^{-1} of squalene content
Production of pure microalgae extracts to modulate the metabolism of human skin and hair follicles [83]	-	Methanol/ ethanol/ ethyl acetate	Achieved an effective treatment using a composition comprising from 0.001 to 35% of dry matter content of an extract of *Monodus* sp.
Method for extraction of lipids in an organic phase and sugars by hydrolysis [84]	Homogenization	Acetone	Results suggest that it is possible to achieve an industrial-scale extraction yield of 96.3% of total lipids in the starting wet algal biomass
Biodiesel production and isolation of several valuable co-products from the marine alkenone-producing microalgae *Isochrysis* [85]	-	n-hexane/ethanol or methanol/ dichloromethane or toluene/acetonitrile	27:8:1 (FAME:alkenones: fucoxanthin) co-production

Pora et al. [82] achieved a maximum squalene content of 6.7 g·L^{-1} from microalgae of the *Thraustochytriales* sp. family by organic solvent extraction (Table 3). This result is up to 2.5-fold other results that have been reported in the literature, which are between 0.9 to 2.46 g·L^{-1} in species such as *Aurantiochytrium* sp. 18W-13a (0.9 and 1.29 g·L^{-1}), extracted by chloroform:methanol (2:1, v/v) [86] and n-hexane containing 10% of chloroform, respectively [87], *Schizochytrium mangrovei* PQ6 (1.01 g·L^{-1}) extracted by n-hexane:diethyl ether:acetic acid (70:30:4, $v/v/v$) [88], *S. limacinum* SR21 (0.9 g·L^{-1} of squalene produced in bioreactor) extracted by chloroform:methanol (2:1, v/v) [89] and *Aurantiochytrium* sp. BR-MP4-A1(2.46 g·L^{-1}) extracted with hexane (3 cycles) after saponification with 10% KOH (w/w) and 75% ethanol (v/v) [90]. In other granted patent, Massetti et al. [84] results suggest that it is possible to achieve 96.3% of lipids extraction yield from *Nannochloropsis* sp. at industrial scale. Their invention comprises the use of acetone, a preferable organic solvent when compared to common solvents such as hexane and chloroform, due its lower toxicity. Despite the use of organic solvent, the perspective of almost 96% of extraction yield is quite expressive. This value was predicted from an experiment that resulted in 281 g of lipids, which means 27.94% of volatile solids present in the starting wet biomass (66.72% in which 29% is related to lipids). Du et al. [91] maximized the lipid extraction yield in an oleaginous microalgae *Neochloris oleabundans*. In this work, the authors achieved a maximum lipid extraction yield of 61.3% dry weight in fresh water stressed (nitrogen limitation) cultivations after multistage extractions (four extractions) with n-ethylbutylamine as solvent. Moreover, some publications have reported less than 50% of lipid yield in different microalgae species using more recent technology based on supercritical CO_2 (Sc-CO_2) extraction, 20% *Nannochloropsis oculata* (Sc-CO_2, 450 bar), 40% *Chlorella vulgaris* (Sc-CO_2 and ethanol 10% v/v, 250 bar) [92], 11.1% *Tetraselmis* sp. (Sc-CO_2, 300 bar) [93] and 12.1% *Nannochloropsis* sp. (Sc-CO_2, 550 bar) [94].

Classical organic solvent-based techniques for microalgae biocompounds extraction and recovery, mostly lipids, include the Bligh and Dyer (B&D) method [95], Folch method [96] and Soxhlet method [97]. The B&D and Folch methods are quite similar techniques widely used for lipid extraction. Folch is a two-step method, which firstly employs chloroform/methanol (2:1 ratio) as solvents to homogenize lipids. The homogenate is filtered and lipids separated in a second step, which comprises washing of the crude oil extract, with at least 5-fold volume of water. It is a time-consuming technique that requires large amounts of solvent. In order to solve the disadvantages of the Folch method, the B&D examined the chloroform-methanol-water phase diagram, proposing a single-step method. Briefly, the homogenate obtained in the chloroform/methanol mixture must be a monophasic system, forming a bi-phase when mixed with water. The chloroform phase contains the lipids and the other phase is composed by methanol, water and non-lipid components.

Soxhlet is a classical technique proposed by Franz von Soxhlet in 1879, that comprises a Soxhlet apparatus in which lipids are extracted from biomass by exhausting reflux in organic solvent, commonly n-hexane or petroleum ether. Despite being a simple and scalable technology, it also is time consuming and requires a large solvent amount. Beyond that, the extraction yield is low, probably due to inefficient polar lipid extraction [98].

Despite being well-established techniques, conventional solvents are toxic and not aligned with environmental and human health concerns. In this sense, Breil et al. [99] studied several potential solvents to substitute chloroform (e.g., 2-methyltetrahydrofuran (MeTHF), cyclopentyl methyl ether (CPME), ethyl acetate (EtOAc), ethyl lactate, dimethyl carbonate (DMC), p-cymene, d-limonene and α–pinene) and methanol (e.g., ethanol and isopropanol) in B&D and Folch methods for lipid extraction. The classical chloroform/methanol/water system still is more selective than the tested solvents. However, their study suggests that the combination of ethyl acetate/ethanol, followed by the addition of water enriched with KCl and ethyl acetate is a suitable option. In addition, Gorgich et al. [100] increased lipid recovery using methyl tert-butyl ether (MTBE), which is less hazardous than classical solvents (e.g., chloroform and hexane), in three microalgae species compared to modified B&D method (chloroform:methanol:water, 1:2:0.8 (v/v)). The highest yield and also the most expressive difference between methods were observed in lipid extraction from *Chlorella zofingiensis*, resulting in 34.06% by MTBE and 26.68% by the B&D method. On the other hand, Ruecker et al. [101], US patent n°10329515B2, described a solventless lipid-extraction process based on alkaline treatment for cell disruption with potassium hydroxide (KOH), followed by sequential aqueous phase separation, achieving process performance and bioproduct quality comparable to the traditional solvent extraction with hexane.

3.2. Alternative Solvents Extraction

By contrast with conventional solvents, alternative solvents are supposed to have lower environmental, safety and health impacts [102]. Some examples of alternative solvents are bio-based solvents such as terpenes, ionic liquids, deep eutectic solvents, and liquid polymers. One of the most well studied liquid polymers is polyethylene glycol (PEG), a biodegradable polymer widely used in aqueous two-phase systems [103]. Focusing on recent developments, this section will discuss the use of ionic liquids and deep eutectic solvents.

Ionic liquids (ILs) are organic salts with melting points of 100 °C or below. However, not all ILs have "green" characteristics and must be avoided. For a better understanding Bubalo et al. [104] present a general description of the potential environmental risks of ILs. These substances are usually composed of an organic cation and an anion that can be organic or inorganic. ILs have great thermal stability, solvating power and non-volatile characteristics due to their high enthalpies of vaporization ($\Delta_{vap}H$). There are several possible combinations between cations and anions for ILs synthesis; it is estimated up to 10^{18} combinations.

This feature significantly broadens the ILs applicability and it has attracted great interest from the modern scientific community, especially for extraction and separation of biocompounds [103,105]. Table 4 shows some examples of ILs used as assistant for microalgal components extraction.

Table 4. Ionic liquids used for microalgae biomolecules extraction.

Microalgae Species	Ionic Liquid	Abbreviation	Target Component
Neochloris oleoabundans [106]	1-butyl-3-methylimidazolium tetrafluoroborate 1-butyl-3-methylimidazolium methyl sulfate 1-butyl-3-methylimidazolium dicyanamide 1-butyl-3- methylimidazolium chloride	[BMIM][BF_4] [BMIM][$MeSO_4$] [BMIM][DCN] [BMIM][Cl]	Lipids
Chlorella vulgaris and *Spirulina platensis* [107]	1-ethyl-3-methylimidazolium acetate choline L-arginate choline glycinate choline L-lysinate choline L-phenyl-alaninate	[Emim][OAc] [Ch][ARG] [Ch][GLY] [Ch][LYS] [Ch][PHE]	Carbohydrates and lipids
Haematococcus pluvialis [35]	ethanolammonium caproate diethanolammonium caproate triethanolammonium caproate	[EAC] [DEAC] [TEAC]	Astaxanthin
Scenedesmus sp. [108]	triethylammonium hydrogen sulfate 1-butylpyridinium hydrogen sulfate 1-butylpyridinium dihydrogen phosphate	[$HNEt_3$][HSO_4] [BPy][HSO_4] [BPy][$H2PO_4$]	Lipids
Chlorella vulgaris [36]	1-octyl-3-methylimidazolium bis(trifluoromethanesulfonyl) imide 1-octyl-3-methylimidazolium acetate	[Omim][NTf2] [Omim][OAc]	Lipids

ILs and deep eutectic solvents (DES) have several similar properties, such as thermal stability, conductivity and low volatility, but differ in the sources and chemical processes involved in their synthesis. DES present a cheaper and easier synthesis and a range source of potential starting materials, being capable of overcoming some ILs' limitations like high cost, purification steps and toxicity [109]. DES are formed by hydrogen bond interactions between a hydrogen bond acceptor (e.g., choline chlorine or choline acetate) and a hydrogen bond donor (e.g., carboxylic acids, amides, amines and alcohols), presenting a final melting point lower than each independent component used in their synthesis. There are also the so-called natural deep eutectic solvents (NADES), when the compounds are primary metabolites (e.g., organic acids, amino acids, sugar, among others) [110,111]. Table 5 summarizes some examples of deep eutectic solvents applied for lipids and polyphenolic compounds extraction from microalgae.

Table 5. Deep eutectic solvents (DES) used to extract microalgae-based compounds.

Microalgae Species	DES	Hydrogen Bond Acceptor	Hydrogen Bond Donor	Monomer Ratio (mol:mol)	Target Product	Yield (%)
Chlorella vulgaris [112]	Ch-Gly	Choline chloride	Glycerol	1:2	Polyphenolic compounds	5.27 [a]
	Ch-EG		Ethylene glycol	1:4		7.19 [a]
	Ch-PDO		1-3-propanediol	1:4		9.19 [a]
	Ch-BDO		1-4-butanediol	1:4		9.87 [a]
Chlorella sp. and *Chlorococcum* sp. [113]	Ch-Fa	Choline chloride	Formic acid	1:3	FAME	9.12/9.00
	Ch-Aa		Acetic acid	1:3		13.91/11.5
	Ch-Oa		Oxalic acid	1:1		10.35/9.40
	Ch-Pa		Propanedioic acid	1:1		10.53/9.52
Chlorella sp. [110]	aCh-O *	Choline chloride	Oxalic acid	1:2	FAME	16.41 [b]
	aCH-EG *	Choline chloride	Ethylene glycol	1:2		15.32 [b]
	aU-A *	Urea	Acetamide	1:2		10.53 [b]

* aqueous, [a] mg GAE.g^{-1} biomass, [b] corresponding to mg linoleic acid (C18:2) g biomass. GAE: gallic acid equivalent. FAME: fatty acid methyl esters.

3.3. Supercritical Fluid Extraction

Supercritical fluid extraction (SFE) has been increasingly employed, especially in the food and pharmaceutical industries, as a green and contamination-free extractive method [114]. SFE combines extraction and separation by strictly controlling process parameters such as temperature, pressure, flow rate and processing time. Therefore, this selective capacity allows products with greater purity and yields to be obtained [115].

Substances in conditions of temperature and pressure above its critical point are called supercritical fluids (SCFs), which combine properties of liquids and gases in an effective way. In this state, supercritical fluids show some interesting properties, such as lower viscosity than liquids and ability to better dissolve substances than gases and with more diffusive power. In addition, SCFs can work for some difficult or impossible extractions using organic solvents. Knez et al. [116] made a critical review about SCFs as solvents for the future, concluding that from an environmental point of view, SCFs are a feasible alternative to conventional organic solvents, capable of providing products with high purity and low energy consumption, once the high pressure technologies are advancing.

Concerning the extraction of microalgae biomolecules, SFE has been widely used mainly for recovery of polyunsaturated fatty acids (PUFAs), pigments and vitamin, in which the most commonly employed solvent is sc-CO_2, pure or associated with some other co-solvent. Yang et al. [117] developed a method based on a solid matrix-supported sc-CO_2 for γ-linolenic acid (GLA) extraction from *Arthrospira platensis*, reaching more than 34.5% GLA of the total fatty acids and 98% of extraction efficiency. On the other hand, Feller et al. [118] compared sc-CO_2 with subcritical n-butane extraction of ω-3 and ω-6 fatty acids in *Phaeodactylum tricornutum*, *Nannochloropsis oculata* and *Porphyridium cruentum*, demonstrating that in all cases subcritical n-butane extracted more PUFAs, ω-3 and ω-6, than the other solvent. In fact, they realized that sc CO_2 is more selective for saturated fatty acids (SFA) extraction. However, sc-CO_2 also extracted more total carotenoids in two of the three tested species.

Another way to improve extraction performance is by combining supercritical fluid with a co-solvent. In this regards, Chronopoulou et al. [119] evaluated sc-CO_2 extraction of several biomolecules (alpha-tocopherol, phylloquinone, gamma-tocopherol, retinol, canthaxanthin, phytofluene and lutein) from *Tetradesmus obliquus* microalgae. In this work, the authors obtained some satisfactory results by using limonene and methanol as sc-CO_2 co-solvents. For example, the use of these co-solvents enabled the extraction of canthaxanthin and lutein that was not possible under only sc-CO_2 conditions, achieving the maximum extraction yield of 16.14% and 1.25%, respectively. Table 6 summarizes the yields and operating conditions for the supercritical extraction of different components from various microalgae species.

Table 6. Supercritical extraction of different components from microalgae species.

Microalgae Species	Solvent(s)	T (°C)	P (bar)	Flow Rate (g min^{-1})	t (min)	Target Compound	Extraction Yield (%)
Chlorella saccharophila [59]	sc-CO_2	73.0	241	3 SLPM	60	FAME	17.60
Spirulina platensis [120]	sc-CO_2 + EtOH	55.0	78.6	52.83 [a]	75	Vitamin E	8.08
Spirulina platensis [121]	sc-CO_2 + EtOH	55.0	80	52.83 [a]	75	Pigments	7.94
Nannochloropsis sp. [94]	sc-CO_2	75.0	550	14.48	100	Lipids	12.08 *
Tetradesmus obliquus [119]	sc-CO_2 + MeOH	60.0	0.25	-	90	Phylloquinone	302.74
		50.0	0.35			Canthaxanthin	16.14
						Lutein	1.25
		40.0				γ-tocopherol	137.43
			0.30			α-tocopherol	200.15
						Retinol	543
	sc-CO_2 + limonene					Phytofluene	142.65
							(mg g^{-1} dw)
Scenedesmus almeriensis [122]	sc-CO_2	65.0	0.55	14.48	120	Lutein	2.97 **
			0.40	7.24		Lipids	15.02
Nannochloropsis gaditana [123]	sc-CO_2	65.0	250	7.24	100	EPA	11.50
Nannochloropsis oculata [118]	sc-CO_2	40.0	300	0.02 [a]	120	Lipids	15.60
Phaeodacthylum tricornutum [118]							14.70
Porphyridium cruentum [118]							4.50

SLPM: Standard liter per minute. EtOH: ethanol. MeOH: methanol. FAME: fatty acid methyl esters. EPA: eicopentaenoic acid. [a] Calculated according to data available in the referenced paper. * 19.5% purity; ** approximately 34% purity.

Ferreira et al. [124] studied the environmental and economic aspects of a biorefinery from *Nannochloropsis* sp., comparing the use of Soxhlet extraction (with n-hexane) and SFE for oil and other co-products extraction. The study showed that despite consuming more energy (262 MJ·MJprod^{-1}) and generate more CO_2 emissions (17 g·MJprod^{-1}) than Sohxlet extraction (220 M·MJprod^{-1} and 14 g·MJprod^{-1}, respectively) in oil production, SFE is more economically feasible once this technology allows the co-extraction of high-value pigments. Thus, SFE presented a final cost (e.g., cultivation and downstream processes) of 365 €·kg$_{oil}$$^{-1}$ while the Soxhlet extraction costs were estimated in 661 €·kg$_{oil}$$^{-1}$. Moreover, there are many specifications currently available for supercritical fluid extraction (Table 7), proving the advancement of this technology in the market.

Table 7. Examples of supercritical fluids extraction equipment available.

Company	Scale	Specifications
ExtrateX Supercritical Fluid Innovation [a]	Laboratory	Volume: 50–1000 mL; CO_2 pump: 0–100 g min^{-1} Pressure: 0–100 bar; Temperature: up to 150 °C Co-solvent pump: 24 mL min^{-1} at 400 bar
	Small production	Volume: 5–20 L; CO_2 pump: 0–24 or 0–40 kg h^{-1} Pressure: 0–350 bar; Temperature: 0–150 °C Co-solvent pump: 100 mL min^{-1} at 350 bar
	Large production	Volume: 25–100 L; CO_2 pump: 0–150 or 0–300 kg h^{-1} Pressure: 0–350 bar; Temperature: 0–150 °C Co-solvent pump: 30 L h^{-1} at 350 bar
Joda Technology Co. (15 specifications available) [b]	Small to large production	Pressure: 400–500 bar; Flow rate: 60–16,000 L h^{-1} Area: 10–700 m^2; Height: 2.5–16 m Capacity of raw material: 10–20 kg or less to 5–12 ton
Tradematt (Henan) Industry Co., Ltd. [c]	Small production	Extractor: 10 L × 2, ≤50 MPa; Separator: 5 L × 2/3, ≤16 MPa Temperature: ≤85 °C; Power: 20 kW CO_2 flow rate: 100 L h^{-1}
	Large production	Extractor: 25 L × 2, ≤40 MPa Separator: 15 L × 2/3, ≤16 MPa Temperature: ≤85 °C; Power: 30 kW CO_2 flow rate: 300 L.h^{-1}
		Volume: 700 L × 3; Pressure: 320–500 bar Flowrate: 4000 L h^{-1}; Area: 500 m^2 Capacity of raw material: 1.5–3.5 ton
		Volume: 1500 L × 3; Pressure: 320–500 bar Flowrate: 800 L h^{-1}; Area: 600 m^2 Capacity of raw material: 3.5–8 ton
Green Mill Supercritical [d]	Small production	CO_2 pump: 100–500 gm min^{-1}, pressure: 7500 psia Extractor: 7.5 L; Separator: 3.3 L Power requirements: 200 V, 60 Hz, 1 phase, ~67 A maximum and 14.7 kW maximum

Websites: [a] www.extratex-sfi.com; [b] www.joda-tech.com; [c] www.cntradematt.com; [d] www.greenmillsupercritical.com.

4. Analysis Method

The chromatographic methods use the differences in the characteristics of molecules in a mixture to separate them, according to properties such as adsorption (liquid–solid), partition (liquid–solid), affinity and molecular weights. There are several types of chromatography and the effectiveness of each one will depend on the nature of the molecule to be separated [125].

4.1. Supercritical Fluid Chromatography

Similar to other chromatographic techniques, supercritical fluid chromatography (SFC) is capable of detecting and separating the components of a given sample. In addition, SFC combines some advantages of both gas chromatography (GC) and high-performance liquid chromatography (HPLC) and can be particularly useful for substances that decompose at high temperatures in GC or have some undetectable functional groups in HPLC systems [126]. Abrahamsson et al. [127] used the combination of SFC and SFE for carotenoid extraction and separation from *Scenedesmus* sp., being able to maximize the extraction of all compounds, except violaxanthin, by adding 10% ethanol as co-solvent to sc-CO_2. These authors obtained 72.9 μg·astaxanthin, 59.9 μg·β-carotene, 436.1 μg·lutein, 670.8 μg·neoxanthin and 89.6 μg·zeaxanthin per gram of freeze dried algae, in 60 min of extraction at 300 bar, 60 °C and 2 mL·min^{-1} of total flow. Additionally, they achieved a repeatability of 2.4% relative standard deviation (RSD) and intermediate-precision of 6.4% RSD, in extracts of SFE using co-solvent. These results are quite similar to those reported for carotenoids, obtained through the conventional technique of HPLC analysis [128].

4.2. Column Chromatography

Column chromatography is solid–liquid adsorption chromatography. This technique consists of a column packed with an adsorbent, commonly alumina (Al_2O_3) or silica gel (SiO_2), forming the stationary phase. The sample carried by the mobile phase passes through the column and the molecules are separated based on its polarity [129]. Silica gel is the most common adsorbent used in stationary phase for column chromatography, although for some applications alumina has been shown to be more suitable. For a better understanding of the particularities of each compound (i.e., silica gel and alumina) this subject was reviewed by Nawrocki [130,131] and Claessens and van Straten [132]. Anyhow, silica gel has been successfully reported for microalgae bio-oil purification, including compounds such as tridocosahexaenoyl glycerol (tri-DHA) [133], mono and polyunsaturated fatty acids [134] from the commercial oil DHASCO® extracted from *Crypthecodinium cohnii*, γ-linolenic acid from *S. platensis* [135], eicosapentaenoic and arachidonic acids (EPA and AA) from *Porphyridium cruentum* [136] and carotenoids such as fucoxanthin from *Nitzschia* sp. [137].

Hydroxyapatite Chromatography

Hydroxyapatite is a calcium phosphate commonly used in chromatographic columns for protein and DNA separation. The adsorption of proteins on hydroxyapatite involves anionic and cationic exchanges and, in order to facilitate this process, the proteins are generally eluted by increasing phosphate gradient [138]. In this sense, several authors have reported the use of hydroxyapatite chromatography for phycobiliproteins separation in various microalgae and red algae species, such as *Rhodella violacea* [139], *Arthrospira maxima* [140], *Corallina elongata* [141], *A. platensis* [142,143] and *Porphyra yezoensis* Ueda [144]. Other reported applications of hydroxyapatite chromatography separations involve recovery of enzyme from *Pavlova pinguis* [145] and phosphoprotein from transgenic *Chlamydomonas reinhardtii* [146].

4.3. Gel Permeation Chromatography

Gel permeation chromatography (GPC) is a size-exclusion chromatography widely used for macromolecules separation. In this technique, the hydrophobic stationary phase is composed of inert particles with small pores. The sample is continuously injected into the chromatographic column, carried by the mobile phase that is commonly an organic solvent. The smaller the size of the molecule, the longer the retention time in the column, as long as these small molecules permeate the pores of the stationary phase. Larger molecules, in contrast, pass straight through the column, presenting shorter retention times. Thus, GPC gives the molecular weight distribution of a sample [125,147]. For microalgae, this technique is particularly useful in the characterization of polysaccharides [148,149] that can reach levels in the range of 8–64% DW [150] and polyhydroxyalkanoates [151].

4.4. Ion-Exchange Chromatography

The ion-exchange chromatography (IEX) uses the difference between the charge properties of the molecules to promote their separations. Therefore, it is used to separate biomolecules with ionizable groups, such as proteins, nucleic acids, peptides, among others. Higher selectivity can be achieved by manipulating some process parameters such as pH, which is capable of altering the surface net charge of charged biomolecules. It is worth mentioning that the correlation between surface net charge and pH is unique for each protein, allowing high levels of specificity in this application [152,153].

Chen et al. [154] developed a single-step C-phycocyanin (C-PC) extraction from *S. platensis* using stirred fluidized bed ion exchange chromatography. The biomass was pre-treated with ultrasonication and from 10% dry weight of the disrupted cells the group reached 59.4% of C-PC recovery yield with 3.0 of purity factor (A_{615nm}/A_{280nm}). However, by combining ultrafiltration with IEX as a strategy to enhance purity of B-phycoerythrin from *Porphyridium cruentum*, Tang et al. [155] achieved a purification factor (A_{545nm}/A_{280nm}) of 5.1, which is considered of analytical grade, with 68.5% yield.

4.5. Affinity Chromatography

Affinity chromatography relies on very specific reversible interactions to separate complex mixtures. The target molecule carried by the mobile phase binds to the ligand immobilized in the stationary phase by affinity. The recovery of these molecules is accomplished by further washing and elution steps. In order to select the most appropriate matrix to compose the stationary phase, besides the specific interaction of the target molecule, some other characteristics should be taken into account, such as high physical and chemical stability, low adsorption of undesirable molecules and good properties to maintain the mobile phase flow throughout the whole process [156].

4.6. Thin-Layer Chromatography

In thin-layer chromatography (TLC), it is obligatory for the mobile phase to be a liquid, but the stationary phase, the 'thin-layer', can be solid or a solid with a certain amount of liquid [157]. This technique is widely used for lipid and phospholipids separation [158–160], although PUFAs oxidation still occurs due to the long period of oxygen contact [161]. TLC has also been reported for fucoxanthinol purification [162], separation of final products of biocatalytic transformation [163], antioxidants [164] and for biodiesel characterization [165].

4.7. High-Performance Liquid Chromatography

Through the use of high pressure, which forces the solvent to pass through the closed column, packed with small particles, high-performance liquid chromatography (HPLC) can achieve high separation resolution. The smaller the particle size that compose the stationary phase, the greater the separation efficiency, however, high pressure is required. The HPLC system consists of a mobile phase reservoir, mobile phase pumping system, sample injection system, chromatographic column and detection, control, acquisition and data recording system [166]. In microalgae-based biocomponents, HPLC has been extensively reported for carotenoids [128,167–169], fatty acid composition [170,171] and phenolic acids profile [172]. Recently, Hussain et al. [173] published the results of their research discussing the use of chiral and achiral ionic liquids (ILs) in HPLC to enantiomeric separation, an important field for biopharmaceuticals. Although the use of chiral ILs is not a fully-elucidated technology, it has great potential to replace the conventional chiral phases used in HPLC. Unlike IL-based phases, the conventional ones have several limitations, such as instability at high temperatures, low solubility, high ultraviolet (UV) absorptivity and high cost [105,174]. Figure 3 shows the most suitable mode of the HPLC method according to biomolecule properties.

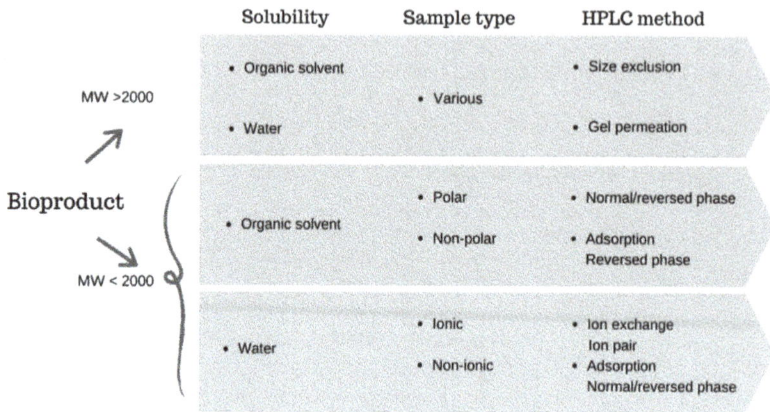

Figure 3. Selection of high-performance liquid chromatography (HPLC) methods for biomolecule separation according to bioproduct properties. MW: molecular weight [129].

4.8. Counter Current Chromatography

In addition to HPLC, another liquid chromatography technique is countercurrent chromatography (CCC), in which both mobile and stationary phases are liquid. The principle of this technique is based on the difference in the affinity of the solutes of a mixture for these phases, and, unlike what the name suggests, there is no countercurrent in this process [175].

Recently, Fábryová et al. [176] developed a new protocol integrated with multi-injection high-performance countercurrent chromatography (HPCCC) for lutein extraction in *Chlorella vulgaris* biomass. The extract with the highest lutein concentration was obtained in the lower phase of a biphasic solvent system composed of n-heptane, ethanol and water (5:4:1.5, $v/v/v$), which was used for both extraction and the mobile phase for HPCCC lutein isolation (3.20 mg·g^{-1} DW). In 2 g of this *C. vulgaris* extract, 60 mg lutein with 92% of purity was obtained, which was subsequently taken to gel permeation chromatography resulting in 50 mg of lutein with 97% purity. Earlier, Cheel et al. [177] reported the yield of 4 mg of nostotrebin 6 (NOS-6) with 99% of purity from 100 mg of crude extract obtained from *Nostoc* sp. NOS-6 has been described as an inhibitor of acetylcholinesterase and butyrylcholinesterase, enzymes with great relevance in studies to treat Alzheimer's disease.

Another CCC-based methodology is the high-speed counter-current chromatography (HSCCC), which is based on hydrodynamic equilibrium. In this technique, while one phase is pumped through the column, the other liquid phase is retained by centrifugal force. This model has high separation efficiency and furthermore, circumvents the sample adsorption problem presented by other chromatographic methods using a solid matrix [178].

HSCCC is mostly reported for plants, fruits or herbal biocomponents-based extraction [179]. However, for microalgae-based biomolecules, HSCCC has been reported for separation of polysaccharides [180], squalene [181], carotenoids [182,183] and lipopeptides [184], in this last case associated with HPLC.

4.9. Gas Chromatography

Gas chromatography (GC) is widely used for detecting and separating volatile molecules. In this system, the stationary phase can be composed of a solid (gas-adsorption) or a non-volatile liquid absorbed on an inert powder (gas–liquid partition) whereas the sample is carried through the column by an inert gas. Thus, the molecules are separated due to their different migration velocity in the chromatographic column [185]. Table 8 shows the applicable chromatographic method according to the type of biomolecule.

Table 8. Comparison of gas chromatography (GC) applicability and the main chromatographic methods for biomolecules separation. Adapted from Ahuja [129] (GC: gas chromatography, HPLC: high-performance liquid chromatography, IEX: ion-exchange chromatography, TLC: thin-layer chromatography).

Biomolecule Type	Chromatographic Methods					
	GC	HPLC	IEX	TLC	Size-Exclusion	Adsorption
Hydrophilic		X	X	X	X	
Hydrophobic	X	X		X	X	X
Ionic		X	X	X		
Non-ionic	X	X		X	X	X
Volatile	X					
Non-volatile		X	X	X	X	X
Simple			X	X		
Complex	X	X			X	X

In microalgae-based products, GC is mostly reported as a methodology for fatty acid methyl esters (FAME) separation in lipid extracts [31]. Over time, several variants of conventional GC were developed in order to improve separation resolution, and broaden the field of application. Examples include the two-dimensional GC, which is composed of two subsequent columns containing different stationary phases [186] that can be performed as GC–GC, when only selected fractions are re-injected in the second column, or GC×GC, when the sample is continuously re-injected [187]. Another example is the pyrolysis gas chromatography mass spectrometry (Py–GC/MS), in which the components are separated on a fused silica column and detected by mass spectrometer that can work in continuous or pulse modes, according to heating mechanism [188]. Py–GC/MS presents some advantages over other techniques, such as Fourier-transform infrared (FTIR) and thermogravimetric analysis (TGA), including the detection of products with low molar weight, without heating rates limitation and providing more complete information about gaseous products. Thus, Py–GC/MS has been used in the analysis of microalgae pyrolysis products aiming biofuel production [189].

5. Separation and Purification Methods

After the extraction, the biomolecules are usually mixed with either solvent or combined in a single phase, making it necessary to separate the compounds of interest from those that are of least interest or to remove impurities, which lower their value or introduce some type of toxicity. Thus, it is generally necessary to apply separation methods to purify the extracted compounds. These include electrophoresis, membrane separation, ultracentrifugation, etc., whose characteristics, advantages and disadvantages will be presented.

5.1. Electrophoresis

Electrophoresis uses the charge of the molecules to make them migrate through an electric field. This technique is commonly used for proteins and other macromolecules separation. Unlike GPC, where all particles are forced through the column, in electrophoresis, smaller particles migrate through the column pores, while larger particles remain almost immobile. Moreover, high-resolution separations can be achieved by two-dimensional electrophoresis (2DE), which was previously reported to solve more than 1000 proteins from *Escherichia coli* bacterium in a single experiment [190]. The advantage of this approach is that in the first moment, the molecules are separated based on their charges and then according to their molar masses [191]. In the literature it is reported that electrophoresis and

2DE have been used in water-soluble and insoluble proteins resolution from microalgae, such as *Porphyridium cruentum* [192], *Haematococcus pluvialis* [193] and *Chlorella protothecoides* [194]. Alternatively, Lipkens et al. [195], US patent nº 9556411B2, described a method based on ultrasound and acoustophoresis for *Dunaliella salina*'s harvesting, lipid extraction and separation. The process comprises three sequential chambers; in each one a specific ultrasonic transducer forms acoustic waves perpendicular to the flow in order to accomplish harvesting, lipid extraction and lipid separation, respectively.

5.2. Membrane Separation Processes

Membrane processes allow separation of different compounds based on their size or charge (e.g., electromembrane). In conventional membrane filtration, while molecules larger than the pore size of membranes are retained, the lower molecules pass through the membrane pores, regardless of their nature, that is, without selectivity.

5.2.1. Ultrafiltration

Ultrafiltration (UF) is widely used for macromolecules concentration and buffer exchange. For example, by using reduced pore size membranes (i.e., 1–100 nm) and mild temperature conditions, high retention of proteins and enzymes can be achieved. However, unlike chromatographic methods, UF has no selectivity and has the disadvantages of being susceptible to fouling and absorption of target compounds during filtration. On the other hand, it is a low-cost technique and easy to scale up for commercial applications [196].

Chaiklahan et al. [197] evaluated the purification of a crude C-PC extracted from *Spirulina* sp. in two steps. First, performing microfiltration with 5 and 0.8/0.2 μm pore size membranes at 150 and 100 mL·min^{-1}, respectively, and then, membrane ultrafiltration with 50 kDa MWCO (Molecular Weight Cut Off) at 69 kPa and 75 mL·min^{-1} flowrate. Under these conditions, the purity ratio increased from 0.54 ± 0.19 (before UF) to 1.07 ± 0.01, being considered as food grade.

Still in a pigment purification approach, Gómez-Loredo et al. [198] studied ultrafiltration as a complementary method to alcohol-salt aqueous two-phase systems (ATPS) for fucoxanthin (i.e., carotenoid) purification, achieving approximately 16% reduction in impurities. In addition, they noted that by increasing the concentration of ethanol in the biphasic system, more contaminants were trapped in the membrane. Thus, in this case, fouling turns out to be a desirable event.

5.2.2. Electromembrane Filtration

Electromembrane filtration (EF) is an alternative to conventional membrane filtration. The great advantage of EF is to promote separation by charge and molecular weight without requiring pressure, being more selective than the conventional membranes [199]. Kim et al. [200] also proved electromembrane filtration as a useful tool for harvesting *Chlorella* sp. KR-1, improving fourfold (6.47) the concentration factor (CF) when electricity was applied. Meanwhile, the maximum CF achieved was 1.32 and 1.56, without electricity and by commercial polyvinylidene (PVDF) membrane, respectively.

5.3. Ultracentrifugation

The differentiated sedimentation of molecules, according to their densities and under the influence of gravity (g-force), is the basic principle of particles separation by centrifugation. In laboratory scale, the g-force of ultracentrifugation can reach values of up to 200,000× *g*, being able, for example, to remove impurities (e.g., chlorophyll, cell debris) in an extract of C-phycocyanin [201] and to estimate molecular weight of enzymes and proteins from microalgae [202]. However, for this last purpose ultracentrifugation is already outdated. Also, it is an energy-intensive procedure.

5.4. Aqueous Two-Phase Systems

In aqueous two-phase systems (ATPS) the phases are generally formed by the combination of polymer/polymer or polymer/salt as phase-forming components. However, the combination of alcohol/salt has also been widely used due to the easy of alcohol removal by evaporation. This technique can be used to purify several biomolecules such as pigments, proteins, enzymes, amino acids, antibodies, etc. [203]. The principle of the ATPS method is the mass transfer mechanisms involved in the equilibrium and phase separation of a mixture. The phase diagram of an ATPS is distinct and provides the optimal operating conditions (e.g., temperature, pH and concentration of the phase-forming components). Additionally, the physicochemical properties and concentration of the crude sample, as well as tie line length (the length of the line joining the composition of the two phases in a phase diagram), phase volume ratio of the two phases and the presence of additives can also affect the partitioning effect of target molecules in liquid biphasic systems and must be taken into account [204]. A variant of this technique that has been widely used to purify molecules from microalgal biomass is the liquid biphasic flotation (LBF). This technique combines ATPS with solvent sublation (SS), which is based on the adsorption of molecules in bubbles surface in aqueous phase. Briefly, the target component is carried by bubbles to the top of the immiscible phase composed of the organic solvent. Then, the bubbles rupture, releasing the target molecules to be recovered in the upper zone of the column [203].

Table 9 shows some recent research about purification of microalgal biomolecules by LBF. Chew et al. [205] achieved almost 94% recovery of C-PC in *Spirulina platensis* using PEG 4000 and $(NH4)_2SO_4$. Meanwhile, Chia et al. [68], members of the same group, improved recovery in 1% by replacing salt with K_2HPO_4 and optimizing the pretreatment parameters (20% power, 50% duty cycle and 7 min of irradiation time). However, the most interesting result was that the purity factor of C-PC almost quadrupled with these adaptations, increasing from 1.63 to 6.17. The purity factor of C-PC extracts is generally given by the ratio A_{615nm}/A_{280nm} (absorbance at 615 and 280 nm), which correlates the presence of C-phycocyanin and other contaminating proteins. Purity factors above 4.0 are considered as analytical grade and can value more than US$ 15 per mg [5]. Furthermore, Safaei et al. [206] reported four-step C-PC purification from *Limnothrix* sp. NS01, but achieving only 5.26 of purity. The steps included treatments with chitosan, activated charcoal, precipitation with ammonium sulfate and IEX. The genus *Limnothrix* includes filamentous cyanobacteria as well as *Spirulina*, which reinforces the importance of the results obtained by Chia et al. [68].

Table 9. Purification of microalgal biomolecules by liquid biphasic flotation.

Microalgae Species	Pretreatment	Type of Polymers	Type of Salts	Type of Alcohol	Bioproduct	Recovery (%)	Purity Factor
Chlorella sorokiniana [207]	Ultrasonication	-	$(NH4)_2SO_4$	2-propanol	Protein	53.16	-
Haematococcus pluvialis [208]	Mechanical cell disruption	-	Na_2CO_3	2-propanol	Astaxanthin	104.28	-
Spirulina platensis [205]	Ultrasonication	PEG 4000	$(NH4)_2SO_4$	-	C-PC	93.50	1.63
Spirulina platensis [68]	Sonication	PEG 4000	K_2HPO_4	-	C-PC	94.89	6.17

PEG: polyethylene glycol. C-PC: C-phycocyanin.

As shown in Table 9, LBF is always associated with a previous pretreatment step. However, it is possible to integrate electric cell disruption and LBF in just one step (i.e., extraction followed by purification), the so-called liquid biphasic electric flotation (LBEF) [209]. Koyande et al. [210] were able to increase separation efficiency (nearly 10%) and protein recovery (nearly 20%) by replacing LBF with LBEF in protein extraction from *C. vulgaris*.

Nevertheless, other adaptations of the methodology have been explored, such as the use of ionic liquids (ILs) [211] or the combination of non-ionic surfactants and ionic liquids [212] in the biphasic systems formation.

5.5. Three-Phase Partitioning

Three-phase partitioning (TPP) is a simple method for protein purification. Briefly, t-butanol and ammonium sulfate are added to an aqueous solution containing proteins. The ammonium sulfate promotes proteins precipitation and after mild centrifugation, proteins are concentrated in an intermediate zone between two immiscible phases: t-butanol on top, and water in the bottom [213]. t-butanol is usually miscible in water, but in the presence of a certain amount of salt, such as ammonium sulfate, it becomes immiscible. In TPP, t-butanol increases protein buoyancy by acting as a co-precipitating agent and assisting in the triphasic formation [214]. Recently, Chia et al. [9] used a combination of TPP and LBF (method described in 5.4), the liquid triphasic flotation system (LTF), for protein recovery from *Chlorella vulgaris* biomass. In this last approach, LTF, the three phases are formed by a mixture of microalgal biomass, salt solution and t-butanol. The proteins are concentrated in the intermediate zone as well as in TPP and the air bubbles, also used in LBF, facilitate its extraction, reducing process time. In this work, they achieved approximately 90% of recovery and above 60% of separation efficiency.

5.6. Ammonium Sulfate Precipitation

The ammonium sulfate precipitation is a very well-established method applied to protein purification. By adding a salt into an aqueous solution containing the proteins, the salt dissociates and the water that once offered a great solvating power to proteins, solvates the dissociated ions preferably. Thus, the charges of the protein molecules tend to interact more, forming aggregates and precipitating, an effect called salting out. In addition, ammonium sulfate is usually chosen for this purpose due to its high water solubility and low cost [215]. Silva et al. [216] performed an experimental design to optimize some parameters (i.e., $(NH_4)_2SO_4$ concentration, volume and pH) in an ammonium sulfate precipitation protocol for C-PC purification from *Spirulina platensis*. The highest C-PC purity factor achieved was 0.88 at 0–20%/20–50% $(NH_4)_2SO_4$ saturation, 0.52 (resuspension volume/initial volume) and pH 7.0, with 83.8% of recovery. However, this methodology is generally used as a pre-purification step and it is often associated with chromatographic methods [217,218] or membrane dialysis [219,220]. Muensean and Kim [221] tested the purification of β-glucosidase, from *Trichoderma citrinoviride* produced by *Chlorella vulgaris*, in three steps. First, they performed the ammonium sulfate precipitation and then passed the sample through a DEAE-Sepharose FF column. Finally, the enzymatic-concentrated sample was loaded onto a Sephacryl S–100 HR column. As expected, yield decreases with each step, but the specific activity of the enzyme and its purification increased more than 30-fold (5.53 to 168.74 $U \cdot mg^{-1}$ and 4.55 to 138.72, respectively). Later, Chen et al. [222] evaluated the purification of C-PC from *S. platensis* by ammonium sulfate precipitation, ion exchange chromatography and the combination of both. The best recovery was obtained with precipitation only ($91 \pm 2\%$), while the highest purity was obtained by combining both techniques (4.33 ± 0.10), in which the purity achieved with ammonium sulfate was 1.92 ± 0.06 and with chromatography 3.70 ± 0.06.

Table 10 summarizes the main advantages and disadvantages of several separation and purification methods described in Section 4. Among all advantages and disadvantages of each method, the chromatographic methods are still widely used at the laboratory scale. On the other hand, technologies based on multi-phase fractioning (e.g., aqueous two-phase and three-phase partitioning) despite being scalable demand large amounts of chemicals, generating environmental issues.

Table 10. Main advantages and disadvantages of several methods for biomolecules separation and purification.

Separation/Purification Method	Advantages	Disadvantages	References
Supercritical fluid chromatography	- Lower viscosity, higher diffusivity and high efficiency - Fast time analysis and lower pressure drop compared to liquid chromatography - Can separate thermally labile compounds - Can use almost all HPLC and GC detectors	- Polarity of the mobile phase is limiting factor (supercritical fluids is incapable of solubilizing highly polar solutes)	[223,224]
Gel permeation chromatography (GPC)	- Rapid analysis time - Requires small amounts of sample - Suitable for separation of molecules with high MW (>2000)	- Requires previous steps of sample preparation - GPC can only resolve solutes with molecular size differing in 10% at least. - High equipment cost	[147,223]
Ion-exchange chromatography	- Applicable for several molecules - High purity levels with high efficiency	- Does not provide direct information of stationary phase surface's events (response conditioned to the interaction between solute, eluent and active sites of resin)	[225]
Thin-layer chromatography	- Short time and simple operation/cheap technology - Can be used for direct analysis of crude samples - Performed at room temperature	- Susceptible to polyunsaturated fatty acids oxidation (open system) - Low separation efficiency	[129,160,161,226]
High-performance chromatography	- Quantitative analysis with accuracy and precision - Great reproducibility and selectivity - Resolution of complex separations - Not limited by volatility or thermal stability of the sample - Sample recovery is relatively simple	- Lower separation efficiency than capillary GC - No universal detector - Many operating parameters (more complex operation)	[129,227]

Table 10. *Cont.*

Separation/Purification Method	Advantages	Disadvantages	References
Gas chromatography	- High resolution of components with short time analysis - Capable of analyze low amounts of sample (μL) - Reliable, relatively simple and low cost comparing to other other chromatographic technologies	- Limited to volatile compounds - Commonly needs to be associated with mass spectroscopy - Not suitable for most polar, thermolabile or high molecular weight compounds - Not suitable for large-scale	[129,186]
Ultrafiltration	- Mild temperature conditions - Less energy consumption than other membrane technologies - High retention of macromolecules	- Conventional UF have a limitation of separation of solutes with 10-fold difference in size - Susceptible to membrane fouling	[196,198]
Electromembrane filtration	- More selective and with reduced fouling formation compared to traditional membrane filtration - Does not require pressure	- Requires additional energy	[199,200]
Ultracentrifugation	- Allows the separation of small biomolecules	- Time consuming and sample loss - Expensive equipment that requires regular maintenance - Incapable of processing various samples simultaneously	[228]
Aqueous two-phase systems	- Low energy consumption and short processing time - High extraction efficiency - Simple operation and feasible for large-scale	- Requires large amounts of polymer in large scale - Environmental issues even recycling the phase-forming polymer	[203,205,211]

Table 10. *Cont.*

Separation/Purification Method	Advantages	Disadvantages	References
Three-phase partitioning	- Simple operation with short processing time - Can be performed at room temperature - Mild conditions avoid protein denaturation - Purify and concentrate proteins (unlike chromatographic methods, that dilute them) - Achieve high purity levels - Can be used directly in cultures containing cell debris without previous pretreatment - Non expensive and feasible for large-scale	- Requires organic solvents - Environmental concerns	[9,213]
Ammonium sulfate precipitation	- Low-cost technology with simple operation - Feasible for large-scale	- Commonly used as a pre-purification method, frequently followed by chromatographic methods or membrane filtration to achieve high purity levels	[216,218–220]

MW: molecular weight.

6. Conclusions

Microalgal biomass is a rich source of biomolecules with applications in several industry sectors, such as food, pharmaceuticals and cosmetics. However, like any biotechnological process, it is necessary to establish strategies to make it economically viable. Recent developments using supercritical fluids, ionic liquid and deep eutectic solvents have shown potential substitutes for traditional extractive methods based on organic solvents. Furthermore, physicochemical treatments, such as supercritical fluid extraction, ultrasound-assisted extraction and pulsed electric fields are promising technologies for large-scale implementation, presenting high efficiency and producing extracts with a high purity level. In particular, the use of PEF treatments can be applied in both upstream, as a growth stimulator, and downstream processes to enhance extraction yield. Concerning biomolecule separation and purification, chromatographic methods are widely applied in laboratory scale, although, improvements in non-chromatographic methods based on multiphase systems have been extending their application and turned them scalable. However, the demand for large amounts of polymer or organic solvent is still the biggest drawback of these methods. Thus, greener technologies that combine high efficiency, selectivity and low energy demand, such as supercritical fluid extraction, are the preferable techniques applied in large-scale microalgae processes. Moreover, the recent developments applied to microalgal biomass have demonstrated not only the high value of microalgal biomass as feedstock for new biorefineries, but also the viability of a new bio-based economy that is more environmentally friendly and more sustainable. Despite the development of competitive bioproducts on the market still being a challenge and requiring great dedication from research centers, the transition from chemicals to bio-based products is an unquestionable trend worldwide.

Author Contributions: Conceptualization, P.S.C., W.G.M.J., T.M.M., N.S.C. and A.A.M.; data curation, P.S.C., W.G.M.J., T.M.M. and N.S.C.; writing—original draft preparation, P.S.C.; writing—review and editing, P.S.C., W.G.M.J., T.M.M., N.S.C. and A.A.M.; visualization, P.S.C., T.M.M. and N.S.C.; supervision, T.M.M., N.S.C. and A.A.M.; project administration, T.M.M., N.S.C.; funding acquisition, T.M.M., N.S.C. All authors have read and agreed to the published version of the manuscript.

Funding: This work was financially supported by Base Funding—UIDB/00511/2020 of Laboratory for Process Engineering, Environment, Biotechnology and Energy—LEPABE—funded by National funds through the FCT/MCTES (PIDDAC); Base Funding—UIDB/04730/2020 of Center for Innovation in Engineering and Industrial Technology, CIETI—funded by national funds through the FCT/MCTES (PIDDAC); project IF/01093/2014/CP1249/CT0003 funded by national funds through FCT/MCTES; project "EXTRATOTECA—Microalgae Extracts for High Value Products"—POCI-01-0247-FEDER-033784, funded by FEDER funds through COMPETE2020—Programa Operacional Competitividade e Internacionalização (POCI) and by national funds (PIDDAC) through FCT/MCTES.

Acknowledgments: António Martins thanks FCT (Fundação para a Ciência e Tecnologia) for funding through program DL 57/2016—Norma transitória. Wilson Júnior thanks European Union's Horizon 2020 research and innovation programme under the Marie Skłodowska-Curie grant agreement number 867473.

Conflicts of Interest: The authors declare no conflict of interest.

References

1. Morais Junior, W.G.; Gorgich, M.; Corrêa, P.S.; Martins, A.A.; Mata, T.M.; Caetano, N.S. Microalgae for biotechnological applications: Cultivation, harvesting and biomass processing. *Aquaculture* **2020**, *528*, 735562. [CrossRef]
2. Mata, T.M.; Martins, A.A.; Caetano, N.S. Microalgae for biodiesel production and other applications: A review. *Renew. Sustain. Energy Rev.* **2010**, *14*, 217–232. [CrossRef]
3. Mata, T.M.; Martins, A.A.; Caetano, N.S. Microalgae processing for biodiesel production. In *Advances in Biodiesel Production: Processes and Technologies*, 1st ed.; Luque, R., Melero, J.A., Eds.; Woodhead Publishing: Cambridge, UK; Elsevier Ltd.: Amsterdam, The Netherlands, 2012; pp. 204–231.
4. Mata, T.M.; Martins, A.A.; Sikdar, S.K.; Costa, C.A.V.; Caetano, N.S. Sustainability considerations about microalgae for biodiesel production. In *Advanced Biofuels and Bioproducts*, 1st ed.; Lee, J.W., Ed.; Springer: New York, NY, USA, 2013; pp. 745–757. [CrossRef]

5. Santos, R.R.; Corrêa, P.S.; Dantas, F.M.L.; Teixeira, C.M.L.L. Evaluation of the Co-production of Total Carotenoids, C-phycocyanin and Polyhydroxyalkanoates by *Arthrospira platensis*. *Bioresour. Technol. Rep.* **2019**, *7*, 100226. [CrossRef]
6. Park, S.; Nguyen, T.H.T.; Jin, E. Improving Lipid Production by Strain Development in Microalgae: Strategies, Challenges and Perspectives. *Bioresour. Technol.* **2019**, *292*, 121953. [CrossRef]
7. Geresh, S.; Adin, I.; Yarmolinsky, E.; Karpasas, M. Characterization of the Extracellular Polysaccharide of *Porphyridium* sp.: Molecular Weight Determination and Rheological Properties. *Carbohydr. Polym.* **2002**, *50*, 183–189. [CrossRef]
8. Shrivastav, A.; Mishra, S.K.; Mishra, S. Polydroxyalkanoate (PHA) Synthesis by *Spirulina subsalsa* from Gujarat Coast of India. *Int. J. Biol. Macromol.* **2010**, *46*, 255–260. [CrossRef] [PubMed]
9. Chia, S.R.; Mak, K.Y.; Khaw, Y.J.; Suhaidi, N.; Chew, K.W.; Show, P.L. An efficient and rapid Method to Extract and Purify Protein–Liquid Triphasic Flotation System. *Bioresour. Technol.* **2019**, *294*, 122158. [CrossRef] [PubMed]
10. Edelmann, M.; Chamlagain, S.A.B.; Kariluoto, S.; Piironen, V. Riboflavin, Niacin, Folate and Vitamin B12 in Commercial Microalgae Powders. *J. Food Compos. Anal.* **2019**, *82*, 103226. [CrossRef]
11. Liu, L.; Pohnert, G.; Wei, D. Extracellular metabolites from industrial microalgae and their biotechnological potential. *Mar. Drugs* **2016**, *14*, 191. [CrossRef] [PubMed]
12. Chew, K.W.; Yap, J.Y.; Show, P.L.; Suan, N.H.; Juan, J.C.; Ling, T.C.; Lee, D.; Chang, J. Microalgae biorefinery: High value products perspectives. *Bioresour. Technol.* **2017**, *229*, 53–62. [CrossRef] [PubMed]
13. Alhattab, M.; Kermanshahi-Pour, A.; Brooks, M.S. Microalgae disruption techniques for product recovery: Influence of cell wall composition. *J. Appl. Phycol.* **2019**, *31*, 61–88. [CrossRef]
14. Agboola, J.O.; Teuling, E.; Wierenga, P.A.; Gruppen, H.; Schrama, J.W. Cell wall disruption: An effective strategy to improve the nutritive quality of microalgae in African catfish (*Clarias gariepinus*). *Aquac. Nutr.* **2019**, *25*, 783–797. [CrossRef]
15. Larrosa, A.P.Q.; Camara, A.S.; Moura, J.M.; Pinto, L.A.A. *Spirulina* sp. Biomass Dried/disrupted by Different Methods and their Application in Biofilms Production. *Food Sci. Biotechnol.* **2018**, *27*, 1659–1665. [CrossRef] [PubMed]
16. Gim, H.; Kim, S.W. Optimization of Cell Disruption and Transesterification of Lipids from *Botryococcus braunii* LB572. *Biotechnol. Bioprocess Eng.* **2018**, *23*, 550–556. [CrossRef]
17. Martínez, J.M.; Gojkovic, Z.; Ferro, L.; Maza, M.; Álvarez, I.; Raso, J.; Funk, C. Use of Pulsed Electric Field Permeabilization to Extract Astaxanthin from the Nordic Microalgae *Haematococcus pluvialis*. *Bioresour. Technol.* **2019**, *289*, 121694. [CrossRef]
18. Rito-Palomares, M. Bioseparation: The limiting step in bioprocess development. *J. Chem. Technol. Biotechnol.* **2008**, *83*, 115–116. [CrossRef]
19. Yu, L.; Wu, F.; Chen, G. Next-Generation Industrial Biotechnology–Transforming the Current Industrial Biotechnology into Competitive Processes. *Biotechnol. J.* **2019**, *14*, 1800437. [CrossRef]
20. Wang, M.; Chen, S.; Zhou, W.; Yuan, W.; Wang, D. Algal cell lysis by bacteria: A review and comparison to conventional methods. *Algal Res.* **2020**, *46*, 101794. [CrossRef]
21. Günerken, E.; D'Hondt, E.; Eppink, M.H.M.; Garcia-Gonzalez, L.; Elst, K.; Wijffels, R.H. Cell disruption for microalgae biorefineries. *Biotech. Adv.* **2015**, *33*, 243–260. [CrossRef]
22. Toursel, B.; Delannoy, F.; Patinier, S. Optimised Method for Breaking Chlorella Walls by Mechanical Crushing. U.S. Patent 10465159B2, 5 November 2019.
23. D'Hondt, E.D.; Martín-Juárez, J.; Kasperoviciene, J.; Koreiviene, J.; Sulcius, S.; Elst, K.; Bastiaens, L. Cell Disruption Technologies. In *Microalgae-Based Biofuels and Bioproducts–From Feedstock Cultivation to End-Products*; Gonzalez-Fernandez, C., Muñoz, R., Eds.; Woodhead Publishing Series in Energy: Duxford, UK, 2017; pp. 133–154. [CrossRef]
24. Balasubramanian, R.K.; Doan, T.T.Y.; Obbard, J.P. Factors affecting cellular lipid extraction from marine microalgae. *Chem. Eng. J.* **2013**, *215–216*, 929–936. [CrossRef]
25. Guedes, A.C.; Amaro, H.M.; Gião, M.S.; Malcata, F.X. Optimization of ABTS radical cation assay specifically for determination of antioxidant capacity of intracellular extracts of microalgae and cyanobacteria. *Food Chem.* **2013**, *138*, 638–643. [CrossRef]
26. González-Delgado, A.D.; Kafarov, V. Microalgae based biorefinery: Evaluation of oil extraction methods in terms of efficiency, costs, toxicity and energy in lab-scale. *Rev. Ion.* **2013**, *26*, 29–37.
27. Gomes, T.A.; Zanette, C.M.; Spier, M.R. An overview of cell disruption methods for intracellular biomolecules recovery. *Prep. Biochem. Biotech.* **2020**, *50*, 635–654. [CrossRef]
28. Bernaerts, T.M.M.; Greysen, L.; Foubert, I.; Hendrickx, M.E.; Loey, A.M.V. Evaluating microalgal cell disruption upon ultra high pressure homogenization. *Algal Res.* **2019**, *42*, 101616. [CrossRef]
29. Elain, A.; Nkounkou, C.; Fellic, M.; Donnart, K. Green extraction of polysaccharides from *Arthrospira platensis* using high pressure homogenization. *J. Appl. Phycol.* **2020**, *32*, 1719–1727. [CrossRef]
30. Piasecka, A.; Krzemińska, I.; Tys, J. Physical methods of microalgal biomass pretreatment. *Int. Agrophys.* **2014**, *38*, 341–348. [CrossRef]
31. Zghaibi, N.; Omar, R.; Kamal, S.M.M.; Biak, D.R.A.; Harun, R. Microwave-Assisted Brine Extraction for Enhancement of the Quantity and Quality of Lipid Production from Microalgae *Nannochloropsis* sp. *Molecules* **2019**, *24*, 3581. [CrossRef] [PubMed]
32. Lee, J.; Yoo, C.; Jun, S.; Ahn, C.; Oh, H. Comparison of several methods for effective lipid extraction from microalgae. *Bioresour. Technol.* **2010**, *101*, S75–S77. [CrossRef]
33. Dai, Y.; Chen, K.; Chen, C. Study of the microwave lipid extraction from microalgae for biodiesel production. *Chem. Eng. J.* **2014**, *250*, 267–273. [CrossRef]

34. Viner, K.J.; Champagne, P.; Jessop, P.G. Comparison of cell disruption techniques prior to lipid extraction from *Scenedesmus* sp. slurries for biodiesel production using liquid CO_2. *Green Chem.* **2018**, *20*, 4330. [CrossRef]
35. Fan, Y.; Niu, Z.; Xu, C.; Yang, L.; Chen, F.; Zhang, H. Biocompatible protic ionic liquids-based microwave-assisted liquid-solid extraction of astaxanthin from *Haematococcus pluvialis*. *Ind. Crop. Prod.* **2019**, *141*, 111809. [CrossRef]
36. Krishnan, S.; Ghani, N.A.; Aminuddin, N.F.; Quraishi, K.S.; Azman, N.S.; Cravotto, G.; Leveque, J. Microwave-assisted lipid extraction from *Chlorella vulgaris* in water with 0.5%–2.5% of imidazolium based ionic liquid as additive. *Renew. Energ.* **2020**, *149*, 244–252. [CrossRef]
37. Adam, F.; Abert-Vian, M.; Peltier, G.; Chemat, F. "Solvent-free" ultrasound-assisted extraction of lipids from fresh microalgae cells: A green, clean and scalable process. *Bioresour. Technol.* **2012**, *114*, 457–465. [CrossRef] [PubMed]
38. Martínez, J.M.; Delso, C.; Álvarez, I.; Raso, J. Pulsed electric field-assisted extraction of valuable compounds from microorganisms. *Compr. Rev. Food Sci. Saf.* **2020**, *19*, 530–552. [CrossRef] [PubMed]
39. Lafarga, T. Cultured microalgae and compounds derived thereof for food applications: Strain selection and cultivation, drying, and processing strategies. *Food Rev. Int.* **2020**, *36*, 559–583. [CrossRef]
40. Toepfl, S.; Heinz, V.; Knorr, D. Applications of pulsed electric fields technology for food industry. In *Pulsed Electric Fields Technology for the Food Industry*; Raso, J., Heinz, V., Eds.; Food Engineering Series; Springer: Boston, MA, USA, 2006; pp. 197–221. [CrossRef]
41. Arshad, R.N.; Abdul-Malek, Z.; Munir, A.; Buntat, Z.; Ahmad, M.H.; Jusoh, Y.M.M.; Bekhit, A.E.; Roobab, U.; Manzoor, M.F.; Aadil, R.M. Electrical systems for pulsed electric field applications in the food industry: An engineering perspective. *Trends Food Sci. Technol.* **2020**, *104*, 1–13. [CrossRef]
42. Lam, G.P.; Postma, P.R.; Fernandes, D.A.; Timmermans, R.A.H.; Vermuë, M.H.; Barbosa, M.J.; Eppink, M.H.M.; Wijffels, R.H.; Olivieri, G. Pulsed electric field for protein release of the microalgae *Chlorella vulgaris* and *Neochloris oleoabundans*. *Algal Res.* **2017**, *24*, 181–187. [CrossRef]
43. Käferböck, A.; Smetana, S.; Vos, R.; Schwarz, C.; Toepfl, S.; Parniakov, O. Sustainable extraction of valuable components from *Spirulina* assisted by pulsed electric fields technology. *Algal Res.* **2020**, *48*, 101914. [CrossRef]
44. Zhang, R.; Lebovka, N.; Marchal, L.; Vorobiev, E.; Grimi, N. Pulsed electric energy and ultrasonication assisted green solvent extraction of bio-molecules from different microalgal species. *Innov. Food Sci. Emerg. Technol.* **2020**, *62*, 102358. [CrossRef]
45. McMillan, J.R.; Watson, I.A.; Ali, M.; Jaafar, W. Evaluation and comparison of algal cell disruption methods: Microwave, waterbath, blender, ultrasonic and laser treatment. *Appl. Energy* **2013**, *103*, 128–134. [CrossRef]
46. Lorente, E.; Farriol, X.; Salvadó, J. Steam explosion as a fractionation step in biofuel production. *Fuel Process Technol.* **2015**, *131*, 93–98. [CrossRef]
47. Lorente, E.; Hapońska, M.; Clavero, E.; Torras, C.; Salvadó, J. Microalgae fractionation using steam explosion, dynamic and tangential cross-flow membrane filtration. *Bioresour. Technol.* **2017**, *237*, 3–10. [CrossRef] [PubMed]
48. Nurra, C.; Torras, C.; Clavero, E.; Ríos, S.; Rey, M.; Lorente, E.; Farriol, X.; Salvadó, J. Biorefinery concept in a microalgae pilot plant. Culturing, dynamic filtration and steam explosion fractionation. *Bioresour. Technol.* **2014**, *163*, 136–142. [CrossRef] [PubMed]
49. Al hattab, M.; Ghaly, A. Microalgae oil extraction pre-treatment methods: Critical review and comparative analysis. *J. Fundam. Renew. Energy Appl.* **2015**, *5*, 172. [CrossRef]
50. Kapoore, R.V.; Butler, T.O.; Pandhal, J.; Vaidyanathan, S. Microwave-assisted extraction for microalgae: From biofuels to biorefinery. *Biology* **2018**, *7*, 18. [CrossRef]
51. Singh, R.; Kumar, A.; Sharma, Y.C. Biodiesel Synthesis from Microalgae (*Anabaena* PCC 7120) by Using Barium Titanium Oxide (Ba_2TiO_4) Solid Base Catalyst. *Bioresour. Technol.* **2019**, *287*, 121357. [CrossRef]
52. Castro, Y.A.; Ellis, J.T.; Miller, C.D.; Sims, R.C. Optimization of Wastewater Microalgae Saccharification Using Dilute Acid Hydrolysis for Acetone, Butanol, and Ethanol Fermentation. *Appl. Energy* **2015**, *140*, 14–19. [CrossRef]
53. Juárez, J.M.; Hernando, A.L.; Torre, R.M.; Lanza, S.B.; Rodríguez, S.B. Saccharification of Microalgae Biomass Obtained from Wastewater Treatment by Enzymatic Hydrolysis. Effect of Alkaline-Peroxide Pretreatment. *Bioresour. Technol.* **2016**, *218*, 265–271. [CrossRef]
54. Huang, W.; Kim, J. Cationic Surfactant-Based Method for Simultaneous Harvesting and Cell Disruption of Microalgal Biomass. *Bioresour. Technol.* **2013**, *149*, 579–581. [CrossRef]
55. Liang, K.; Zhang, Q.; Cong, W. Enzyme-Assisted Aqueous Extraction of Lipid from Microalgae. *J. Agric. Food Chem.* **2012**, *60*, 11771–11776. [CrossRef]
56. Zhang, Y.; Kong, X.; Wang, Z.; Sun, Y.; Zhu, S.; Li, L.; Lv, P. Optimization of Enzymatic Hydrolysis for Effective Lipid Extraction from Microalgae *Scenedesmus* sp. *Renew. Energy* **2018**, *125*, 1049–1057. [CrossRef]
57. Zhou, Y.; Lai, Y.S.; Eustance, E.; Rittmann, B.E. Promoting *Synechocystis* sp. PCC 6803 Harvesting by Cationic Surfactants: Alkyl-chain Length and Dose Control the Release of Extracellular Polymeric Substances and Biomass Aggregation. *ACS Sustain. Chem. Eng.* **2018**, *7*, 2127–2133. [CrossRef]
58. Lai, Y.S.; Zhou, Y.; Martarella, R.; Wang, Z.; Rittmann, B.E. Synergistic Integration of C12-C16 Cationic Surfactants for Flocculation and Lipid Extraction from *Chlorella* Biomass. *ACS Sustain. Chem. Eng.* **2017**, *5*, 752–757. [CrossRef]
59. Alhattab, M.; Kermanshahi-pour, A.; Brooks, S. Dispersed Air Flotation of *Chlorella saccharophila* and subsequent Extraction of Lipids–Effect of Supercritical CO_2 Extraction Parameters as Surfactant Pretreatment. *Biomass Bioenerg.* **2019**, *127*, 105297. [CrossRef]

50. Rakesh, S.; Dhar, D.W.; Prasanna, R.; Saxena, A.K.; Saha, S.; Shukla, M.; Sharma, K. Cell disruption methods for improving lipid extraction efficiency in unicellular microalgae. *Eng. Life Sci.* **2015**, *15*, 443–447. [CrossRef]
51. González-González, L.M.; Astals, S.; Pratt, S.; Jensen, P.D.; Schenk, P.M. Impact of osmotic shock pre-treatment on microalgae lipid extraction and subsequent methane production. *Bioresour. Technol. Rep.* **2019**, *7*, 100214. [CrossRef]
52. López, B.H.; Morales, J.M.V. Extraction of antioxidants from *Haematococcus pluvialis* by osmotic shock with sucrose, for use in the preparation of syrup for jams. In Proceedings of the XIV International Engineering Congress (CONIIN), Queretaro, Mexico, 14–19 May 2018; pp. 1–5. [CrossRef]
53. Koyande, A.K.; Tanzil, V.; Dharan, H.M.; Subramaniam, M.; Robert, R.N.; Lau, P.; Khoiroh, I.; Show, P. Integration of osmotic shock assisted liquid biphasic system for protein extraction from microalgae *Chlorella vulgaris*. *Biochem. Eng. J.* **2020**, *157*, 107532. [CrossRef]
54. Heo, Y.M.; Lee, H.; Lee, C.; Kang, J.; Ahn, J.; Lee, Y.M.; Kang, K.; Choi, Y.; Kim, J. An integrative process for obtaining lipids and glucose from *Chlorella vulgaris* biomass with a single treatment of cell disruption. *Algal Res.* **2017**, *27*, 286–294. [CrossRef]
55. Mercer, P.; Armenta, R.E. Developments in oil extraction from microalgae. *Eur. J. Lipid Sci. Technol.* **2011**, *113*, 539–547. [CrossRef]
56. Zheng, Y.; Xiao, R.; Roberts, M. Polymer-enhanced enzymatic microalgal cell disruption for lipid and sugar recovery. *Algal Res.* **2016**, *14*, 100–108. [CrossRef]
57. Bai, M.; Han, C.; Lu, W.; Wan, H.; Chang, J.; Chen, C.; Chang, H. Method for Manufacturing an Active Substance for Inducing Self-Lysis in Microalga Cells, Active Substance Obtained Therefrom, and Method for Inducing Self-Lysis in Microalga Cells. U.S. Patent 10196600B2, 5 February 2019.
58. Chia, S.R.; Chew, K.W.; Show, P.L.; Xia, A.; Ho, S.; Lim, J.W. *Spirulina platensis* based biorefinery for the production of value-added products for food and pharmaceutical applications. *Bioresour. Technol.* **2019**, *289*, 121727. [CrossRef] [PubMed]
59. Van Eykelenburg, C. *Spirulina platensis*: Morphology and Ultrastructure. Ph.D. Thesis, Delft University, Delft, The Netherlands, 1980. Available online: https://repository.tudelft.nl/islandora/object/uuid%3A744c0474-507a-42c7-b29b-4139f7cb4f22 (accessed on 20 September 2020).
70. Gouveia, J.D.G. The Effects of Light Quality on the Morphology and Hydrocarbon Production of Botryococcus braunii. Master's Thesis, University of Exeter, Exeter, UK, September 2010. Available online: https://core.ac.uk/download/pdf/12826395.pdf (accessed on 20 September 2020).
71. Wang, S.; Hu, Q.; Sommerfeld, M.; Chen, F. Cell wall proteomics of the green alga *Haematococcus pluvialis* (Chlorophyceae). *Proteomics* **2004**, *4*, 692–708. [CrossRef] [PubMed]
72. Yuan, T.; Li, X.; Xiao, S.; Guo, Y.; Zhou, W.; Xu, J.; Yuan, Z. Microalgae Pretreatment with Liquid Hot Water to Enhance Enzymatic Hydrolysis Efficiency. *Bioresour. Technol.* **2016**, *220*, 530–536. [CrossRef] [PubMed]
73. Zhou, J.; Zhang, F.; Meng, H.; Bao, G.; Zhang, Y.; Li, Y. Development of a Silicon Carbide Disruption Method Enables Efficient Extraction of Proteins from Cyanobacterium *Synechocystis* sp. PCC 6803. *Process Biochem.* **2014**, *49*, 2199–2202. [CrossRef]
74. Woitzik, D.; Weckesser, J.; Jürgens, U. Isolation and characterization of cell wall components of the unicellular cyanobacterium *Synechococcus* sp. PCC 6307. *J. Gen. Microbiol.* **1988**, *134*, 619–627. [CrossRef]
75. Gille, A.; Hollenbach, R.; Trautmann, A.; Posten, C.; Brivida, K. Effect of Sonication on Bioaccessibility and Cellular Uptake of Carotenoids from Preparations of Photoautotrophic *Phaeodactylum tricornutum*. *Food Res. Int.* **2019**, *118*, 40–48. [CrossRef]
76. Serive, B.; Kaas, R.; Bérard, J.; Pasquet, V.; Picot, L.; Cadoret, J. Selection and Optimisation of a Method for Efficient Metabolites Extraction from Microalgae. *Bioresour. Technol.* **2012**, *124*, 311–320. [CrossRef]
77. De Martino, A.; Meichenin, A.; Shi, J.; Pan, K.; Bowler, C. Genetic and Phenotypic Characterization of *Phaeodactylum tricornutum* (Bacillariophyceae) Accessions. *J. Phycol.* **2007**, *43*, 992–1009. [CrossRef]
78. Ramos, G.J.P.; Bicudo, C.E.M.; Neto, A.G.; Moura, C.W.N. *Monoraphidium* and *Ankistrodesmus* (Chlorophyceae, Chlorophyta) from Pantanal dos Marimbus, Chapada Diamantina, Bahia State, Brazil. *Hoehnea* **2012**, *39*, 421–434. [CrossRef]
79. Imam, S.H.; Snell, W.J. The *Chlamydomonas* cell wall degrading enzyme, lysine, acts on two substrates within the framework of the wall. *J. Cell Biol.* **1988**, *106*, 2211–2221. [CrossRef]
80. Wilson, D.P. The triradiate and other forms of *Nitzschia closterium* (Ehrenberg) Wm. Smith, forma *minutissima* of Allen and Nelson. *J. Mar. Biol. Assoc. UK* **1946**, *26*, 235–270. [CrossRef]
81. Patel, A.; Mikes, F.; Matsakas, L. An overview of current pretreatment methods used to improve lipid extraction from oleaginous microorganisms. *Molecules* **2018**, *23*, 1562. [CrossRef] [PubMed]
82. Pora, B.; Qian, Y.; Caulier, B.; Comini, S.; Looten, P.; Segueilha, L. Method for the Preparation and Extraction of Squalene from Microalgae. U.S. Patent 10087467B2, 2 October 2018.
83. Zanella, L.; Pertile, P.; Massironi, M.; Massironi, M.; Caviola, E. Extracts of Microalgae and their Applications. U.S. Patent 9974819B2, 22 May 2018.
84. Massetti, F.; Capuano, F.; Medici, R.; Miglio, R. Process for the Extraction of Lipids and Sugars from Algal Biomass. U.S. Patent 10336965B2, 2 July 2019.
85. Reddy, C.M.; O'Neil, G.W. Use of Marine Algae for co-Producing Alkenones, Alkenone Derivatives and Co-Products. U.S. Patent 10208321B2, 19 February 2019.
86. Nakazawa, A.; Matsuura, H.; Kose, R.; Kato, S.; Honda, D.; Inouye, I.; Kaya, K.; Watanabe, M. Optimization of culture conditions of the thraustochytrid *Aurantiochytrium* sp. strain 18W-13a for squalene production. *Bioresour. Technol.* **2012**, *109*, 287–291. [CrossRef] [PubMed]

87. Kaya, K.; Nakazawa, A.; Matsuura, H.; Honda, D.; Inouye, I.; Watanabe, M. Thraustochytrid *Aurantiochytrium* sp. 18W-13a accumulates high amounts of squalene. *Biosci. Biotechnol. Biochem.* **2011**, *75*, 2246–2248. [CrossRef] [PubMed]
88. Hoang, M.H.; Ha, N.C.; Thom, L.T.; Tam, L.T.; Anh, H.T.L.; Thu, N.T.H.; Hong, D.D. Extraction of squalene as value-added product from the residual biomass of *Schizochytrium mangrovei* PQ6 during biodiesel producing process. *J. Biosci. Bioeng.* **2014**, *118*, 632–639. [CrossRef]
89. Patel, A.; Liefeldt, S.; Rova, U.; Christakopoulos, P.; Matsakas, L. Co-production of DHA and squalene by Thraustochytrid from forest biomass. *Sci. Rep.* **2020**, *10*, 1992. [CrossRef]
90. Li, Q.; Chen, G.; Fan, K.; Lu, F.; Aki, T.; Jiang, Y. Screening and Characterization of Squalene-producing Thraustochytrids from Hong Kong Mangroves. *J. Agric. Food Chem.* **2009**, *57*, 4267–4272. [CrossRef]
91. Du, Y.; Schuur, B.; Brilman, D.W.F. Maximizing lipid yield in *Neochloris oleoabundans* algae extraction by stressing and using multiple extraction stages with n-ethlylbutylamine as switchable solvent. *Ind. Eng. Chem. Res.* **2017**, *56*, 8073–8080. [CrossRef]
92. Obeid, S.; Beaufils, N.; Camy, S.; Takache, H.; Ismail, A.; Pontalier, P. Supercritical carbon dioxide extraction and fractionation of lipids from freeze-dried microalgae *Nannochloropsis oculata* and *Chlorella vulgaris*. *Algal Res.* **2019**, *34*, 49–56. [CrossRef]
93. Hernández, D.; Solana, M.; Riaño, B.; García-González, M.C.; Bertucco, A. Biofuels from microalgae: Lipid extraction and methane production from residual biomass in a biorefinery approach. *Bioresour. Technol.* **2014**, *170*, 370–378. [CrossRef]
94. Leone, G.P.; Balducchi, R.; Mehariya, S.; Martino, M.; Larocca, V.; Di Sanzo, G.; Lovine, A.; Casella, P.; Marino, T.; Karatza, D.; et al. Selective Extraction of ω-3 Fatty Acids from *Nanochloropsis* sp. Using Supercritical CO_2 Extraction. *Molecules* **2019**, *24*, 2406. [CrossRef] [PubMed]
95. Bligh, E.G.; Dyer, W.J. A rapid method of total lipid extraction and purification. *Can. J. Biochem. Physiol.* **1959**, *37*, 911–917. [CrossRef] [PubMed]
96. Folch, J.; Lees, M.; Stanley, G.H.S. A simple method for the isolation and purification of total lipids from animal tissues. *J. Biol. Chem.* **1956**, *226*, 497–509.
97. Kumar, R.R.; Rao, P.H.; Arumugan, M. Lipid extraction methods from microalgae: A comprehensive review. Front. *Energy Res.* **2015**, *2*, 1–9. [CrossRef]
98. Guldhe, A.; Singh, B.; Ansari, F.A.; Sharma, Y.; Bux, F. Extraction and conversion of microalgal lipids. In *Algae Biotechnology*; Bux, F., Chisti, Y., Eds.; Green Energy and Technology; Springer: New York, NY, USA, 2016; pp. 91–110. [CrossRef]
99. Breil, C.; Vian, M.A.; Zemb, T.; Kunz, W.; Chemat, F. "Bligh and Dyer" and Folch methods for solid-liquid-liquid extraction of lipids from microorganisms. Comprehension of solvatation mechanisms and towards substitution with alternatives solvents. *Int. J. Mol. Sci.* **2017**, *18*, 708. [CrossRef] [PubMed]
100. Gorgich, M.; Mata, T.M.; Martins, A.A.; Branco-Vieira, M.; Caetano, N.S. Comparison of different lipid extraction procedures applied to three microalgal species. *Energy Rep.* **2020**, *6*, 477–482. [CrossRef]
101. Ruecker, C.M.; Adu-Peasah, S.P.; Engelhardt, B.S.; Veeder, G.T. Soventless Extraction Process. U.S. Patent 10329515B2, 25 June 2019.
102. Doble, M.; Kruthiventi, A.M. Alternate solvents. In *Green Chemistry and Engineering*; Doble, M., Kruthiventi, A.M., Eds.; Academic Press: New York, NY, USA, 2007; pp. 93–104. [CrossRef]
103. Clarke, C.J.; Tu, W.; Levers, O.; Bröhl, A.; Hallett, J.P. Green and sustainable solvents in chemical processes. *Chem. Rev.* **2018**, *118*, 747–800. [CrossRef]
104. Bubalo, M.C.; Radoševic, K.; Redovniković, I.R.; Halambek, J.; Srček, V.G. A brief overview of the potential environmental hazards of ionic liquids. *Ecotox Environ. Saf.* **2014**, *99*, 1–12. [CrossRef]
105. Ho, T.D.; Zhang, C.; Hantao, L.W.; Anderson, J.L. Ionic Liquids in Analytical Chemistry: Fundamentals, Advances, and Perspectives. *Anal. Chem.* **2014**, *86*, 262–285. [CrossRef]
106. Zhou, W.; Wang, Z.; Alam, M.A.; Xu, J.; Zhu, S.; Yuan, Z.; Huo, S.; Guo, Y.; Qin, L.; Ma, L. Repeated utilization of ionic liquid to extract lipid from algal biomass. *Int. J. Polym. Sci.* **2019**, *9209210*, 1–7. [CrossRef]
107. To, T.Q.; Procter, K.; Simmons, B.A.; Subasshchandrabose, S.; Atkin, R. Low cost ionic liquid-water mixtures for effective extraction of carbohydrate and lipid from algae. *Faraday Discuss* **2018**, *206*, 93. [CrossRef] [PubMed]
108. Yu, Z.; Chen, X.; Xia, S. The mechanism of lipids extraction from wet microalgae *Scenedesmus* sp. by ionic liquid assisted subcritical water. *J. Ocean Univ. China* **2016**, *15*, 549–552. [CrossRef]
109. El Archkar, T.; Fourmentin, S.; Greige-Gerges, H. Deep eutectic solvents: An overview on their interactions with water and biochemical compounds. *J. Mol. Liq.* **2019**, *288*, 111028. [CrossRef]
110. Lu, W.; Alam, M.A.; Pan, Y.; Wu, J.; Wang, Z.; Yuan, Z. A new approach of microalgal biomass pretreatment using deep eutectic solvents for enhance lipid recovery for biodiesel production. *Bioresour. Technol.* **2016**, *218*, 123–128. [CrossRef]
111. Jablonský, M.; Škulcová, A.; Šima, J. Use of deep eutectic solvents in polymer chemisty–A review. *Molecules* **2019**, *24*, 3978. [CrossRef]
112. Mahmood, W.M.A.W.; Lorwirachsutee, A.; Theodoropoulos, C.; Conzalez-Miguel, M. Polyol-based deep eutectic solvents for extraction of natural polyphenolic antioxidants from *Chlorella vulgaris*. *ACS Sustain. Chem. Eng.* **2019**, *7*, 5018–5026. [CrossRef]
113. Pan, Y.; Alam, M.A.; Wang, Z.; Huang, D.; Hu, K.; Chen, H.; Yuan, Z. One-step production of biodiesel from wet and unbroken microalgae biomass using deep eutectic solvent. *Bioresour. Technol.* **2017**, *238*, 157–163. [CrossRef]
114. Manjare, S.D.; Dhingra, K. Supercritical Fluids in Separation and Purification: A Review. *Mater Sci. Energy Technol.* **2019**, *2*, 463–484. [CrossRef]

115. Johner, J.C.F.; Meireles, M.A.A. Construction of a Supercritical Fluid Extraction (SFE) Equipment: Validation Using Annato and Fennel and Extract Analysis by Thin Layer Chromatography Couple to Image. *Food Sci. Technol.* **2016**, *36*, 210–247. [CrossRef]
116. Knez, Ž.; Pantić, M.; Cör, D.; Novak, Z.; Hrnčič, M.K. Are Supercritical Fluids Solvents for the Future? *Chem. Eng. Process* **2019**, *141*, 107532. [CrossRef]
117. Yang, X.; Li, Y.; Li, Y.; Ye, D.; Yuan, L.; Sun, Y.; Han, D.; Hu, Q. Solid Matrix-Supported Supercritical CO_2 Enhances Extraction of γ-Linolenic Acid from Cyanobacterium *Arthrospira (Spirulina) platensis* and Bioactivity Evaluation of the Molecule in Zebrafish. *Mar. Drugs* **2019**, *17*, 203. [CrossRef] [PubMed]
118. Feller, R.; Matos, A.P.; Mazzutti, S.; Moecke, E.H.S.; Tres, M.V.; Derner, R.B.; Oliveira, J.V.; Junior, A.F. Polyunsaturated ω-3 and ω-6 Fatty Acids, Total Carotenoids and Antioxidant Activity of Three Marine Microalgae Extracts Obtained by Supercritical CO_2 and Subcritical n-butane. *J. Supercrit. Fluids* **2018**, *133*, 437–443. [CrossRef]
119. Chronopoulou, L.; Dal Bosco, C.; Di Caprio, F.; Prosini, L.; Gentili, A.; Pagnanelli, F.; Palocci, C. Extraction of Carotenoids and Fat-Soluble Vitamins from *Tetradesmus obliquus* Microalgae: An Optimized Approach by Using Supercritical CO_2. *Molecules* **2019**, *24*, 2581. [CrossRef]
120. Mendiola, J.A.; García-Martínez, D.; Rupérez, F.J.; Martínez-Álvarez, P.J.; Reglero, G.; Cifuentes, A.; Barbas, C.; Ibañez, E.; Señoráns, F.J. Enrichment of Vitamin E from *Spirulina platensis* Microalga by SFE. *J. Supercrit. Fluids* **2008**, *43*, 484–489. [CrossRef]
121. Mendiola, J.A.; Marín, F.R.; Hernández, S.F.; Arredondo, B.O.; Señoráns, F.J.; Ibañez, E.; Reglero, G. Characterization via Liquid Chromatography Coupled to Diode Array Detector and Tandem Mass Spectrometry of Supercritical Fluid Antioxidant Extracts of *Spirulina platensis* Microalga. *J. Sep. Sci.* **2005**, *28*, 1031–1038. [CrossRef]
122. Mehariya, S.; Lovine, A.; Di Sanzo, G.; Larocca, V.; Martino, M.; Leone, G.P.; Casella, P.; Karatza, D.; Marino, T.; Musmarra, D.; et al. Supercritical Fluid Extraction of Lutein from *Scenedesmus almeriensis*. *Molecules* **2019**, *24*, 1324. [CrossRef]
123. Molino, A.; Martino, M.; Larocca, V.; Di Sanzo, G.; Spagnoletta, A.; Marino, T.; Karatza, D.; Lovine, A.; Mehariya, S. Eicosapentaenoic Acid Extraction from *Nannochloropsis gaditana* Using Carbon Dioxide at Supercritical Conditions. *Mar. Drugs* **2019**, *17*, 132. [CrossRef]
124. Ferreira, A.F.; Ribeiro, L.A.; Batista, A.P.; Marques, P.A.S.S.; Nobre, B.P.; Palavra, A.M.F.; Silva, P.P.; Gouveia, L.; Silva, C. A biorefinery from *Nannochloropsis* sp. microalga–Energy and CO_2 emission and economic analyses. *Bioresour. Technol.* **2013**, *138*, 235–244. [CrossRef]
125. Coskun, O. Separation Techniques: Chromatography. *North Clin. Istanbul.* **2016**, *3*, 156–160. [CrossRef]
126. Rajan, P.M.V.; Barron, A.R. Physical Methods in Chemistry and Nano Science. Openstax CNX™, 337–392. Available online: http://cnx.org/content/col10699/1.25/ (accessed on 25 September 2019).
127. Abrahamsson, V.; Rodriguez-Meizoso, I.; Turner, C. Determination of Carotenoids in Microalgae Using Supercritical Fluid Extraction and Chromatography. *J. Chromatogr. A* **2012**, *1250*, 63–68. [CrossRef]
128. Aluç, Y.; Kankilic, G.B.; Tüzün, I. Determination of Carotenoids in Two Algae Species from the Saline Water of Kapulukaya Reservoir by HPLC. *J. Liq. Chromatogr. R. T* **2018**, *41*, 93–100. [CrossRef]
129. Ahuja, S. Chromatographic methods. In *Chromatography and Separation Science*; Academic Press: New York, NY, USA, 2003; pp. 81–100.
130. Nawrocki, J. Silica Surface Controversies, Strong Adsorption Sites, their Blockage and Removal. Part I. *J. Chromatogr.* **1991**, *31*, 177. [CrossRef]
131. Nawrocki, J. Silica Surface Controversies, Strong Adsorption Sites, their Blockage and Removal. Part II. *J. Chromatogr.* **1991**, *31*, 193. [CrossRef]
132. Claessens, H.A.; van Straten, M.A. Review on the Chemical and Thermal Stability of Stationary Phases for Reversed-Phase Liquid Chromatography. *J. Chromatogr. A* **2004**, *1060*, 23–41. [CrossRef]
133. Venegas-Venegas, E.; Guil-Guerrero, J.L.; Rincón-Cervera, M.A.; Ramos-Bueno, R.P. Tricosahexaenoyl glycerol Purification from Docosahexanoic Acid Single Cell Oil (DHASCO). *Eur. J. Lipid Sci. Technol.* **2014**, *116*, 266–271. [CrossRef]
134. González-Fernández, M.J.; Ramos-Bueno, R.P.; Rodríguez-García, I.; Guil-Guerrero, J.L. Purification Process for MUFA- and PUFA- based Monoacylglycerols from Edible Oils. *Biochimie* **2017**, *139*, 107–114. [CrossRef]
135. Ronda, S.R.; Lele, S. Argentated Silica Gel Chromatography for Separation of γ-linolenic Acid from Microalgae. *Biosci. Biotech. Res. Asia* **2007**, *4*, 621–626.
136. Guil-Guerrero, J.L.; Belarbi, E.; Rebolloso-Fuentes, M.M. Eicosapentaenoic and arachidonic acids purification from the red microalga *Porphyridium cruentum*. *Bioseparation* **2001**, *9*, 299–306. [CrossRef]
137. Wu, H.; Fu, X.; Cao, W.; Xiang, W.; Hou, Y.; Ma, J.; Wang, Y.; Fan, C. Induction of Apoptosis in Human Glioma Cells by Fucoxanthin via Triggering of ROS-Mediated Oxidative Damage and Regulation of MAPKs and PI3K-AKT Pathways. *J. Agric. Food Chem.* **2019**, *67*, 2212–2219. [CrossRef]
138. Schröder, E.; Jönsson, T.; Poole, L. Hydroxyapatite chromatography: Altering the phosphate-dependent elution profile of protein as a function of pH. *Anal. Biochem.* **2003**, *313*, 176–178. [CrossRef]
139. Reuter, W.; Nickel, C.; Wehmeyer, W. Isolation of Allophycocyanin B from *Rhodella violacea* Results in a Model of the Core from Hemidiscoidal Phycobilisomes of Rhodophyceae. *FEBS Lett.* **1990**, *273*, 155–158. [CrossRef]
140. Gómez-Lojero, C.; Pérez-Gómez, B.; Krogmann, D.W.; Peña-Diaz, A. The Tricylindrical Core of the Phycobilisome of the Cyanobacterium *Arthrospira (Spirulina) maxima*. *Int. J. Biochem. Cell Bid.* **1997**, *29*, 959–970. [CrossRef]

141. Rossano, R.; Ungaro, N.; D'Ambrosio, A.; Liuzzi, G.M.; Riccio, P. Extracting and Purifying R-phycoerythrin from Mediterranean Red Algae *Corallina elongata* Ellis & Solander. *J. Biotechnol.* **2002**, *101*, 289–293. [CrossRef]
142. Niu, J.; Wang, G.; Lin, X.; Zhou, B. Large-scale recovery of C-phycocyanin from *Spirulina platensis* using expanded bed adsorption chromatography. *J. Chromatogr. B* **2006**, *850*, 267–276. [CrossRef] [PubMed]
143. Yang, F.; Wong, K.; Yang, Y.; Li, X.; Jiang, J.; Zheng, W.; Wu, H.; Chen, T. Purification and in vitro antioxidant activities of tellurium-containing phycobiliproteins from tellurium-enriched *Spirulina platensis*. *Drug Des. Dev. Ther.* **2014**, *8*, 1789–1800. [CrossRef]
144. Cai, C.; Wu, L.; Li, C.; He, P.; Li, J.; Zhou, J. Purification, Crystallization and Preliminary X-ray Analysis of Phycocyanin and Phycoerythrin from *Porphyra yezoensis* Ueda. *Acta Cryst.* **2011**, *F67*, 579–583. [CrossRef]
145. Ohsawa, N.; Tsujita, M.; Morikawa, S.; Itoh, N. Purification and Characterization of a Monohalomethane-producing Enzyme S-adenosyl-L-methionine: Halide Ion Methyltransferase from a Marine Microalga, *Pavlova pinguis*. *Biosci. Biotechnol. Biochem.* **2001**, *65*, 2397–2404. [CrossRef]
146. Ravi, A.; Guo, S.; Rasala, B.; Tran, M.; Mayfield, S.; Nikolov, Z.L. Separation Options for Phosphorylated Osteopontin from Transgenic Microalgae *Chlamydomonas reinhardtii*. *Int. J. Mol. Sci.* **2018**, *19*, 585. [CrossRef]
147. Williams, T. Gel Permeation Chromatography: A Review. *J. Mater. Sci.* **1970**, *5*, 811–820. [CrossRef]
148. Zhang, J.; Liu, L.; Ren, Y.; Chen, F. Characterization of Exopolysaccharides Produced by Microalgae with Antitumor Activity on Human Colon Cancer Cells. *Int. J. Biol. Macromol.* **2019**, *128*, 761–767. [CrossRef]
149. Yang, X.; Liu, M.; Qi, B.; Li, L.; Deng, J.; Hu, X.; Wu, Y.; Hao, S. Extraction, Purification and Partial Characterizations of Polysaccharides from *Gracilaria lemaneiformis*. *Adv. Mat. Res.* **2014**, *881–883*, 776–780. [CrossRef]
150. Muhamad, I.I.; Zulkifli, N.; Selvakumaran, S.; Lazim, N.A.M. Bioactive Algal-Derived Polysaccharides: Multi-Functionalization, Therapeutic Potential and Biomedical Applications. *Curr. Pharm. Des.* **2019**, *25*, 1147–1162. [CrossRef] [PubMed]
151. de Morais, M.G.; Stillings, C.; Dersch, R.; Rudisile, M.; Pranke, P.; Costa, J.A.V.; Wendorff, J. Extraction of Poly(3-hydroxybutyrate) from *Spirulina* LEB 18 for Developing Nanofibers. *Polímeros* **2015**, *25*, 161–167. [CrossRef]
152. GE Healthcare. Ion Exchange Chromatography & Chromatofocusing. Principles and Methods. 2010. Available online: https://www.gelifesciences.com/en/us/solutions/protein-research/knowledge-center/protein-purification-methods/ion-exchange-chromatography (accessed on 25 September 2019).
153. Owczarek, B.; Gerszberg, A.; Hnatuszko-Konka, K. A Brief Reminder of Systems of Production and Chromatography-Based Recovery of Recombinant Protein Biopharmaceuticals. *BioMed. Res. Int.* **2019**, 1–13. [CrossRef]
154. Chen, K.; Wang, S.S.; Show, P.; Hsu, S.; Chang, Y. Rapid and Efficient Recovery of C-Phycocyanin from Highly Turbid *Spirulina platensis* Algae Using Stirred Fluidized Bed Ion Exchange Chromatography. *Sep. Purif. Technol.* **2019**, *209*, 636–645. [CrossRef]
155. Tang, Z.; Zhao, J.; Ju, B.; Li, W.; Wen, S.; Pu, Y.; Qin, S. One-step chromatographic procedure for purification of B-phycoerythrin from *Porphyridium cruentum*. *Protein Expr. Purif.* **2016**, *123*, 70–74. [CrossRef]
156. Urh, M.; Simpson, D.; Zhao, K. Affinity Chromatography: General Methods. *Methods Enzymol.* **2009**, *463*, 417–438. [CrossRef]
157. Mangold, H.K. *Dietary Fats and Health*; Perkins, E.G., Visek, W.J., Eds.; American Oil Chemists' Society: Champaign, IL, USA, 1983; pp. 110–123.
158. Yang, M.; Meng, Y.; Chu, Y.; Fan, Y.; Cao, X.; Xue, S.; Chi, Z. Triacylglycerol accumulates exclusively outside the chloroplast in shortterm nitrogen-deprived *Chlamydomonas reinhardtii*. BBA-Mol. *Cell Biol. Lipids* **2018**, *1863*, 1478–1487. [CrossRef]
159. Shen, P.; Wang, H.; Pan, Y.; Meng, Y.; Wu, P.; Xue, S. Identification of Characteristic Fatty Acids to Quantify Triacylglycerols in Microalgae. *Front. Plant. Sci.* **2016**, *7*, 162. [CrossRef]
160. Dyńska-Kukulska, K.; Ciesielski, W. Methods of extraction and thin-layer chromatography determination of phospholipids in biological samples. *Rev. Anal. Chem.* **2012**, *31*, 43–56. [CrossRef]
161. Lage, S.; Gentili, F.G. Quantification and Characterisation of Fatty Acid Methyl Esters in Microalgae: Comparison of Pretreatment and Purification Methods. *Bioresour. Technol.* **2018**, *257*, 121–128. [CrossRef] [PubMed]
162. Sun, P.; Wong, C.; Li, Y.; He, Y.; Mao, X.; Wu, T.; Ren, Y.; Chen, F. A Novel Strategy for Isolation and Purification of Fucoxanthinol and Fucoxanthin from the Diatom *Nitzschia laevis*. *Food Chem.* **2019**, *277*, 566–572. [CrossRef] [PubMed]
163. Żyszka-Haberecht, B.; Poliwoda, A.; Lipok, J. Structural constraints in cyanobacteria-mediated whole-cell biotransformation of methoxylated and methylated derivatives of 2′-hydroxychalcone. *J. Biotechnol.* **2019**, *293*, 36–46. [CrossRef] [PubMed]
164. Jaime, L.; Mendiola, J.A.; Herrero, M.; Soler-Rivas, C.; Santoyo, S.; Señorans, F.J.; Cifuentes, A.; Ibáñez, E. Separation and characterization of antioxidants from *Spirulina platensis* microalga combining pressurized liquid extraction, TLC, and HPLC-DAD. *J. Sep. Sci.* **2005**, *28*, 2111–2119. [CrossRef]
165. de Jesus, S.S.; Ferreira, G.F.; Maciel, M.R.W.; Filho, R.M. Biodiesel Purification by Column Chromatography and Liquid-Liquid Extraction Using Green Solvents. *Fuel* **2019**, *235*, 1123–1130. [CrossRef]
166. Harris, D.C. *Quantitative Chemical Analysis*, 8th ed.; W. H. Freeman and Company: New York, NY, USA, 2007.
167. Cerón-Garcia, M.C.; González-López, C.V.; Camacho-Rodríguez, J.; López-Rosales, L.; García-Camacho, F.; Molina-Grima, E. Maximizing Carotenoid Extraction from Microalgae Used as Food Additives and Determined by Liquid Chromatography (HPLC). *Food Chem.* **2018**, *257*, 316–324. [CrossRef]
168. Soares, A.T.; Júnior, J.G.M.; Lopes, R.G.; Derner, R.B.; Filho, N.R.A. Improvement of the Extraction Process for High Commercial Value Pigments from *Desmodesmus* sp. Microalgae. *J. Braz. Chem. Soc.* **2016**, *27*, 1083–1093. [CrossRef]

169. Morowvat, M.H.; Ghasemi, Y. Developing a Robust Method for Quantification of β-Carotene in *Dunaliella salina* Biomass Using HPLC Method. *Int. J. Pharm. Clin. Res.* **2016**, *8*, 1423–1428.
170. Soares, A.T.; da Costa, D.C.; Vieira, A.A.H.; Filho, N.R.A. Analysis of Major Fatty Acid Composition of Freshwater Microalgae. *Heliyon* **2019**, *5*, e01529. [CrossRef]
171. Jones, J.; Manning, S.; Montoya, M.; Keller, K.; Poenie, M. Extraction of Algal Lipids and Their Analysis by HPLC and Mass Spectrometry. *J. Am. Oil Chem. Soc.* **2012**, *89*, 1371–1381. [CrossRef]
172. Jerez-Martel, I.; García-Poza, S.; Rodríguez-Martel, G.; Rico, M.; Afonso-Olivares, C.; Gómez-Pinchetti, J.L. Phenolic Profile and Antioxidant Activity of Crude Extracts from Microalgae and Cyanobacteria Strains. *J. Food Qual.* **2017**, 1–8. [CrossRef]
173. Hussain, A.; AlAjmi, M.F.; Hussain, I.; Ali, I. Future of Ionic Liquids for Chiral Separations in High-Performance Liquid Chromatography and Capillary Electrophoresis. *Crit. Rev. Anal. Chem.* **2019**, *49*, 289–305. [CrossRef] [PubMed]
174. Ventura, S.P.M.; Nobre, B.P.; Ertekin, F.; Hayes, M.; Garciá-Vaquero, M.; Vieira, F.; Koc, M.; Gouveia, L.; Aires-Barros, M.R.; Palavra, A.M.F. Extraction of Value-Added Compounds from Microalgae. In *Microalgae-Based Biofuels and Bioproducts*; Gonzalez-Fernandez, C., Muñoz, R., Eds.; Woodhead Publishing: Cambridge, UK, 2017. [CrossRef]
175. Berthod, A.; Ruiz-Angel, M.; Hassoun, M. Countercurrent Chromatography. *Extrus. Elution* **2007**, 1–8. [CrossRef]
176. Fábryová, T.; Cheel, J.; Kubáč, D.; Hrouzek, P.; Vu, D.L.; Tůmová, L.; Kopecký, J. Purification of Lutein from the Green Microalgae *Chlorella vulgaris* by Integrated use a New Extraction Protocol and a Multi-injection High Performance Counter-Current Chromatography (HPCCC). *Algal Res.* **2019**, *41*, 101574. [CrossRef]
177. Cheel, J.; Kučerová, P.; Garrard, I.; Ignatova, S.; Hrouzek, P.; Kopecký, J. Two-Step Separation of Nostotrebin 6 from Cultivated Soil Cyanobacterium (*Nostoc* sp.) by High Performance Countercurrent Chromatography. *Molecules* **2014**, *19*, 8773–8787. [CrossRef] [PubMed]
178. Khan, B.M.; Liu, Y. High Speed Counter Current Chromatography: Overview of Solvent-system and Elution-mode. *J. Liq. Chromatogr. Relat. Technol.* **2018**, *41*, 629–636. [CrossRef]
179. Li, H.; Chen, F. Separation and Purification of Epimedin A, B, C, and Icariin from the Medicinal Herb *Epimedium brevicornum* Maxim by Dual-Mode HSCCC. *J. Chromatogr. Sci.* **2009**, *47*, 337–340. [CrossRef]
180. Wu, X.; Li, R.; Zhao, Y.; Liu, Y. Separation of Polysaccharides from *Spirulina platensis* by HSCCC with Ethanol-Ammonium Sulfate ATPS and Their Antioxidant Activities. *Carbohydr. Polym.* **2017**, *173*, 465–472. [CrossRef]
181. Lu, H.; Jiang, Y.; Chen, F. Preparative Separation and Purification of Squalene from the Microalga *Thraustochytrium* ATCC 26185 by High-Speed Counter-Current Chromatography. *J. Chromatogr. A* **2003**, *994*, 37–43. [CrossRef]
182. Chen, F.; Li, H.; Wong, R.N.; Ji, B.; Jiang, Y. Isolation and Purification of the Bioactive Carotenoid Zeaxanthin from the Microalga *Microcystis aeruginosa* by High-Speed Counter-Current Chromatography. *J. Chromatogr. A* **2005**, *1064*, 183–186. [CrossRef]
183. Li, H.; Fan, K.; Chen, F. Isolation and Purification of Canthaxanthin from the Microalga *Chlorella zofingiensis* by High-Speed Counter-Current Chromatography. *J. Sep. Sci.* **2006**, *29*, 699–703. [CrossRef] [PubMed]
184. Cheel, J.; Urajová, P.; Hájek, J.; Hrouzek, P.; Kuzma, M.; Bouju, E.; Faure, K.; Kopecký, J. Separation of Cyclic Lipopeptide Puwainaphycins from Cyanobacteria by Countercurrent Chromatography Combined with Polymeric Resins and HPLC. *Anal. Bioanal. Chem.* **2017**, *409*, 917–930. [CrossRef] [PubMed]
185. Rose, B.A. Gas Chromatography and its Analytical Applications. A Review. *Analyst* **1959**, *84*, 574b–595. [CrossRef]
186. Biedermann, M.; Grob, K. Advantages of Comprehensive Two-Dimensional Gas Chromatography for Comprehensive Analysis of Potential Migrants from Food Contact Materials. *Anal. Chim. Acta* **2019**, *1057*, 11–17. [CrossRef] [PubMed]
187. Lebanov, L.; Tedone, L.; Kaykhaii, M.; Linford, M.R.; Paull, B. Multidimensional Gas Chromatography in Essential Oil Analysis. Part 1: Technical Developments. *Chromatographia* **2019**, *82*, 377–398. [CrossRef]
188. Kusch, P. Application of Pyrolysis-Gas Chromatography/Mass Spectrometry (Py-GC/MS). *Compr. Anal. Chem.* **2017**, *75*, 169–207. [CrossRef]
189. Liu, C.; Duan, X.; Chen, Q.; Chao, C.; Lu, Z.; Lai, Q.; Megharaj, M. Investigations on Pyrolysis of Microalgae *Diplosphaera* sp. MM1 by TG-FTIR and Py-GC/MS: Products and Kinetics. *Bioresour. Technol.* **2019**, *294*, 122126. [CrossRef] [PubMed]
190. Berg, J.M.; Tymoczko, J.L.; Stryer, L. The Purification of Proteins Is an Essential First Step in Understanding Their Function. In *Biochemistry*, 5th ed.; W. H. Freeman and Company: New York, NY, USA, 2002. Available online: https://www.ncbi.nlm.nih.gov/books/NBK22410 (accessed on 25 September 2019).
191. Smith, R. Two-Dimensional Electrophoresis: An Overview. In *Methods in Molecular Biology, Two-Dimensional Electrophoresis Protocols*; Tyther, R., Sheehan, D., Eds.; Humana Press: Totowa, NJ, USA, 2009. [CrossRef]
192. Tran, T.; Lafarge, C.; Pradelles, R.; Perrier-Cornet, J.; Cayot, N.; Loupiac, C. Effect of High Hydrostatic Pressure on the Structure of the Soluble Protein Fraction in *Porphyridium cruentum* Extracts. *Innov. Food Sci. Emerg. Technol.* **2019**, *58*, 102226. [CrossRef]
193. Wang, S.; Hu, Q.; Sommerfeld, M.; Chen, F. An optimized protocol for isolation of soluble proteins from microalgae for two-dimensional gel electrophoresis analysis. *J. Appl. Phycol.* **2003**, *15*, 485–496. [CrossRef]
194. Dai, L.; Reichert, C.L.; Hinrichs, J.; Weiss, J. Acid Hydrolysis Behavior of Insoluble Protein-Rich Fraction Extracted from *Chlorella protothecoides*. *Colloids Surf. A* **2019**, *569*, 129–136. [CrossRef]
195. Lipkens, B.; Mitchell, E.; Carmichael, J.; Mealey, D.; Dionne, J. Ultrasound and Acoustophoresis for Collection and Processing of Oleaginous Microorganisms. U.S. Patent 9556411B2, 31 January 2017.
196. Saxena, A.; Tripathi, B.; Kumar, M.; Shahi, V.K. Membrane-based techniques for the separation and purification of proteins: An overview. *Adv. Colloid Interface Sci.* **2009**, *145*, 1–22. [CrossRef] [PubMed]

197. Chaiklahan, R.; Chirasuwan, N.; Loha, V.; Tia, S.; Bunnag, B. Separation and Purification of Phycocyanin from *Spirulina* sp. Using a Membrane Process. *Bioresour. Technol.* **2011**, *102*, 7159–7164. [CrossRef] [PubMed]
198. Gómez-Loredo, A.; Gonzále-Valdez, J.; Rito-Palomares, M. Insights on the Downstream of Fucoxanthin, a microalgal carotenoid, from an aqueous two-phase system stream exploiting ultrafiltration. *J. Appl. Phycol.* **2015**, *27*, 1517–1523. [CrossRef]
199. Barkia, I.; Saari, N.; Manning, S.R. Microalgae for High-Value Products towards Human Health and Nutrition. *Mar. Drugs* **2019**, *17*, 304. [CrossRef]
200. Kim, D.; Hwang, T.; Oh, Y.; Han, J. Harvesting *Chlorella* sp. KR-1 Using Cross-Flow Electro-Filtration. *Algal Res.* **2014**, *6*, 170–174. [CrossRef]
201. Furuki, T.; Maeda, S.; Imajo, S.; Hiroi, T.; Amaya, T.; Hirokawa, T.; Ito, K.; Nozawa, H. Rapid and Selective Extraction of Phycocyanin from *Spirulina platensis* with Ultrasonic Cell Disruption. *J. Appl. Phycol.* **2003**, *15*, 319–324. [CrossRef]
202. Lien, T.; Schreiner, Ø.; Steine, M. Purification of a Derepressible Arylsulfatase from *Chlamydomonas reinhardti*. Properties of the Enzyme in Intact Cells and in Purified State. *Biochim. Biophys. Acta* **1975**, *384*, 168–179. [CrossRef]
203. Mathiazakan, P.; Shing, S.Y.; Ying, S.S.; Kek, H.K.; Tang, M.S.Y.; Show, P.L.; Ooi, C.; Ling, T.C. Pilot-scale Aqueous Two-phase Floatation for Direct Recovery of Lipase Derived from *Burkholderia cepacia* Strain ST8. *Sep. Purif. Technol.* **2016**, *171*, 206–213. [CrossRef]
204. Leong, H.Y.; Chang, C.; Lim, J.W.; Show, P.L.; Lin, D.; Chang, J. Liquid Biphasic Systems for Oil-rich Algae Bioproducts Processing. *Sustainability* **2019**, *11*, 4682. [CrossRef]
205. Chew, K.W.; Chia, S.R.; Krishnamoorthy, R.; Tao, Y.; Chu, D.; Show, P.L. Liquid biphasic flotation for the purification of C-phycocyanin from *Spirulina platensis* microalga. *Bioresour. Technol.* **2019**, *288*, 121219. [CrossRef]
206. Safaei, M.; Maleki, H.; Soleimanpour, H.; Norouzy, A.; Zahiri, H.S.; Vali, H.; Noghabi, K.A. Development of a Novel Method for the Purification of C-Phycocyanin Pigment from a Local Cyanobacterial Strain *Limnothrix* sp. NS01 and Evaluation of Its Anticancer Properties. *Sci. Rep.* **2019**, *9*, 9474. [CrossRef] [PubMed]
207. Phong, W.N.; Show, P.L.; Teh, W.H.; Teh, T.X.; Lim, H.M.Y.; Nazri, N.S.B.; Tan, J.; Ling, T.C. Proteins Recovey from Wet Microalgae Using Liquid Biphasic Flotation (LBF). *Bioresour. Technol.* **2017**, *244*, 1329–1336. [CrossRef] [PubMed]
208. Khoo, K.S.; Chew, K.; Ooi, C.W.; Ong, H.C.; Ling, T.C.; Show, P.L. Extraction of Natural Astaxanthin from *Haematococcus pluvialis* Using Liquid Biphasic Flotation System. *Bioresour. Technol.* **2019**, *290*, 121794. [CrossRef]
209. Sankaran, R.; Show, P.L.; Cheng, Y.; Tao, Y.; Ao, X.; Nguyen, T.D.P.; Quyen, D.V. Integration Process for Protein Extraction from Microalgae Using Liquid Biphasic Eletric Flotation (LBEF) System. *Mol. Biotechnol.* **2018**, *60*, 749–761. [CrossRef] [PubMed]
210. Koyande, A.K.; Chew, K.W.; Lim, C.; Lee, S.Y.; Lam, M.K.; Show, P. Optimization of Protein Extraction from *Chlorella vulgaris* Via Nobel Sugaring-out Assisted Liquid Biphasic Electric Flotation System. *Eng. Life Sci.* **2019**, 1–10. [CrossRef]
211. Pei, Y.C.; Li, Z.Y.; Liu, L.; Wang, J.J.; Wang, H.Y. Selective Separation of Protein and Saccharides by Ionic Liquids Aqueous Two-phase Systems. *Sci. China Chem.* **2010**, *53*, 1554–1560. [CrossRef]
212. Álvarez, M.S.; Longo, M.A.; Deive, F.J.; Rodríguez, A. Non-ionic surfactants and ionic liquids are a suitable combination for aqueous two-phase systems. *Fluid Phase Equilibr.* **2019**, *502*, 112302. [CrossRef]
213. Rachana, C.R.; Lyju, J.V. Three Phase Partitioning-A Novel Protein Purification Method. *Int. J. Chem. Tech. Res.* **2014**, *6*, 3467–3472.
214. Dennison, C.; Lovrien, R. Three Phase Partitioning: Concentration and Purification of Proteins. *Protein Expr. Purif* **1997**, *11*, 149–161. [CrossRef]
215. Duong-Ly, K.C.; Gabelli, S.B. Salting Out of Proteins Using Ammonium Sulfate Precipitation. *Methods Enzymol.* **2014**, *541*, 85–94. [CrossRef]
216. Silva, L.A.; Kuhn, K.R.; Moraes, C.C.; Burket, A.V.; Kalil, S.J. Experimental Design as a Tool for Optimization of C-Phycocyanin Purification by Precipitation from *Spirulina platensis*. *J. Braz. Chem. Soc.* **2009**, *20*, 5–12. [CrossRef]
217. Herrera, A.; Boussiba, S.; Napoleone, V.; Hohlberg, A. Recovery of C-Phycocyanin from the Cyanobacterium *Spirulina maxima*. *J. Appl. Phycol.* **1989**, *1*, 325–331. [CrossRef]
218. Básaca-Loya, G.A.; Valdez, M.A.; Enríquez-Guevara, E.A.; Gutiérrez-Millán, L.E.; Burboa, M.G. Extraction and purification of B-Phycoerythrin from the red microalga *Rhodosorus Marinus*. *Cienc Mar.* **2009**, *35*, 359–368. [CrossRef]
219. Thengodkar, R.R.M.; Sivakami, S. Degradation of Chlorpyrifos by an Alkaline Phosphatase from the Cyanobacterium *Spirulina platensis*. *Biodegradation* **2010**, *21*, 637–644. [CrossRef] [PubMed]
220. Julianti, E.; Singgih, M.; Mulyani, L.N. Optimization of extraction method and characterization of phycocyanin pigment from *Spirulina platensis*. *J. Math Fund. Sci.* **2019**, *51*, 168–176. [CrossRef]
221. Muensean, K.; Kim, S.M. Purification and Characterization of β-Glucosidase Produced by *Trichoderma Citrinoviride* Cultivated on Microalga *Chlorella vulgaris*. *Appl. Biochem. Microbiol.* **2015**, *51*, 102–107. [CrossRef]
222. Chen, C.; Kao, P.; Tan, C.H.; Show, P.L.; Cheah, W.Y.; Lee, W.; Ling, T.C.; Chang, J. Using an Innovative PH-Stat CO_2 Feeding Strategy to Enhance Cell Growth and C-Phycocyanin Production from *Spirulina platensis*. *Biochem. Eng. J.* **2016**, *112*, 78–85. [CrossRef]
223. Robards, K.; Haddad, P.R.; Jackson, P.E. Supercritical Fluid Chromatography. In *Principles and Practice of Modern Chromatography Methods*; Robards, K., Haddad, P.R., Jackson, P.E., Eds.; Elsevier Academic Press: San Diego, CA, USA, 2004; pp. 381–405. [CrossRef]
224. Yuan, H.; Olesik, S.V. Supercritical fluid chromatography. In *Encyclopedia of Analytical Chemistry*; John Wiley & Sons, Ltd.: New York, NY, USA, 2006. [CrossRef]

225. Acikara, Ö.B. Ion-exchange chromatography and its applications. In *Column Chromatography*; Martin, D.F., Martin, B.B., Eds.; IntechOpen: Rijeka, Croatia, 2013. [CrossRef]
226. Cheng, S.; Huang, M.; Shiea, J. Thin layer chromatography/mass spectrometry. *J. Chromatogr. A* **2011**, *1218*, 2700–2711. [CrossRef]
227. Dong, M.W. *Modern HPLC for Practicing Scientists*; John Wiley & Sons: New York, NY, USA, 2006. [CrossRef]
228. Niu, Z.; Pang, R.T.K.; Liu, W.; Li, W.; Cheng, R.; Yeung, W.S.B. Polymer-based precipitation preserves biological activities of extracellular vesicles from an endometrical cell line. *PLoS ONE* **2017**, *12*, e0186534. [CrossRef]

processes

MDPI

Article

Innovative and Green Extraction Techniques for the Optimal Recovery of Phytochemicals from Saudi Date Fruit Flesh

Kashif Ghafoor [1,*], Md. Zaidul Islam Sarker [2], Fahad Y. Al-Juhaimi [1], Elfadil E. Babiker [1], Mohammed S. Alkaltham [1], Abdullah K. Almubarak [1] and Isam A. Mohamed Ahmed [1]

1 Department of Food Science and Nutrition, College of Food and Agricultural Sciences, King Saud University, Riyadh 11451, Saudi Arabia
2 Food Science Program, Cooperative Research, Extension and Education Services, Northern Marianas College, Saipan, MP 96950, USA
* Correspondence: kghafoor@ksu.edu.sa; Tel.: +966-11-4691951

Abstract: Saudi Arabia is one of the major producers of date (*Phoenix dactylifera*) fruit. Date fruit flesh is considered a healthy food due to the presence of natural antioxidants. Green and innovative supercritical fluid (SFE, 52.5 °C temperature, 27.50 MPa pressure, 5 mL CO_2/min flow rate) and subcritical (SubCO_2, 250 extraction cycles, 29 °C temperature, 6.8 MPa, 12 h, ethanol solvent) extraction techniques were used to produce flesh extracts from four Saudi date fruits (Sukari (SKFE), Ambara (AMFE), Majdool (MJFE) and Sagai (SGFE)), and extracts prepared using 6 h Soxhlet extraction at 70 °C for 16 h using n-hexane as solvent, were taken as control. SFE produced the highest ($p < 0.05$) extract yields, whereas the SubCO_2 method recovered significantly higher ($p < 0.05$) amounts of phytochemicals. Total phenolics (186.37–447.31 mg GAE/100 g), total flavonoids (82.12–215.28 mg QE/100 g), total anthocyanins (0.41–1.34 mg/100 g), and total carotenoid (1.24–2.85 mg BCE/100 g) were quantified in all the flesh extracts. The biological properties evaluation showed that flesh extracts had high antioxidant (17.79–45.08 µg AAE/mL), antiradical (191.36–34.66 µg/mL DPPH IC_{50}), ferric-reducing (2.18–5.01 mmol TE/100 g) and ABTS-scavenging (444.75–883.96 µmol TE/100 g) activities. SubCO_2 was the best technique and Majdool the best date variety, in terms of both phytochemicals and biological properties.

Keywords: *Phoenix dactylifera*; date fruit flesh; date varieties; supercritical CO_2; subcritical CO_2; phytochemicals; biological properties

Citation: Ghafoor, K.; Sarker, M.Z.I.; Al-Juhaimi, F.Y.; Babiker, E.E.; Alkaltham, M.S.; Almubarak, A.K.; Ahmed, I.A.M. Innovative and Green Extraction Techniques for the Optimal Recovery of Phytochemicals from Saudi Date Fruit Flesh. *Processes* **2022**, *10*, 2224. https://doi.org/10.3390/pr10112224

Academic Editors: Zongbi Bao and Qianqian Xu

Received: 9 October 2022
Accepted: 25 October 2022
Published: 29 October 2022

1. Introduction

For a long time, the date palm (*Phoenix dactylifera* L.) has been recognized as one of the most valuable fruiting plants in the Arabian Peninsula [1,2]. Saudi Arabia ranks among the top date fruit producers with continuously increasing exports [3]. Date fruits are excellent dietary sources of energy and fibre and are sometimes considered one of the staple foods of the region. Different date palm products, such as fruits, pollens, leaves, seeds and trunks, are considered to have medicinal properties [4]. The date fruit can be defined as a fleshy pericarp containing one seed [5]. The date fruit is rich in carbohydrates (mainly sucrose, glucose and fructose), vitamins (especially vitamin B complex), fiber, minerals, proteins and many other natural products and phytochemicals, including phenolics, carotenoids, phytosterols and flavonoids. These can be used in the promotion and protection of human health due to their antioxidant and anti-inflammatory properties [6]. In general, date fruits are a good source of minerals and amino acids, and they have been used for the treatment of chronic illnesses and diseases [7]. It has been reported that the consumption of date fruit is associated with a reduction of high blood pressure and oxidative stress. Furthermore, date fruits also aid in the treatment of health problems such as diabetes, cancer and atherosclerosis [8]. The presence of phytochemicals such as anthocyanins, flavonoids, tannins, phenolic compounds and water-soluble vitamins adds additional health benefits to these fruits, many of which still need to be explored [6,9].

The nutraceutical industry is one of the important future industries which may target the recovery, isolation and further utilization of various health-promoting components from various types of fruits, vegetables and agri-products. The extraction of phytochemicals from plant materials is an important area of research, as their utilization and biological/functional properties depend on extraction yield and quality of the extract and selectivity of the method [10]. The generally used solvent-based methods are nonselective, although in many cases they have high gravimetric yields [11]. In general, studies investigating date fruit as a source of nutrients and phytochemicals have involved the application of conventionally used organic solvents such as methanol [5,9]. There is a scarcity of reports involving the use of innovative and green techniques to study functional compounds and their properties in date fruit flesh. Supercritical fluid extraction (SFE) using supercritical CO_2 is a green and innovative method for the recovery of natural antioxidants for food and pharmaceutical uses, as regulatory requirements discourage use of toxic organic solvents [12]. CO_2, being environment friendly, safe, nontoxic, noncarcinogenic, nonflammable and cheap, is often used as a supercritical fluid, although the initial costs of SFE setup are quite high. CO_2 can be used for a wide range of chemical and biochemical extraction processes [11]. Application of subcritical CO_2 in subcritical CO_2 ($SubCO_2$) extraction is also increasing, as the apparatus used in this method is generally identical to the Soxhlet method and may be less costly in comparison to SFE systems [13,14]. In both states, the selectivity of CO_2 during extraction processes can controlled or adjusted through temperature, cosolvent and pressure to obtain extract fractions, particularly rich in specific phytochemicals [11–15]. These two techniques are also generally referred to as green methods as they work at lower temperatures and the extracts obtained are considered safe for food applications [11,13,14].

The objectives of the present study included the application of conventional (Soxhlet) and green/innovative (SFE and $SubCO_2$) extraction methods for the recovery of different phytochemicals, such as total phenolic compounds, total flavonoids, total anthocyanins and total carotenoids from fruit samples of four (Sukari date flesh, Ambara date flesh, Majdool date flesh and Sagai date flesh) different cultivars of date grown in Saudi Arabia. The SFE process was optimized using response surface methodology and regression analysis techniques. The biological activities of these flesh extracts were also studied using different in vitro antioxidant activity assays.

2. Materials and Methods

2.1. Date Flesh Sample Preparation

Sukari, Ambara, Majdool and Sagai date fruit (Figure 1), cultivated in Saudi Arabia and harvested during date season, were obtained from a date fruit market in Riyadh, Saudi Arabia during 2020. Date flesh was manually removed from the fruit and dried under vacuum at 50 °C. The dried flesh was made into a powder form using a blender (Panasonic, Shah Alam, Malaysia). Analytical grade chemicals from Sigma-Aldrich (St. Louis, MO, USA) were used in this work.

Figure 1. Date Fruits from Four Date Cultivars grown in Saudi Arabia.

2.2. Moisture Content Determination

A halogen moisture analyzer-HB43 (Mettler-Toledo GmbH, Giessen, Germany) was used for the determination of moisture in date flesh samples following a Ca 2d- 25 AOAC official method [16]. The instrument was fixed at 105 °C, followed by sample evaluation. Date flesh sample (5 g) was placed on an aluminum plate in the sample handler for 5 min, followed by increase of its temperature with the help of halogen light. The samples' moisture percentage was stored in an automatic recorder.

2.3. Date Flesh Antioxidants' Extraction

2.3.1. Conventional Extraction

The conventional method employed to extract date flesh (25 g grounded sample) bioactive compounds (25 g) was carried out in Whatman® cellulose extraction thimbles (28 × 100 mm) and using n-hexane (250 mL) as solvent in a Soxhlet system. The extraction was carried out for 10–12 h at 70 °C. The dried extract was obtained by evaporating the n-hexane (containing extract) at 50 °C using a rotary evaporator (Eyela, Tokyo Rikakikai Co., Tokyo, Japan). The extracts were stored in airtight, light-proof containers at −20 °C. The calculation of the extract yield was based on following equation:

$$Yield\ (\%) = \frac{Weight\ of\ extract}{Weight\ of\ ground\ sample} \times 100 \quad (1)$$

2.3.2. Supercritical Fluid CO_2 Extraction (SFE)

The SFE unit (Jasco Corporation, Tokyo, Japan) used in this study consisted of two pumps: a CO_2 pressurizing pump and cosolvent recovery pump. The cosolvent used was 30% ethanol and the CO_2 cylinder pump was also equipped with a cooling system, whereas the pump head was cooled using a mixture (50:50) of ethylene glycol and deionized water. The process variables included extraction temperature (X_1) (35–70 °C) CO_2 pressure (X_2)

(15–40 MPa), and CO_2 flow rate (X_3) (0.5–5 mL/min). The extraction trials were based on a central composite rotatable experimental design [11,13,17]. A total of at least 18 trials were carried out for optimization of the extraction. In each run, a 20 g sample was placed in the extraction vessel and CO_2 was pressurized into the vessel to carry out bioactive compounds' extraction at different combinations of process variables. The extracts were collected at backpressure regulator in a Schott bottle followed by vacuum drying. The percentage yields of the extracts were estimated according to Equation (1). Extract samples thus prepared were packaged in light-proof bottles, sealed to prevent air penetration, and stored at −20 °C. In order to optimize the SFE extraction process for the extract yield, a predictive model was designed as follows:

$$Y = b_0 + b_1X_1 + b_2X_2 + b_3X_3 + b_{11}X_1^2 + b_{22}X_2^2 + b_{33}X_3^2 + b_{12}X_1X_2 + b_{13}X_1X_3 + b_{23}X_2X_3 \quad (2)$$

The response variable Y denoted the extract yield. b_0 is the regression coefficient for the offset term. The regression coefficients were also denoted using b_2 and b_3 (linear terms); b_{11}, b_{22} and b_{33} (quadratic terms) and b_{12}, b_{13} and b_{23} (interaction terms). The process variables of SFE temperature, pressure and CO_2 flow rate were represented by X_1, X_2 and X_3 in the above model, respectively.

2.3.3. Subcritical CO_2 Extraction (SubCO_2)

The extraction of bioactive compounds from date flesh was also carried out using SubCO_2 methods, as reported earlier [14,18,19]. The extraction process was accomplished by treating 150 g powdered date flesh with 75 mL 95% absolute alcohol (cosolvent) followed by placement of the sample in the extraction vessels. The extraction cycles (250 cycles), pressure (6.8 MPa) and temperatures at heating (29 °C) and cooling controls were set and achieved, followed by the flow of the liquid CO_2 (6 kg) into the extraction vessel. The extract and remaining raw sample after extraction were recovered through system depressurization; the extract was dried under reduced pressure and stored at −20 °C. The extract yield was calculated according to Equation (1).

2.4. Date Flesh Bioactive Compounds

2.4.1. Analysis of Total Phenolic Contents (TPC)

The TPC determination was based on the Folin–Ciocalteu (FC) method [20]. A 10 mg vacuum-dried extract, obtained from three different techniques, was dissolved in 100 mL methanol. A sample of 200 µL from the extract in methanol, or a standard solution (gallic acid in methanol) having different concentrations, was added to 400 µL of FC reagent, followed by the addition of deionized water to make a total volume in test tube as 4.6 mL. One mL of 10% sodium carbonate was mixed with the contents of test tubes after 10 min, followed by further room temperature incubation for 2 h. The absorbance values were recorded at 765 nm using a spectrophotometer (Shimadzu, Kyoto, Japan). TPC results were expressed as mg gallic acid equivalent (GAE) per 100 g of date flesh.

2.4.2. Analysis of Total Flavonoids Contents (TFC)

The TFC were evaluated using a colorimetric method [21]. A sample of 1 mL of extract (prepared in TPC method) was taken in a test tube sodium nitrite (5%, 0.3 mL), and aluminum chloride (10%, 0.3 mL) solutions were added. The contents of reaction mixture were kept at room temperature for 5 min, after which 2 mL of sodium hydroxide (1 M) was added and the volume of the contents increased to 10 mL using deionized water. The absorbance values were recorded at 510 nm, and quercetin was used was used as a standard compound. The results for TFC were presented as mg of quercetin equivalent (QE)/100 g of date flesh.

2.4.3. Analysis of Total Anthocyanins

The evaluation of total anthocyanin [22] in date flesh extract samples was carried out by taking 1 mL of extract sample (from TPC method) and mixing it with 10 mL of

5% ethanol. The samples were then centrifuged at 1800× *g* for 5 min and 200 μL of the supernatant was mixed with 3.8 mL of HCl (1 M). The contents were then incubated for 3 h at room temperature. The absorbance (A) values were recorded at 520 nm using HCl as a blank. The standard used was malvidin-3-glucoside, which was dissolved in methanol and prepared following the same procedure as that used for the extract sample. Absorbance values (B) of standard solution were recorded. The total anthocyanins were presented as mg/100 g of date flesh and initially calculated according to the following formula:

$$Total\ anthocyanins\ \left(mg\ g^{-1}\right) = \frac{A \times dilution\ factor \times 1000}{B} \quad (3)$$

2.4.4. Analysis of Total Carotenoids

A previously described method [23], in which β-carotene was used as a standard, was used for the determination of total carotenoids in date flesh extracts. The dried extract was mixed in 500 μL of 5% NaCl by mixing at vortex, followed by centrifugation at 3000× *g* for 10 min. *n*-hexane was used to dilute the samples (supernatants) before the measurement of absorbance values at 460 nm. Different concentrations of the standard were used to obtain a standard curve. The total carotenoids were quantified from the standard curve and sample absorbance values as mg of β-carotene equivalent (BCE) per 100 g of date flesh.

2.5. Biological Properties of Date Flesh Extracts

2.5.1. Phosphomolybdenum Complex Method for the Antioxidant Activity

A previously described phosphomolybdenum complex (PC) method [24] was used for the estimation of the antioxidant activity of date flesh extracts obtained using different techniques. A 100 μg sample of date flesh extract from each technique was solubilized in 1 mL of methanol. From this solution, a 400 μL sample of this solution was mixed with 4 mL of PC solution that contained 0.6 M sulphuric acid, 2 mM sodium phosphate and 4 mM ammonium molybdate. The blank consisted of 4 mL PC solution and of reagent solution and 400 μL of methanol. Blank and sample tubes were capped and placed in water bath to incubate at 95 °C for 1 h. Followed by cooling under tap water, samples were measured for absorbance at 695 nm. The ascorbic acid calibration curve was prepared following the above procedure for comparison and measurement of antioxidant activity.

2.5.2. Free Radical Scavenging Activity

The radical scavenging activity of date flesh extracts was measured using 1-diphenyl-2-picrylhydrazyl (DPPH) method [25]. Briefly, 1 mL sample (1 mg of extract of sample was dissolved in 100 mL methanol) was made to react with 2 mL of DPPH solution (prepared mixing 1 mg of DPPH reagent in 100 mL of methanol). After 5 min incubation at room temperature, the absorbance values of reaction mixture were estimated at 517 nm. The radical scavenging activities were expressed as IC_{50} values, where a lower value reflected higher ability of the extract to scavenge DPPH free radicals.

2.5.3. Ferric Reducing Antioxidant Power (FRAP)

A method described by Benzie and Szeto [26] was used for the evaluation of FRAP of date flesh extracts, after some modifications. The FRAP reagent used in this assay consisted of acetate buffer (300 mM and pH 3.6); ferric chloride hexahydrate (20 mM in water) and TPTZ (10 mM in 40 mM HCl). Date flesh extract was dissolved in 3 mL of FRAP solution, making the concentration of the extract 0.02–1 mg/mL. Once incubated at room temperature for 30 min, the absorbance values of the reaction mixture were recorded at 593 nm. The trolox (6-hydroxy-2,5,7,8-tetramethylchroman-2-carboxylic acid) was employed as a standard for calibration curve preparation. The FRAP activity was expressed as millimole trolox equivalent (mmolTE)/100 g.

2.5.4. Cation Radical Scavenging Assay

A 2, 2- azinobis, 3- ethylbenzothiazoline, 6- sulfonic acid diammonium salt (ABTS) was used to measure the cation radical scavenging activity of date flesh extracts. A previously explained method [27] was used for this assay, in which ABTS (7 mM) was reacted with potassium persulfate (2.45 mM) under dark conditions for 12–16 h at ambient temperatures. Diluted mixture (ethanol for dilution) was evaluated for absorbance using a spectrophotometer and a value of 0.700 ± 0.01 at 734 nm was achieved for the mixture. Samples from extracts (10–100 µg/mL ethanol) were mixed with 3 mL of ABTS reagent and incubated at 23 °C for 6 min, after which the absorbance values were recorded. Trolox was used for constructing the calibration curve and the results were expressed as µmol TE/100 g.

2.6. Statistical Analyses

The analytical and extraction methods were properly replicated (at least triplicates) and the data were expressed as means ± SD. Regression analysis and analysis of variance (ANOVA) methods were applied to the obtained data using a statistical analysis system (SAS, version 9.4) software and the significance was defined as $p < 0.05$.

3. Results and Discussion

3.1. Date Flesh Extraction Using Supercritical Fluid Technique, Optimization and Modelling of the Method

Date flesh samples from four different date fruit verities, namely Sukari, Ambara, Majdool and Sagai (as shown in Figure 1), were estimated for moisture contents. These were found to be 21.71 ± 0.97, 28.35 ± 1.29, 28.42 ± 1.02 and 26.93 ± 1.10%, respectively. Considering the possibility that SFE, using CO_2 as supercritical fluid, could yield extracts of superior quality with higher bioactive contents from date fruit flesh in comparison to other conventional methods [11], an experiment involving 18 different sets of process variables (temperature (X_1), pressure (X_2) and CO_2 flow rate (X_3)) was designed (Table 1). The extract yield was selected as response variable (Y). The extracts from flesh of Sukari, Ambara, Majdool and Sagai dates were denoted as SKFE, AMFE, MJFE and SGFE, respectively. A similar approach was previously followed and reported on using date seed from the same varieties [18].

Table 1. Date flesh extraction experimental design for supercritical fluid extraction (SFE).

Run	SFE Process Variables			Response Variable (Yield %)			
	Temperature °C (X_1)	Pressure MPa (X_2)	CO_2 Flow mL/min (X_3)	Sukari Flesh (Y_1)	Ambara Flesh (Y_2)	Majdool Flesh (Y_3)	Sagai Flesh (Y_4)
1	62.91	20.07	4.09	4.85 ± 0.58	4.32 ± 0.41	6.11 ± 0.42	6.14 ± 0.47
2	62.91	34.93	4.09	4.74 ± 0.62	4.61 ± 0.18	6.23 ± 0.35	5.61 ± 0.25
3	52.50	27.50	2.75	7.62 ± 0.75	9.94 ± 0.71	8.64 ± 0.42	8.74 ± 0.92
4	52.50	40.00	2.75	7.61 ± 0.51	10.01 ± 0.68	10.83 ± 0.78	9.18 ± 0.35
5	52.50	27.50	2.75	8.62 ± 0.72	9.59 ± 0.45	9.59 ± 0.92	9.72 ± 0.45
6	52.50	27.50	2.75	9.81 ± 0.55	9.71 ± 0.64	9.72 ± 0.34	9.45 ± 0.62
7	52.50	27.50	0.50	5.93 ± 0.66	8.94 ± 0.63	6.24 ± 0.47	5.62 ± 0.374
8	62.91	20.07	1.41	4.96 ± 0.34	4.72 ± 0.24	5.45 ± 0.24	4.45 ± 0.28
9	52.50	27.50	2.75	8.79 ± 0.49	9.70 ± 0.57	9.84 ± 0.44	8.78 ± 0.34
10	62.91	34.93	1.41	5.28 ± 0.64	4.91 ± 0.72	6.53 ± 0.36	5.65 ± 0.62
11	70.00	27.50	2.75	2.10 ± 0.21	1.2 ± 0.49	1.1 ± 0.08	1.54 ± 0.09
12	52.50	15.00	2.75	7.60 ± 0.48	9.09 ± 0.45	8.75 ± 0.78	6.86 ± 0.52
13	42.09	20.07	1.41	5.91 ± 0.35	5.87 ± 0.35	6.84 ± 0.35	5.80 ± 0.37
14	42.09	34.93	4.09	8.97 ± 0.43	6.90 ± 0.40	7.61 ± 0.38	7.72 ± 0.62
15	42.09	20.07	4.09	8.55 ± 0.68	6.77 ± 0.27	10.98 ± 0.62	10.78 ± 0.66
16	42.09	34.93	1.41	5.92 ± 0.71	6.53 ± 0.42	6.24 ± 0.47	6.74 ± 0.28
17	35.00	27.50	2.75	5.86 ± 0.19	3.32 ± 0.12	3.66 ± 0.19	3.41 ± 0.18
18	52.50	27.50	5.00	11.13 ± 1.65	10.23 ± 0.46	11.86 ± 0.89	10.90 ± 0.95

Values shown here are means ± standard deviations.

SFE process from different date varieties were invariably affected by the selected process parameters in terms of the total extract yield. In the case of SKFE, the regression analyses of the collected data revealed that all SFE parameters had significant effects ($p < 0.05$) on the yield of SKFE and the total model was mainly affected by linear and quadratic terms. The total model was highly significant ($p = 0.001$), and a full model can be presented using Equation (4). The R^2 value for this model was 0.9291.

$$Y_{SKFE} = -48.641575 + 1.767577X_1 + 0.508405X_2 + 4.228512X_3 - 0.016315X_1^2 - 0.0008774X_2^2 - 0.0882784X_3^2 \\ -0.000356X_1X_2 - 0.056812X_1X_3 - 0.000251X_2X_3 \quad (4)$$

Preparation of AMFE using SFE method was significantly ($p < 0.05$) affected by the temperature of the extraction, whereas pressure and fluid flow rate seemed to be nonsignificant. The total model was significant ($p < 0.05$); however, only quadratic terms showed significant effects. A full model for AMFE can be presented using Equation (5) with R^2 value of 0.9458.

$$Y_{AMFE} = -72.441290 + 2.839923X_1 + 0.434073X_2 + 2.317291X_3 - 0.027195X_1^2 - 0.006639X_2^2 - 0.198606X_3^2 \\ -0.000501X_1X_2 - 0.017653X_1X_3 - 0.005399X_2X_3 \quad (5)$$

MJFE was prepared from Majdool fruit flesh using the designed experiment. Temperature and CO_2 flow rate were significant ($p < 0.05$) process variables. The linear and quadratic interaction during regression analysis revealed significant impact. The total model (R^2 value: 0.9419), as presented using Equation (6), was also significant ($p < 0.05$).

$$Y_{MJFE} = -48.828114 + 2.224780X_1 - 0.437084X_2 + 4.937134X_3 - 0.022949X_1^2 + 0.002444X_2^2 - 0.070720X_3^2 \\ +0.008355X_1X_2 - 0.046149X_1X_3 - 0.046830X_2X_3 \quad (6)$$

The SFE model (Equation (7)) for extract yield (SGFE) can be presented using the following equation (R^2 value: 0.9350), and it was observed that, like the majority of other date flesh extract data, the linear and quadratic terms demonstrated a significant effect on the response variable, i.e., the yield of the extract. Temperature seemed to be the most significant ($p = 0.0002$) process variable during SFE for SGFE.

$$Y_{SGFE} = -56.097246 + 2.078330X_1 + 0.274720X_2 + 5.538783X_3 - 0.020790X_1^2 - 0.005262X_2^2 \\ -0.114781X_3^2 + 0.004509X_1X_2 - 0.038622X_1X_3 - 0.071940X_2X_3 \quad (7)$$

In general, temperature and pressure were the most important process variables during SFE using CO_2, in addition to various other factors (time, cosolvent, CO_2 flow rate, etc.) that could have affected the extraction of natural antioxidants from different types of plant material [14,18,25]. During the extraction of carotenoids and other bioactive compounds, temperature and pressure were observed to be the main variables affecting the extract yield [28]. The relationship of SFE pressure and temperature for the extract yields from flesh of four different date fruits can be presented by response surface plots (Figure 2) obtained using regression analysis data and according to the models presented in Equations (4)–(7). As can be observed from data in Table 1 and Figure 2, the yields of extracts first increased with increasing temperature at fixed pressure and the maximal yields were obtained at an SFE temperature of 52.50 °C and pressure of 27.50 MPs. Regarding CO_2 flow rate, the highest extract yields were obtained when the flow rate was kept at 5 mL/min and this was the highest rate used among all 18 trials. We observed that higher temperatures (60–70 °C) did not seem to have positive effects on the extracts' yields. The SFE extracts prepared during experimental run 18, demonstrating the maximal extract yields for all four types of date fruit flesh, were considered optimal in the current study. The high extract yields in run 18 indicate that the levels of process variables chosen in this run and their combinations proved the best to maximize extract yields from date flesh. However, as the process was optimized only for the global, it may be important that bioactive compound (phenolics, flavonoids, carotenoids etc.) yields be considered a response variable in place of extract yield.

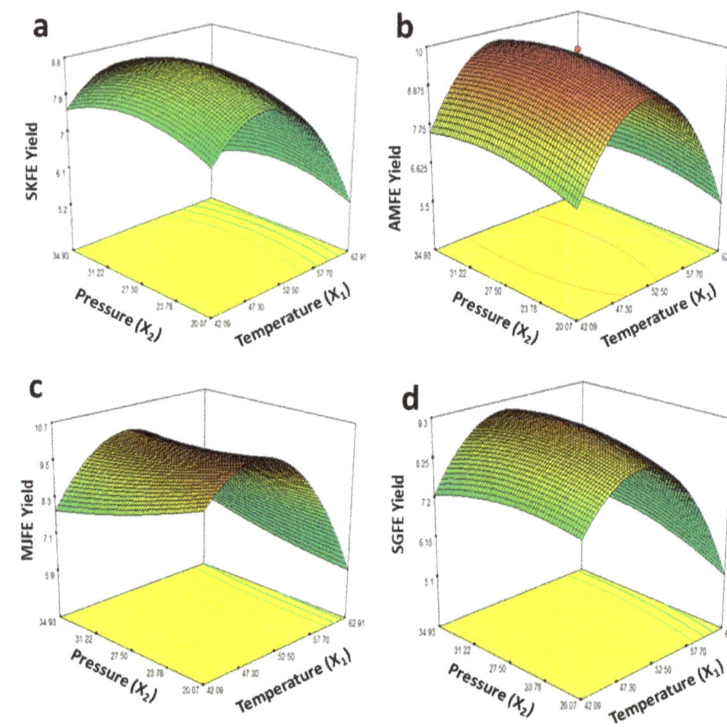

Figure 2. Response surface graphs showing relationship of supercritical fluid (CO_2) extraction (SFE) temperature and pressure (process variables) with extract yields (response variable) from Sukari (SKFE, (**a**)), Ambara (AMFE, (**b**)), Majdool (MJFE, (**c**)) and Sagai (SGFE (**d**)) flesh extracts.

Global extraction yield is an important response variable when it comes to optimization of an extraction process, as it generally directly relates with the yield of individual phytochemicals and natural antioxidants from plant matrices [29].

3.2. Comparison of Different Techniques for the Extraction Yields from Date Flesh

SFE is believed to yield extracts with greater purity and higher biological activity [28]. However, other techniques, such as subcritical CO_2, can also be recommended [18]. The optimized extraction yields, as obtained during run 18 in Table 1, were compared with those obtained from two other techniques i.e., Soxhlet and subcritical CO_2 ($SubCO_2$), as optimized in other studies [18,19]. The results for extraction yields from three different techniques employed in the current study are presented in Table 2.

Table 2. Date flesh yield (%) obtained using Soxhlet, subcritical CO_2 ($SubCO_2$) and supercritical fluid-CO_2 (SFE) extraction methods.

Sample	Extract Yields (%)		
	Soxhlet (70 °C for 12 h, n-Hexane)	$SubCO_2$ (250 Cycles, 29 °C, 6.8 MPa, Ethanol (95%) as Co-Solvent)	SFE (52.5 °C, 27.50 MPa, 5 mL CO_2/min)
Sukari seed	1.92 ± 0.08^{bC}	3.68 ± 0.76^{abB}	11.13 ± 1.65^{abA}
Ambara seed	2.17 ± 0.16^{aC}	4.19 ± 0.57^{aB}	10.23 ± 0.46^{bA}
Majdool seed	0.44 ± 0.05^{dC}	2.75 ± 0.67^{cB}	11.86 ± 0.89^{aA}
Sagai seed	1.46 ± 0.13^{cC}	3.52 ± 0.52^{bB}	10.90 ± 0.95^{abA}

Values shown here are means ± standard deviations (SD). Means with different small letter superscripts in the same column are significantly different ($p < 0.05$). Means within rows with different capital letter superscripts are significantly different ($p < 0.05$).

It was observed that the extract yield obtained using SFE was significantly ($p < 0.05$) higher than the other two methods in each type of date flesh. Moreover, the extract yields of the $SubCO_2$ method were significantly ($p < 0.05$) higher than those using Soxhlet method. In addition to the effects of extraction technique used, significant differences in extract yields were observed within date fruit cultivars within the same technique owing to the possibility of differences in fruit matrices and compositional variations. The beneficial effects of $SubCO_2$ and SFE in the recovery of extracts from all the varieties were quite evident. In the present study, $SubCO_2$ method, an advanced technique based on Soxhlet method using liquid CO_2 and pure ethanol as cosolvent, significantly improved the solvating properties of CO_2 at 29 °C [15,30]. The Soxhlet method, using n-hexane along with higher temperatures (70 °C), may be considered environmentally toxic and energy intensive [31]. In the present study, the high temperature method seemed to be less beneficial (0.44–2.17% yield) for extraction from date flesh, whereas it showed higher extraction yields (93.44–5.66%) from date seed in a previous study [18], proving the significance of plant matrix during extraction. Majdool flesh showed the lowest extract yields in comparison to the other three date flesh types during the Soxhlet and $SubCO_2$ methods. However, Majdool extract yield was the highest when obtained using SFE, proving the importance of selection of the appropriate optimized method for each type of plant material. A general concept that higher temperatures positively affect extraction yields [22,32] was also contradicted here. We inferred that solubility properties of CO_2, both in subcritical (2.75–4.19% yield) and supercritical (10.23–11.86% yield) states, showed profound improvements in date flesh extract recovery at lower temperatures in comparison to the Soxhlet method. Extraction pressure ($SubCO_2$ and SFE systems) also showed positive effects on extraction yields. The application of high pressure during extraction of phytochemicals may cause cell disruptions and structural modification in the plant matrix, increasing the density of solvent and improve the CO_2 solubility strength [11,17,18]. Other plant matrices, such as feijoa leaf [33] and date seed [18], also showed significant improvements in extract yields due to pressure and the use of CO_2 as solvent, as compared to low pressure and organic solvent high temperature methods (Soxhlet). $SubCO_2$ was previously used as pretreatment for enhanced recovery of sugars from date fruit flesh. It is expected to enhance the recovery of bioactive compounds from date flesh, despite the fact that the global extract yield from date flesh is considerably lower than SFE [30]. It was clear that the extraction yield of date flesh using the Soxhlet system was very low compared to green extraction methods (SFE, $SubCO_2$). Furthermore, it makes use of toxic organic solvents, such as n-hexane in the current study, whereas ethanol was used as cosolvent or modifier in SFE and $SubCO_2$ systems. Organic solvents (30% ethanol in SFE and 99.9% ethanol in $SubCO_2$) were used as modifiers/cosolvents in current study. A modifier or cosolvent is used to improve the solubility power of CO_2 for extraction of phytochemicals [13]. Ethanol is generally considered a nontoxic, food pharmaceutical-grade and environment friendly alcohol [11,32].

3.3. Phytochemicals and Their Contents in Date Flesh Extracts

Date flesh is the edible part of date fruit and contains sugars, essential and nonessential amino acids, dietary fiber, vitamins, minerals and phytochemicals, including phenolic antioxidants, tannin-based pigments, carotenoids and epicatechin oligomers [5,6]. In the current study, date flesh extracts were prepared using different techniques from four different date verities. The extracts were evaluated for total phenolic compounds (TPC), total flavonoids (TFC), total anthocyanins (TAC) and total carotenoids contents (TCC). The results for these quantitative analyses are presented in Table 3.

Table 3. Antioxidants and bioactive compounds of date fruit flesh extracts from four date varieties subjected to different extraction methods.

Extraction Technique	Date Extract	Total Phenolics (mgGAE/100 g)	Total Flvonoids (mgQE/100 g)	Total Anthocyanins (mg/100 g)	Total Carotenoids (mgBCE/100 g)
Soxhlet	SKFE	198.24 ± 1.77^{i}	98.67 ± 1.49^{h}	0.41 ± 0.01^{i}	1.33 ± 0.04^{j}
	AMFE	204.37 ± 1.04^{i}	82.12 ± 0.98^{i}	0.62 ± 0.02^{f}	1.40 ± 0.03^{i}
	MJFE	221.06 ± 2.94^{h}	123.94 ± 2.19^{g}	0.52 ± 0.02^{g}	1.24 ± 0.03^{k}
	SGFE	186.37 ± 1.05^{j}	85.83 ± 0.70^{i}	0.48 ± 0.02^{g}	1.48 ± 0.03^{h}
$SubCO_2$	SKFE	426.14 ± 5.30^{b}	199.01 ± 4.04^{b}	0.92 ± 0.02^{e}	2.15 ± 0.06^{g}
	AMFE	316.01 ± 3.34^{f}	157.50 ± 2.24^{e}	1.38 ± 0.03^{a}	2.85 ± 0.04^{a}
	MJFE	447.32 ± 4.43^{a}	215.28 ± 4.53^{a}	0.98 ± 0.03^{d}	2.59 ± 0.03^{c}
	SGFE	335.51 ± 5.72^{e}	166.32 ± 4.07^{d}	1.29 ± 0.05^{b}	2.67 ± 0.03^{b}
SFE CO_2	SKFE	346.35 ± 2.32^{d}	180.31 ± 2.03^{c}	0.98 ± 0.04^{d}	2.26 ± 0.02^{f}
	AMFE	300.77 ± 1.77^{g}	152.88 ± 2.53^{e}	1.34 ± 0.03^{ab}	2.36 ± 0.01^{e}
	MJFE	375.25 ± 5.06^{c}	137.45 ± 2.04^{f}	1.02 ± 0.05^{d}	2.53 ± 0.03^{d}
	SGFE	325.71 ± 1.34^{ef}	123.13 ± 1.98^{g}	1.22 ± 0.03^{c}	2.39 ± 0.05^{e}

Values shown here are means ± SD. $SubCO_2$: Subcritical CO_2 extraction; SFE CO_2: Supercritical fluid extraction; SKFE: Sukari Flesh Extract; AMFE: Ambara Flesh Extract; MJFE: Majdool Flesh Extract; Sagai Flesh Extract. Means with small letters superscripts in columns are significantly different ($p < 0.05$).

3.3.1. Total Phenolics and Total Flavonoids

Among phytochemicals, phenolic compounds are dominant. They are considered non-nutrients, but are very important due to the health benefits associated with their consumption [34]. TPC of date flesh extracts obtained using three different techniques are presented in Table 1. As can be observed, date flesh is a good source of these natural antioxidants, and extraction methods significantly ($p < 0.05$) affected their contents in the extracts from the flesh of Saudi date fruit. TPC ranged between 186.37–447.32 mgGAE/100 g among 12 extract samples, where $SubCO_2$ extracts showed higher contents than in extracts prepared using other methods. Significant ($p < 0.05$) differences in TPC of fruit from different varieties were also observed. Majdool date flesh seemed to be particularly rich in these compounds. Extraction methods seemed to considerably improve the recovery of these compounds from date flesh. SKFE, when prepared using the Soxhlet method, contained 198.24 mgGAE/100 g of TPC, which increased to 426.14 mgGAE/100 g when the $SubCO_2$ method was applied for their recovery from date flesh matrix.

The second most abundant group of phytochemicals detected in date flesh extract was TFC. Similar to phenolics, flavonoids are considered as polyphenols with valuable biological properties that are beneficial for human health [35]. Significant ($p < 0.05$) difference were observed in TFC of date flesh extracts depending on the variety of fruit and the extraction methods used. Majdool seemed to be the richest source, and $SubCO_2$ the best extraction method, in the current study. The minimum (85.83 mgQE/100 g) TFC were observed in SGFE, whereas those of MJFE were the highest (215.28 mgQE/100 g), among the studied samples.

Considering the TPC and TFC of the date flesh extracts, it could be inferred that the $SubCO_2$ technique could be a good process for maximizing their recovery. Other modified processes, such as subcritical water extraction (SCWE), could also be used [36].

Application of SCWE was previously reported for the recovery of phenolic compounds from microwave pretreated date fruit pulp (MW-FP), which significantly improved their recovery in comparison to conventional methods [37]. Both $SubCO_2$ and SFE techniques were found to be more effective for the recovery of phenolic and flavonoids compounds using less energy-extensive methods than Soxhlet. The process used lower extraction temperatures and can be considered a green method for obtaining phytochemical-rich extracts from date flesh [38–40].

3.3.2. Total Anthocyanins and Total Carotenoids

The total anthocyanin (TAC) and total carotenoids contents (TCC) in date fruit flesh extracts from three different extraction methods are shown in Table 3. The detected TAC in date fruit flesh were in the range of 0.41–1.38 mg/100 g. The recovery of TAC was significantly ($p < 0.05$) higher when $SubCO_2$ and SFE techniques were used for their recovery. The varietal differences among TAC were also evident, with Ambara date flesh showing higher contents of anthocyanins. Among TCC results, the $SubCO_2$ AMFE showed the highest contents (2.85 mgBCE/100 g), whereas the lowest TCC (1.24 mgBCE/100 g) was detected in Soxhlet MJFE. Similar to the TAC, TCC of $SubCO_2$ and SFE extracts were significantly higher than those prepared using Soxhlet. Both anthocyanins and carotenoids are important phytochemicals with various reported health benefits. Anthocyanins, pigmented natural antioxidants, have been studied for their antioxidative, anti-inflammatory and cancer-preventive properties using different in vivo and in vitro methods [41,42]. Similar to anthocyanins, carotenoids are also colored compounds or natural pigments with reported health benefits [43]. The TAC and TCC reported in the present study were in accordance with other reports. Fard, Khasab, and Khalas date varieties were observed to contain 0.24−1.52 mg/100 g and 1.31−3.03 mg/100 g of TAC and TCC, respectively [44]. Babova et al. [45] used both supercritical and subcritical CO_2 for the extraction of anthocyanins and other polyphenolic antioxidants from bilberry fruit. A combined process (SFE and $SubCO_2$) was used to improve the selectivity and recovery of phytochemicals from bilberry. This approach resulted in improved extraction selectivity for anthocyanins, cyanidin-3-O-glucoside, and cyanidin-3-O-arabinoside. Other phenolic antioxidants, including delphinidin-3-O-glucoside, ellagic acid pentoside, feruloyl hexoside and several quercetin glycosides, were also selectively extracted.

3.4. Biological Activities of Date Flesh Extracts

The human body undergoes various degenerative phenomena as a result of free radicals. These free radicals are responsible for different deleterious oxidation reactions. Phytochemicals, such as those detected in date fruit flesh extracts, as well as certain vitamins, enzymes and amino acids, help in the eradication of such reactive species or help increase the human body's immunity [46]. The research on in vitro estimation of the ability of different phytochemical and natural extracts to fight against oxidative processes is important. It was observed that extracts from the flesh of date fruit contained large amounts of phytochemicals, hence, their biological activities were determined using different in vitro assays, including phosphomolybdenum complex method (antioxidant), free radical scavenging activity (DPPH), ferric-reducing antioxidant power (FRAP), and ABTS cation radical-scavenging (ABTS). The results are summarized in Table 4. These assays are based on calorimetric principles and involve finding the abilities of phytochemicals to prevent the activities of the oxidation causing agents or reactive species such DPPH radicals, ferric and ABTS ions [46].

Table 4. Biological properties and antioxidant potential of date flesh extracts obtained using different extraction techniques.

Extraction Techniques	Date Extract	Antioxidant (µgAAE/mL)	DPPH IC_{50} (µg/mL)	FRAP (mmolTE/100 g)	ABTS (µmolTE/100 g)
Soxhlet	SKFE	23.90 ± 1.50^{f}	158.41 ± 3.05^{d}	2.41 ± 0.05^{j}	474.64 ± 13.08^{g}
	AMFE	17.79 ± 0.69^{g}	179.48 ± 1.51^{b}	2.57 ± 0.03^{i}	486.37 ± 9.84^{g}
	MJFE	27.59 ± 1.60^{e}	170.09 ± 1.48^{c}	2.93 ± 0.03^{h}	574.22 ± 6.63^{f}
	SGFE	22.44 ± 1.70^{f}	191.36 ± 2.10^{a}	2.18 ± 0.04^{k}	444.75 ± 4.92^{h}
$SubCO_2$	SKFE	38.91 ± 2.89^{bc}	44.65 ± 1.25^{h}	5.01 ± 0.07^{a}	789.04 ± 13.65^{c}
	AMFE	30.24 ± 2.58^{e}	63.15 ± 1.56^{f}	3.79 ± 0.06^{f}	669.10 ± 11.33^{e}
	MJFE	45.08 ± 1.53^{a}	34.66 ± 1.68^{j}	4.66 ± 0.16^{b}	883.96 ± 10.58^{a}
	SGFE	36.72 ± 1.60^{cd}	53.20 ± 1.67^{g}	3.8 ± 0.14^{f}	594.16 ± 7.26^{f}
SFE CO_2	SKFE	29.17 ± 1.69^{e}	86.14 ± 1.56^{e}	3.61 ± 0.03^{g}	715.76 ± 6.24^{d}
	AMFE	35.56 ± 2.51^{d}	46.43 ± 1.33^{h}	3.95 ± 0.04^{e}	655.85 ± 12.31^{e}
	MJFE	40.17 ± 1.01^{b}	40.18 ± 0.44^{i}	4.25 ± 0.04^{c}	848.31 ± 22.16^{b}
	SGFE	34.40 ± 0.73^{d}	54.60 ± 1.45^{g}	4.10 ± 0.06^{d}	584.31 ± 9.87^{f}

Values shown here are means ± SD. $SubCO_2$: Subcritical CO_2 extraction; SFE CO_2: supercritical fluid extraction; Antioxidant: phosphomolybdenum complex antioxidant activity; DPPH: DPPH free radical scavenging activity; FRAP: ferric reducing antioxidant power and ABTS: ABTS cation radical-scavenging; SKFE: Sukari Flesh Extract; AMFE: Ambara Flesh Extract; MJFE: Majdool Flesh Extract; Sagai Flesh Extract. Means values carrying different small letter superscripts in columns are significantly different ($p < 0.05$).

3.4.1. Antioxidant Activity Using Phosphomolybdenum Complex

The data presented in Table show that the antioxidant activity of date flesh extracts was significantly ($p < 0.05$) affected by the extraction technique used. The antioxidant activity ranged from 17.79 in Soxhlet-AMFE to 45.08 µgAAE/mL in $SubCO_2$-MJFE. The $SubCO_2$ method seemed to be the best for SKFE, MJFE and SGFE in terms of their antioxidant activity. The SFE method enhanced the antioxidant activities (35.56 µgAAE/mL) of AMFE in comparison to Soxhlet and $SubCO_2$ methods, where it was 17.79 and 30.24 µgAAE/mL, respectively.

3.4.2. Scavenging Activity against DPPH Radicals

The ability of date flesh extracts to scavenge free DPPH radicals was expressed using IC_{50} values, i.e., the antioxidants' quantity in the extract for 50% reduction in DPPH radicals. A lower IC_{50} value indicated that lesser amounts of antioxidants were needed in free radical scavenging, indicating their higher antiradical ability [18]. Both $SubCO_2$ and SFE extracts had significantly higher ability to scavenge free radicals or lower IC_{50} values than Soxhlet extracts. The lowest IC_{50} value was in 34.66 µg/mL in $SubCO_2$-MJFE followed by 40.18 µg/mL in SFE-MJFE. Under each extraction technique, MJFE showed better antiradical activity then other date flesh extracts obtained by the same method. Furthermore, the ability of date flesh extracts to scavenge DPPH radicals was superior than date seed extracts, as observed in a previous study [18], in which seed extracts had 109.69 µg/mL of IC_{50} value in $SubCO_2$ extracts of Majdool seed and the highest IC_{50} value (353.83 µg/mL), or the weakest potential against DPPH free radicals, was observed in extracts of Sagai date seed. We may infer that both seed and flesh extracts of Majdool dates have good antioxidant and antiradical potential. MJFE also demonstrated the presence of higher amounts of phenolics and flavonoids (Table 3) and it is probable that these phytochemicals contributed to the higher antioxidant potential of these dates.

3.4.3. Reducing Ferric to Ferrous Ions

In the FRAP test, the antioxidants present in plant extracts can transfer an electron, which results in the reduction of ferric ions (Fe^{3+}) to ferrous ions (Fe^{2+}) [46]. The FRAP values of date flesh extracts (Table 4) ranged from 2.18 in Soxhlet-SGFE to 5.01 mmolTE/100 g in $SubCO_2$-SKFE. The highest FRAP value (4.25 mmolTE/100 g) among SFE extracts was shown in the case of MJFE, whereas the FRAP potential of the same extract, when prepared using Soxhlet, was 2.93). Hence, significant ($p < 0.05$) improvements were observed when SFE and $SubCO_2$ were used for extract preparation from date flesh. The FRAP values of flesh extracts of all other dates were also higher in SFE and $SubCO_2$ extracts in comparison to their Soxhlet counterparts. The differences in FRAP values could be related to the date variety; however, extraction method seemed to play a major role. Similar results were obtained in previous studies where both extraction methods and plant variety seemed to affect the FRAP of the extracts [47,48].

3.4.4. Antiradical Activity against ABTS Cations

This test can be referred to as a mixed assay, as it involves the transfer of both electrons and a hydrogen atom from the main reagent, 2,2′-Azinobis-(3-ethylbenzothiazoline-6-sulfonic acid (ABTS), during its reaction with antioxidants present in phytochemical-rich extracts [49,50]. The date flesh extracts prepared in the current study, using different extraction techniques and date varieties, showed substantial ABTS cations scavenging abilities. The $SubCO_2$ and SFE extraction methods showed significance ($p < 0.05$) in the abilities of date flesh extracts to scavenge ABTS cations, when compared with Soxhlet extract. The lowest (444.75 µmolTE/100 g) activity was observed in Soxhlet-SGFE, whereas the highest (883.96 µmolTE/100 g) was observed in $SubCO_2$ MJFE. MJFE showed higher ABTS cation scavenging activity in each extraction method, whereas SGFE showed the lowest values in each extraction category. SKFE also showed good ABTS potential among each extraction category. The use of innovative techniques for the extraction of phytochemicals can increase their recovery which, in turn, results in improvement of their biological potential. Bioactive compound quantities and antiradical activity against ABTS cations were reported to increase in date fruit extracts prepared using the subcritical water extraction method in another study [37].

The biological activities reported here have been previously studied for different fruit extracts. Generally, they are associated with the presence of polyphenolic antioxidants and other phytochemicals as is the case with bilberry fruit extracts prepared using SFE and $SubCO_2$, where the antioxidant properties were closely associated with the presence of phytochemicals such as cyanidin-3-O-glucoside, delphinidin-3-O-glucoside, chlorogenic acid, caffeic acid derivatives, flavonoids, proanthocyanidins, ellagic acid and other phenolic acids [45]. These compounds have been reported to neutralize reactive oxygen species and inhibit their formation. Furthermore, these phytochemicals, and others (such as those reported here in SKFE, AMFE, MJFE and SGFE) may induce DPPH radical scavenging, ferric chelation and ABTS cation scavenging [51]. The phytochemicals present in fruit extracts are rich in phytochemicals which can impart anti-inflammatory effects, thereby preventing cell degeneration, offering treatment of diabetes, and suppressing cancer cell invasion [52–54]. In the current study, the $SubCO_2$ extraction method seemed to be the best for extracting higher amounts of phytochemicals with better bioactivities. Majdool date, which is also known as Mejdool and Mahjool, seemed to have the highest contents of bioactive compounds, with significantly higher antioxidant effects than the remaining studied date varieties. Majdool is a globally-demanded date fruit due to its large size and soft fruit. Current studies have reported it as an excellent source of nutraceuticals [55]. The higher yields of phytochemicals and increased activities in extracts from date flesh using SFE and $SubCO_2$ methods in comparison to Soxhlet (which may extract nonpolar compounds) might be a result of extraction of both polar and nonpolar compounds due to combined temperature and pressure effects (SFE and $SubCO_2$ methods), as reported in other studies [56,57]. Further studies may be carried out for detailed analytical characterization of different bioactive compounds and any other toxic components, if present in the date flesh extracts, using chromatographic analytical techniques.

4. Conclusions

Date fruit is a good source of phytochemicals with valuable antioxidant properties, which may make it suitable for functional food and nutraceutical applications. Such applications could be made more effective and beneficial if these phytochemicals were recovered using safe and innovative extraction techniques. Date fruit flesh from four different date varieties was extracted in order to analyze their bioactive compounds using conventional and modern extraction processes—i.e., Soxhlet and supercritical/subcritical (SFE/SubCO_2) techniques. The extraction yields from modern methods were found to be significantly higher than those from the conventional method. The SFE process was optimized using response surface optimization technique. It was observed that low temperature SFE (57.5 °C) and SubCO_2 (29 °C) methods were more effective in the recovery of phenolics, flavonoids, anthocyanins and carotenoids from date flesh than high temperature (70 °C) Soxhlet methods, as the quantities of all these bioactive compounds were higher when pressure-assisted innovative/green methods were used. The functional qualitiesd of date flesh extracts were also assessed using in vitro procedures for biological/antioxidant properties, i.e., the phosphomolybdenum complex, DPPH, FRAP and ABTS methods. It was observed that date flesh extracts prepared using both SFE and SubCO2 methods were of significantly improved quality in terms of these properties. In addition to these beneficial effects of extraction methods, varietal influences on the qualitative and quantitative aspects of date flesh were also observed. Majdool date flesh extract was richer in bioactive compounds than Sukari, Ambara and Sagai date flesh extracts. As there is generally a correlation among bioactive compound contents and functional properties of the extracts, Majdool flesh extracts showed better antioxidant properties. Further studies based on utilization of these extracts in functional foods could be carried out. A more detailed analytical characterization of date flesh extracts using chromatographic procedures would also help in understanding the selectivity of the extraction techniques and the possible effects of individual compounds on the antioxidant properties of date flesh extracts.

Author Contributions: Conceptualization, methodology, resources, supervision, project administration, funding acquisition, writing-review and editing, K.G.; investigation, writing—original draft preparation, visualization, software, data curation, and validation, M.Z.I.S.; Data curation, resources, F.Y.A.-J.; formal analysis, E.E.B.; M.S.A.; A.K.A. and I.A.M.A. All authors have read and agreed to the published version of the manuscript.

Funding: This project was funded by the National Plan for Science, Technology and Innovation (MAARIFAH), King Abdulaziz City for Science and Technology, Kingdom of Saudi Arabia, award number (15-AGR3527-02).

Institutional Review Board Statement: Not applicable.

Informed Consent Statement: Not applicable.

Data Availability Statement: No new data were created or analyzed in this study. Data sharing is not applicable to this article.

Conflicts of Interest: The authors declare no conflict of interest.

References

1. Ahmed, I.A.; Ahmed, A.W.K.; Robinson, R.K. Chemical composition of date varieties as influenced by the stage repining. *Food Chem.* **1995**, *54*, 305–309. [CrossRef]
2. Ghnimi, S.; Umer, S. Date fruit (*Phoenix dactylifera* L.): An underutilized food seeking industrial valorization. *NFS J.* **2017**, *6*, 1–10. [CrossRef]
3. 25 February 2021/13, Rajab, 1442. Available online: https://saudigazette.com.sa (accessed on 1 October 2022).
4. Abdullahi, M.H.; Garko, M. Medicinal value of date palm (*Phoenix dactylifera* L.). In Proceedings of the Agricultural Society of Nigeria Conference, Nsukka, Nigeria, 11–14 March 2012.
5. Al-Farsi, M.A.; Lee, C.Y. Nutritional and functional properties of dates: A review. *Crit. Rev. Food Sci. Nutr.* **2008**, *48*, 877–887. [CrossRef]
6. Al-Alawi, R.A.; Al-Mashiqri, J.H.; Al-Nadabi, J.S.M.; Al-Shihi, B.I.; Baqi, Y. Date palm tree (*Phoenix dactylifera* L.): Natural products and therapeutic options. *Front. Plant Sci.* **2017**, *8*, 845. [CrossRef] [PubMed]

7. Khalid, S.; Khalid, N.; Khan, R.S.; Ahmed, H.; Ahmad, A. A review on chemistry and pharmacology of Ajwa date fruit and pit. *Trends Food Sci. Tech.* **2017**, *63*, 60–69. [CrossRef]
8. Nasir, M.U.; Hussain, S.; Jabbar, S.; Rahid, F.; Khalid, N.; Mehmood, A. A review on the nutritional content, functional properties and medicinal potential of dates. *Sci. Lett.* **2014**, *3*, 17–22.
9. Saafi, E.B.; El Arem, A.; Issaoui, M.; Hammami, M. Phenolic content and antioxidant activity of four date palm (*Phoenix dactylifera* L.) fruit varieties grown in Tunisia. *Food Sci. Technol.* **2009**, *56*, 2314–2319.
10. Spigno, G.; Tramelli, L.; De-Faveri, D.M. Effects of extraction time, temperature and solvent on concentration and antioxidant activity of grape marc phenolics. *J. Food Eng.* **2008**, *81*, 200–208. [CrossRef]
11. Ghafoor, K.; Al-Juhaimi, F.Y.; Choi, Y.H. Supercritical fluid extraction of phenolic compounds and antioxidants from grape (*Vitis labrusca* B.) seeds. *Plant Foods Hum. Nutr.* **2012**, *67*, 407–414. [CrossRef]
12. Babovic, N.; Djilas, S.; Jadranin, M.; Vajs, V.; Ivanovic, J.; Petrovic, S.; Zizovic, I. Supercritical carbon dioxide extraction of antioxidant fractions from selected *Lamiaceae* herbs and their antioxidant capacity. *Innov. Food Sci. Emerg. Technol.* **2010**, *11*, 98–107. [CrossRef]
13. Ferdosh, S.; Sarker, M.Z.I.; Ab Rahman, N.N.N.; Akand, M.J.H.; Ghafoor, K.; Awang, M.B.; Ab Kadir, M.O. Supercritical carbon dioxide extraction of oil from *Thunnus tonggol* head by optimization of process parameters using response surface methodology. *Kor. J. Chem. Eng.* **2013**, *30*, 1466–1472. [CrossRef]
14. Easmin, S.; Zaidul, I.S.M.; Ghafoor, K.; Sahena, F.; Juliana, M.J.; Jahurul, M.H.A.; AL-Juhaimi, F.Y.; Fauzi, M.B.; Alfi, K. Extraction of α-glucosidase inhibitory compounds from *Phaleria macrocarpa* fruit flesh using solvent, sonication, and subcritical carbon dioxide soxhlet methods. *J. Food Biochem.* **2017**, *41*, e12399. [CrossRef]
15. Leitner, W. Green chemistry: Designed to dissolve. *Nature* **2000**, *405*, 129–130. [CrossRef] [PubMed]
16. AOAC. *Official Method of Analysis*, 5th ed.; Association of Official Agricultural Chemists: Washington, DC, USA, 1998.
17. Ghafoor, K.; Park, J.; Choi, Y.H. Optimization of supercritical carbon dioxide extraction of bioactive compounds from grape peel (*Vitis labrusca* B.) by using response surface methodology. *Innov. Food Sci. Emerg. Technol.* **2010**, *11*, 485–490. [CrossRef]
18. Ghafoor, K.; Sarker, M.Z.I.; Al-Juhaimi, F.Y.; Babiker, E.E.; Alkaltham, M.S.; Almubarak, A.K. Extraction and evaluation of bioactive compounds from date (*Phoenix dactylifera*) seed using supercritical and subcritical CO_2 techniques. *Foods* **2022**, *11*, 1806. [CrossRef] [PubMed]
19. Arumugham, T.; Rambabu, K.; Hasan, S.W.; Show, P.L.; Rinklebe, J.; Banat, F. Supercritical carbon dioxide extraction of plant phytochemicals for biological and environmental applications—A review. *Chemosphere* **2021**, *271*, 129525. [CrossRef]
20. Singleton, V.L.; Rossi, J.A. Colorimetry of total phenolics with phosphomolybdic-phosphotungstic acid reagents. *Am. J. Enol. Vitic.* **1965**, *16*, 144–158.
21. Biglari, F.; AlKarkhi, A.F.; Easa, A.M. Antioxidant activity and phenolic content of various date palm (*Phoenix dactylifera*) fruits from Iran. *Food Chem.* **2008**, *107*, 1636–1641. [CrossRef]
22. Ghafoor, K.; Hui, T.; Choi, Y.H. Optimization of ultrasound-assisted extraction of total anthocyanins from grape peel. *J. Food Biochem.* **2011**, *35*, 735–746. [CrossRef]
23. Ranjith, A.; Kumar, K.S.; Venugopalan, V.V.; Arumughan, C.; Sawhney, R.C.; Singh, V. Fatty acids, tocols, and carotenoids in pulp oil of three sea buckthorn species (*Hippophae rhamnoides, H. salicifolia, and H. tibetana*) grown in the Indian Himalayas. *J. Am. Oil Chem. Soc.* **2006**, *83*, 359–364. [CrossRef]
24. Prieto, P.; Pineda, M.; Aguilar, M. Spectrophotometric quantitation of antioxidant capacity through the formation of a phosphomolybdenum complex: Specific application to the determination of vitamin E. *Anal. Biochem.* **1999**, *269*, 337–341. [CrossRef] [PubMed]
25. Lee, S.K.; Mbwambo, Z.H.; Chung, H.; Luyengi, L.; Gamez, E.J.; Mehta, R.G.; Pezzuto, J.M. Evaluation of the antioxidant potential of natural products. *Comb. Chem. High Throughput Screen.* **1998**, *1*, 35–46. [CrossRef] [PubMed]
26. Benzie, I.F.; Szeto, Y.T. Total antioxidant capacity of teas by the ferric reducing/antioxidant power assay. *J. Agric. Food Chem.* **1999**, *47*, 633–636. [CrossRef] [PubMed]
27. Re, R.; Pellegrini, N.; Proteggente, A.; Pannala, A.; Yang, M.; Rice-Evans, C. Antioxidant activity applying an improved ABTS radical cation decolorization assay. *Free Rad. Biol. Med.* **1999**, *26*, 1231–1237. [CrossRef]
28. Vafaei, N.; Rempel, C.B.; Scanlon, M.G.; Jones, P.J.H.; Eskin, M.N.A. Application of supercritical fluid extraction (SFE) of tocopherols and carotenoids (Hydrophobic Antioxidants) compared to non-SFE methods. *Appl. Chem.* **2022**, *2*, 68–92. [CrossRef]
29. Kashaninejad, M.; Blanco, B.; Benito-Román, O.; Beltrán, S.; Mehdi Niknam, S.; Sanz, M.T. Maximizing the freeze-dried extract yield by considering the solvent retention index: Extraction kinetics and characterization of *Moringa oleifera* leaves extracts. *Food Bioprod. Process.* **2021**, *130*, 132–142. [CrossRef]
30. Arumugham, T.; AlYammahi, J.; Rambabu, K.; Hassan, S.W.; Banat, F. Supercritical CO_2 pretreatment of date fruit biomass for enhanced recovery of fruit sugars. *Sustain. Energy Technol. Assess.* **2022**, *52*, 102231. [CrossRef]
31. Kumar, S.; Prasad, S.R.; Banerjee, R.; Agarwal, D.K.; Kulkarni, K.S.; Ramesh, K.V. Green solvents and technologies for oil extraction from oilseeds. *Chem. Cent. J.* **2017**, *11*, 9. [CrossRef]
32. Ghafoor, K.; Choi, Y.H.; Jeon, J.Y.; Jo, I.H. Optimization of ultrasound-assisted extraction of phenolic compounds, antioxidants and anthocyanins from grape (*Vitis vinifera*) seeds. *J. Agric. Food Chem.* **2009**, *57*, 4988–4994. [CrossRef]
33. Santos, P.H.; Kammers, J.C.; Silva, A.P.; Oliveira, J.V.; Hense, H. Antioxidant and antibacterial compounds from feijoa leaf extracts obtained by pressurized liquid extraction and supercritical fluid extraction. *Food Chem.* **2021**, *344*, 128620. [CrossRef]

34. de la Rosa, L.A.; Moreno-Escamilla, J.O.; Rodrigo-García, J.; Alvarez-Parrilla, E. Phenolic compounds. In *Postharvest Physiology and Biochemistry of Fruits and Vegetables*; Yahia, E.M., Ed.; Woodhead Publishing: Sawston, UK, 2019; pp. 253–271.
35. Shen, N.; Wang, T.; Gan, Q.; Liu, S.; Wang, L.; Jin, B. Plant flavonoids: Classification, distribution, biosynthesis, and antioxidant activity. *Food Chem.* **2022**, *383*, 132531. [CrossRef] [PubMed]
36. Li, B.; Akram, M.; Al-Zuhair, S.; Elnajjar, E.; Munir, M.J. Subcritical water extraction of phenolics, antioxidants and dietary fibres from waste date pits. *J. Environ. Chem. Eng.* **2020**, *8*, 104490. [CrossRef]
37. Rambabu, K.; AlYammahi, J.; Thanigaivelan, A.; Bharath, G.; Sivarajasekar, N.; Velu, S.; Banat, F. Sub-critical water extraction of reducing sugars and phenolic compounds from date palm fruit. *Biomass Conv. Bioref.* **2022**. [CrossRef]
38. Todd, R.; Baroutian, S. A techno-economic comparison of subcritical water, supercritical CO_2 and organic solvent extraction of bioactives from grape marc. *J Clean. Prod.* **2017**, *158*, 349–358. [CrossRef]
39. Mesquita, P.C.; Rodrigues, L.G.G.; Mazzutt, S.; da Silva, M.; Vitali, L.; Lanza, M. Intensified green-based extraction process as a circular economy approach to recover bioactive compounds from soursop seeds (*Annona muricata* L.). *Food Chem. X* **2021**, *12*, 100164. [CrossRef]
40. Vinitha, U.G.; Sathasivam, R.; Muthuraman, M.S.; Park, S.U. Intensification of supercritical fluid in the extraction of flavonoids: A comprehensive review. *Physiol. Mol. Plant Path.* **2022**, *118*, 101815. [CrossRef]
41. Mazza, G.; Kay, C.D. Bioactivity, absorption and metabolism of anthocyanins. In *Recent Advances in Polyphenols Research*; Daayf, F., Lattanzio, V., Eds.; Blackwell Publishing Ltd.: West Sussex, UK, 2008.
42. Padayachee, A.; Netzel, G.; Netzel, M.; Day, L.; Zabaras, D.; Mikkelsen, D.; Gidley, M.J. Binding of polyphenols to plant cell wall analogues—Part 1: Anthocyanins. *Food Chem.* **2012**, *134*, 155–161. [CrossRef]
43. Murillo, E.; Deli, J.; Nagy, V.; Molinar-Toribio, E.; Sandor, V.; Marton, K.; Agocs, A. Carotenoid profile of two capsorubin-rich tropical plants. *J. Food Compos. Anal.* **2021**, *97*, 103798. [CrossRef]
44. Al-Farsi, M.; Alasalvar, C.; Morris, N.; Baron, M.; Shahidi, F. Comparison of antioxidant activity, anthocyanins, carotenoids, and phenolics of three native fresh and sun-dried date (*Phoenix dactylifera* L.) varieties grown in Oman. *J. Agric. Food Chem.* **2005**, *53*, 7592–7599. [CrossRef] [PubMed]
45. Babova, O.; Occhipinti, A.; Capuzzo, A.; Maffei, M.E. Extraction of bilberry (*Vaccinium myrtillus*) antioxidants using supercritical/subcritical CO_2 and ethanol as co-solvent. *J. Supercrit. Fluids* **2016**, *107*, 358–363. [CrossRef]
46. Munteanu, I.G.; Apetrei, C. Analytical methods used in determining antioxidant activity: A review. *Int. J. Mol. Sci.* **2021**, *22*, 3380. [CrossRef]
47. Al Juhaimi, F.; Ozcan, M.M.; Adiamo, O.Q.; Alsawmahi, O.N.; Ghafoor, K.; Babiker, E.E. Effect of date varieties on physico-chemical properties, fatty acid composition, tocopherol contents, and phenolic compounds of some date seed and oils. *J. Food Process. Preserv.* **2018**, *42*, e13584. [CrossRef]
48. Al Juhaimi, F.; Ghafoor, K.; Ozcan, M.M. Physical and chemical properties, antioxidant activity, total phenol and mineral profile of seeds of seven different date fruit (*Phoenix dactylifera* L.) varieties. *Int. J. Food Sci. Nutr.* **2012**, *63*, 84–89. [CrossRef]
49. Hamzah, N.N.; Ferdosh, S.; Sarker, M.Z.I.; Ghafoor, K.; Yunus, K.; Chowdhury, A.J.K.; Bari, N.A.A. Biological activities and extraction technologies of *Pheonix dactylifera*: A review. *Nat. Prod. J.* **2019**, *9*, 3–13.
50. Siddeeg, A.; AlKehayez, N.M.; Abu-Hiamed, H.A.; Al-Sanea, E.A.; AL-Farga, A.M. Mode of action and determination of antioxidant activity in the dietary sources: An overview. *Saudi J. Biol. Sci.* **2021**, *28*, 1633–1644. [CrossRef] [PubMed]
51. Sak, K. Site-specific anticancer effects of dietary flavonoid quercetin. *Nutr. Cancer* **2014**, *66*, 177–193. [CrossRef] [PubMed]
52. Seo, B.N.; Ryu, J.M.; Yun, S.P.; Jeon, J.H.; Park, S.S.; Oh, K.B.; Park, J.K.; Han, H.J. Delphinidin prevents hypoxia-induced mouse embryonic stem cell apoptosis through reduction of intracellular reactive oxygen species-mediated activation of JNK and NF-kappa B, and Akt inhibition. *Apoptosis* **2013**, *18*, 811–824. [CrossRef] [PubMed]
53. Gharib, A.; Faezizadeh, Z.; Godarzee, M. Treatment of diabetes in the mouse model by delphinidin and cyanidin hydrochloride in free and liposomal forms. *Planta Med.* **2013**, *79*, 1599–1604. [CrossRef]
54. Im, N.K.; Jang, W.J.; Jeong, C.H.; Jeong, G.S. Delphinidin suppresses PMA-induced MMP-9 expression by blocking the NF-kappa B activation through MAPK signaling pathways in MCF-7 human breast carcinoma cells. *J. Med. Food* **2014**, *17*, 855–861. [CrossRef]
55. Sirisena, S.; Ng, K.; Ajlouni, S. The emerging Australian date palm industry: Date fruit nutritional and bioactive compounds and valuable processing by-products. *Compr. Rev. Food Sci. Food Saf.* **2015**, *14*, 813–823. [CrossRef]
56. Wrona, O.; Rafińska, K.; Walczak-Skierska, J.; Możeński, C.; Buszewski, B. Extraction and determination of polar bioactive compounds from alfalfa (*Medicago sativa* L.) using supercritical techniques. *Molecules* **2019**, *24*, 4608. [CrossRef] [PubMed]
57. Luong, D.; Sephton, M.A.; Watson, J.S. Subcritical water extraction of organic matter from sedimentary rocks. *Anal. Chim. Acta* **2015**, *879*, 48–57. [CrossRef] [PubMed]

processes MDPI

Article

Carotenoid-Producing Yeasts: Selection of the Best-Performing Strain and the Total Carotenoid Extraction Procedure

Olja Šovljanski [1], Anja Saveljić [1,*], Ana Tomić [1], Vanja Šeregelj [1], Biljana Lončar [1], Dragoljub Cvetković [1], Aleksandra Ranitović [1], Lato Pezo [2], Gordana Ćetković [1], Siniša Markov [1] and Jasna Čanadanović-Brunet [1,*]

[1] Faculty of Technology Novi Sad, University of Novi Sad, Bulevar cara Lazara 1, 21000 Novi Sad, Serbia
[2] Institute of General and Physical Chemistry, University of Belgrade, Studentski trg 12-16, 11000 Belgrade, Serbia
* Correspondence: anja.saveljic@uns.ac.rs (A.S.); jasnab@uns.ac.rs (J.Č.-B.)

Abstract: Yeasts are considered an extraordinary alternative source of natural carotenoids and pigmented terpenoids with multiple applications. Production of carotenoids by yeast fermentation technology has many benefits; it is cost-effective, easily scalable, and safe. The aim of this research is the isolation of yeasts from natural resources and selection of the most potent bioagent for carotenoid production. Additionally, an upgraded carotenoid extraction protocol we established, which implies the testing of four methods for cell lysis (hydrochloric acid treatment, ultrasound treatment, milling treatment, and osmotic pressure treatment), three extraction methods (conventional extraction, ultrasound extraction, and conventional + ultrasound extraction), and three extraction solvents (acetone, isopropanol/methanol (50:50), and ethanol). For the first time, the obtained results were further modeled by an artificial neural network (ANN). Based on the obtained maximal carotenoid yield (253.74 ± 9.74 mg/100 g d.w) for the best-performing *Rhodotorula mucilaginosa*, the optimized extraction procedure involving milling treatment (for cell lysis) and conventional extraction with acetone (for carotenoid extraction) convincingly stood out compared to the other 35 tested protocols. Therefore, the selected carotenoid extraction protocol was verified with respect to its universality for all other yeast isolates, demonstrating its simplicity and effectiveness.

Keywords: carotenoid-producing yeasts; *Rhodotorula*; carotenoid extraction; red-pigmented yeasts; total carotenoid; artificial neural network

Citation: Šovljanski, O.; Saveljić, A.; Tomić, A.; Šeregelj, V.; Lončar, B.; Cvetković, D.; Ranitović, A.; Pezo, L.; Ćetković, G.; Markov, S.; et al. Carotenoid-Producing Yeasts: Selection of the Best-Performing Strain and the Total Carotenoid Extraction Procedure. *Processes* **2022**, *10*, 1699. https://doi.org/10.3390/pr10091699

Academic Editors: Zongbi Bao and Qianqian Xu

Received: 2 August 2022
Accepted: 24 August 2022
Published: 26 August 2022

1. Introduction

Carotenoids belong to a class of terpenoid-pigmented, lipid-soluble compounds and are important biomolecules for human health; in addition to their use as coloring agents, they exhibit anticancer and antioxidant activity, as well as pro-vitamin A function [1]. Their roles have already been established in phytomedicine, chemical, pharmaceutical, cosmetic, food, and feed industries, and demand for their production has only increased. Carotenoids are found in a wide variety of plants and microorganisms. Most of carotenoids are currently extracted either from plants, such as annatto, tomato, carrot, paprika, and grapes, which entails seasonal and geographic variabilities, or are chemically produced, generating hazardous waste [2].

Considering this high demand for carotenoids in various industries, emphasis is also placed on optimization of extraction methods. The chosen method is important because there are many diverse carotenoids with different levels of polarity and physical and chemical barriers in the food matrices, in addition to their sensitivity to excess heat, light, acids, and long extraction times. Non-polar solvents, such as hexane, are usually a good choice for extraction of non-polar carotenes or esterified xanthophylls, whereas polar solvents, such as methanol, ethanol, and acetone, are more suitable for extraction of polar carotenoids [3]. The numerous extraction methods can be classified into four main categories: ultrasound (UAE), pressurized liquid extraction (PLE), supercritical fluid extraction (SFE), and enzymatic extraction (EE). Additionally, certain pretreatments in microbial biotechnology production are required to extract the targeted product, such as cell lysis (if the product is intracellularly located) [4]. Conventional extraction procedures are based on the extractive capabilities of different solvents and involve heating or mixing, such as the classical use of solvents in combination with a vortex apparatus, which is regularly used to mix samples of interest with dilutant, which helps with homogenization of the solution, followed by centrifugation with a supernatant [5]. Ultrasound extraction (UAE) is a modern, rapid, and inexpensive method meant to improve the extraction yield [6]. The main effect of ultrasound is acoustic cavitation, which causes cell rupture and increases the mass transfer of extractants. UAE is performed at relatively low temperatures, allowing for the preservation of thermolabile carotenoids. However, the optimized range of ultrasound intensity is a crucial parameter to obtain the highest yield of carotenoids without a significant increase in temperature causing damage to their structure [4]. In comparison to conventional techniques, UAE is considered to be more sensitive and selective. As carotenoids are intracellular molecules, their extraction is difficult; therefore, UAE should be combined with conventional methods, such as maceration and homogenization [6].

Although chemically synthesized carotenoids still dominate the market, this is starting to change, and considering the interest in green and sustainable industrial engineering, increasing efforts are being devoted to the search for natural sources and methods of carotenoid production. The challenge associated with the mentioned methods for mass production of carotenoids is that they are limited by low volume and high production costs, so there is an obvious need for an alternative production method, such as microbial biotechnological carotenoid biosynthesis. This alternative means of obtaining carotenoids has shown achieved improved yields with the possibility of using agro-industrial waste as cheap substrates and can directly affect the environmental and economic aspects of production. The product can also be obtained relatively quickly, at any time of the year [5], although pretreatments, such as cell lysis, are needed to extract the final product, as carotenoids present with intracellular metabolites [7].

The mentioned group of natural pigments can be produced by a diverse range of microorganisms, although not all are industrially or economically feasible. Therefore, microorganisms with carotenoid-producing characteristics, such as bacteria, archaea, algae, and fungi, are abundant in the ecosystem. Industrial production microorganisms include *Streptomyces chrestomyceticus*, *Blakeslea trispora*, *Phycomyces blakesleeanus*, *Flavobacterium* sp., *Phaffia* sp., *Actinomycetes*, and *Rhodotorula* sp. [8]. In recent years, microalgae have been used as a source for industrial carotenoid production, although there are other promising candidates. For example, yeasts have the potential to produce significant amounts of carotenoids, such as lycopene, β-carotene, astaxanthin, torulene, torularhodin, etc., through fermentation using numerous agricultural products and byproducts. They are mainly represented by the genera *Rhodotorula* sp., *Rhodosporidium* sp., *Sporobolomyces* sp., and *Xanthophylomyces* sp. [9]. These yeast species grow as pigmented colonies, which is why they are known as "red yeasts". Red yeasts have shown good potential as biocatalysts due to their biotransformation of carbon sources into a variety of primary and secondary metabolites. As such, species belonging to genus *Rhodotorula*, among others, have piqued the interest of many researchers with respect to carotenoid production. Their tolerance to inhibitory compounds occurring in natural substrates and high yields of various carotenoids, combined with still quite unexplored metabolic pathways, motivates a search for innovative biotechnological production of carotenoids [2].

In this research, yeasts isolated from natural sources were identified and characterized, and the most potent representative bioagent for carotenoid production was selected based on the widest range of biochemical and physiological characteristics. The main aim of the research was to establish an upgraded carotenoid extraction protocol, which implies testing four different methods for cell lysis (hydrochloric acid treatment, ultrasound treatment, milling treatment, and osmotic pressure treatment), three extraction methods (conventional extraction, ultrasound extraction, and conventional + ultrasound extraction), and three extraction solvents (acetone, isopropanol/methanol (50:50), and ethanol). The obtained results were modelled by an artificial neural network and the defined best protocol for carotenoid extraction from yeast cells was further verified based on its universality for all other yeast strains from the *Rhodotorula* genus.

The innovation of this study is reflected in application of advanced mathematical tools for simplification and rapid evaluation of different experimental steps to obtain the best-performing yeast strain for carotenoid production, optimization of carotenoid extraction protocol, and validation of the universality of the selected extraction operating parameters. In addition to the main advantages of the proposed study, the following steps provide an improved understanding of the differences between the tested extraction protocols for carotenoid-producing yeast cells and utilization of time, energy, and other sources in a microbiological laboratory, representing the first step in further scale-up and utilization of alternative conditions for biotechnological production of carotenoids.

2. Materials and Methods

2.1. Yeast Isolates and Their Characteristics

Isolation was performed at the Laboratory for Testing Food Products (Department of Microbiological Analysis), Faculty of Technology, Novi Sad. Red yeast originates from food-related sources (i.e., food samples, food byproducts, agricultural soil from crops, etc.). In order to isolate red-pigmentated yeasts, a total of more than 100 samples were tested, which can be dived in the following groups: fresh fruit and vegetables (apples, cherries, cucumbers, Jerusalem artichokes, carrots, beets, tomatoes, cherry tomatoes, peppers, potatoes, grapefruits, lettuces, etc.), processed fruits (figs, plums, apricots, pineapples, watermelons, mangos, bananas, oranges, etc.), dairy products (butter, cheese, natural and processed cheese products, yogurt, sour cream, dips, ice cream, frozen desserts made with dairy ingredients, milk powder, etc.), juices and sugar-rich solutions (orange juice, apple juice, carrot juice, beet juice, cherry juice, elder juice, tomato juice, sugar cane, dextrose solutions, high-fructose corn syrup, malt syrup, etc.), and agricultural soil. All food samples were subjected to conventional microbiological analysis for isolation and determination of yeast and aliquots of the processed sample streaked and incubated on Sabouraud maltose agar (SMA) (HiMedia, Mumbai, India). Briefly, a series of dilutions was prepared for each sample, and aliquots were streaked on the mentioned nutrient agar. After incubation (48 h at 30 °C), all pure colonies with the targeted macromorphology (orange or red color) were selected, and the pure red-pigmented cultures were obtained using a new batch of sterile nutrient agar plates.

In order to identify selected red yeast isolates, an API 20C AUX yeast identification system (BioMérieux, Marcy-l'Étoile, Lyon, France) was used according to the manufacturer's recommendation and scientific literature [10–12].

For permanent storage, the selected red-pigmented yeast strains were deposited in the Collection of Microorganisms at the Faculty of Technology, University of Novi Sad, and stored in a deep freezer at −80 °C (with addition of glycerol as cryoprotectant). Prior to use, cultures were incubated in SMA medium at 30 °C for 48 h.

The growth curve was determined during incubation of all yeast isolates in Sabouraud maltose broth (HiMedia, Mumbai, India) for 216 h. The initial concentration of cells was about 4 log CFU/mL (obtained after overnight incubation and preparation of a dilation series) in the nutrient medium, with sampling at 0 h and every 12 h during the first 3 days and every 24 h until the end of incubation period at 30 °C. The total number of living yeast cells (after biomass collection) was determined by the plate counting method using SMA as a nutrient medium. Cell number is expressed as log CFU. In addition, kinetic modelling was performed to determine and predict yeast behavior during incubation (see Section 2.6.1).

In order to define biokinetic zones for growth, yeast isolates were incubated on a solid YPD medium at various temperatures (20, 30, 37, and 44 °C) and in a liquid YPD medium with different pH values (3, 5, 7, and 9) for 120 h with visual observation of growth every

24 h. The absence of growth was categorized as a negative result (0), whereas the presence of growth was categorized as either low (1), medium (2), or high (3). Non-inoculated media were used as controls (blanks). Mathematical analysis was performed to interpret the obtained results according to the methodology reported by Šovljanski et al. [13] (see Section 2.6.2).

The obtained results of assimilation of different carbon sources during API identification were additionally used for mathematical ranking of yeast strains (see Section 2.6.3). For this step of selection of the most improved yeast isolate, isolates were additionally tested in terms of extracellular enzyme production, which can be decisive during further biowaste fermentation. Briefly, lipase activity was detected in tributyrin agar medium (48 h at 30 °C), and a positive reaction was indicated by a halo zone around the colony. The protease activity was tested using skim-milk-supplemented nutrient agar (HiMedia, Mumbai, India) during 48 h of incubation at 30 °C. For both mentioned tests, a positive reaction is indicated by a halo zone around colonies. Cellulase activity was observed after incubation for 3 days at 30 °C on YPD with 20 g L^{-1} carboxymethylcellulose (CMC, Sigma Aldrich, St. Louis, MO, USA). YPD medium contains 10 g L^{-1} yeast extract, 20 g L^{-1} peptone, and 10 g L^{-1} glucose. A positive reaction is indicated by a halo zone following Congo red staining.

2.2. Biomass Production

Yeast cells obtained by multiplying a pure laboratory culture of the mentioned red yeast isolates were used to obtain biomass. After the first cell passage (SMA, 30 °C for 72 h), the obtained biomass was used for the second passage in YP nutrient medium (20 g L^{-1} yeast extract, 10 g L^{-1} glucose, and 10 g L^{-1} peptone). The mentioned nutrient medium and conditions were selected for simultaneous production of yeast biomass and carotenoids according to the recommendation of multiple scientific groups [14–18]. Incubation was performed at 30 °C for 120 h hours in the dark. The obtained biomass was collected and centrifuged twice on an ultracentrifuge at 6000 rpm for 10 min (ROTINA 380R, Hettich, Tuttlingen, Municipality, Germany). Between centrifugations, the supernatant was discarded, and a pellet was lyophilized by a laboratory lyophilizator (Alpha 2–4 LSC, Martin Christ, Osterode, Germany) for 28 h.

2.3. Experimental Design for Extraction of Total Carotenoids from Yeast Cells

In order to define an improved protocol for extraction of total carotenoids from yeast biomass, the experimental design involved three qualitative variables: a method for cell lysis (X1), a carotenoid extraction method (X2), and a solvent extraction method(X3). According to the principle of varying the variables relative to one another, the experimental design involved:

- Methods for cell lysis: chemical (hydrochloric acid treatment and sodium chloride treatment) and mechanical (ultrasound wave treatment and milling) methods;
- Carotenoid extraction methods: conventional extraction (CE), ultrasound extraction (USE), and a combination of conventional and ultrasound extraction (CUSE); and
- Solvent extraction methods: acetone, isopropanol: methanol (50:50), and ethanol extraction.

Carotenoid yield was determined as the output of the experimental procedure using the spectrophotometric method [19] with microtiter plates. Briefly, 200 µL sample aliquots were applied in wells, and absorbances were immediately analyzed at wavelengths of 663, 645, 505, and 453 mm (with the application of an extraction solvent as a blank). Carotenoid yield (mg/100 mg d.w.) was determined based on Equation (1). All analyses were performed in triplicate, and the obtained results are presented as mean values with standard deviations.

$$\text{Carotenoid yield} = 0.216 \cdot A_{663} - 1.22 \cdot A_{654} - 0.304 \cdot A_{505} + 0.452 \cdot A_{453} \quad (1)$$

2.4. Methods for Cell Lysis

Cell lysis by hydrochloric acid was performed using a 4 M acid solution. The biomass-to-acid solution ratio was 1:5, and the mix was subjected to agitation at 150 rpm for 5 min at 35 °C. Cell lysis by ultrasound was performed done using a T 25 ULTRA-TURRAX instrument (IKA Werke, Staufen im Breisgau, Germany). Briefly, 60 g of biomass and 300 mL of NaCl solution were subjected to 4 ultrasonic cycles (50 Hz, 11,000 rpm, 20 °C) for 10 min in dark glass. Cell lysis by osmotic pressure treatment was performed with sodium chloride solution in a 1:5 ratio. The mixture was subjected to a temperature of 40 °C for 24 h. For all three techniques, the obtained precipitate after the centrifugation process (10,000 rpm, 10 min) was double-washed using distilled water to eliminate chemical residues. For milling treatment, biomass was subjected to grinding (1 min, 5000 rpm) with a batch mill tube mill control system (LLG, Meckenheim, Germany).

2.5. Methods for Carotenoid Extraction

The extraction process was performed with conventional and ultrasound methods, as well as a combination of conventional and ultrasound methods. These methods were chosen due to their differing mechanisms for extraction of targeted carotenoids, as well as the possibility of preventing the loss of the bioactivity of the targeted compounds. The conventional method is commonly used for evaluation of other methods; however, with a very high recovery rate of carotenoids, it can be time-consuming costly, using a large amount of solvent [4]. On the other hand, ultrasound, as a greener method, which includes an acoustic cavitation base for extraction, can be applied for many processes, including intracellular metabolite extraction, as well as microbial inactivation [4]. Three solvents were used for each extraction method, as previously mentioned. Briefly, conventional extraction (CE) involved the addition of the solvent in a ratio of 1:2.5 (m/V) and mixture in a vortex for 2 min. After the addition of the same ratio of sample: solvent as for conventional extraction, ultrasound extraction (USE) was performed in an ultrasound bath for 20 min. Consequently, the combination of conventional and ultrasound extractions (CUSE) involved, first, the use of a vortex and then an ultrasonic bath procedure. Following the extraction process, all samples were centrifugated at 8000 rpm for 5 min, and supernatants were separated for further analysis.

In addition to experience with extraction technologies, we followed the recommendations of Mata-Gomez et al. [20], Lopez et al. [21], Michelon et al. [22], and Nemer et al. [23] for the selection and formation of the presented cell lysis and carotenoid extraction methods.

2.6. Statistical Analysis

2.6.1. Growth Kinetics Modelling

In this investigation, time-kill kinetics modelling was implemented by applying the four-parameter sigmoidal computational model, which is convenient for biological systems and described in detail by Brlek et al. [24]. The projected data should be shaped in an S-shaped curve, and the model can be expressed in the form of Equation (2).

$$y(t) = d + \frac{a - d}{1 + \left(\frac{t}{c}\right)^{b}} \quad (2)$$

The yeast isolate number (log CFU/mL) during incubation (hours) is denoted as $y(t)$, whereas the regression coefficients are denoted as follows: a, minimum of the experimentally obtained values (at t = 0); d, the maximally acquired value (at $t = \infty$); c, the inflection point (the point between a and d); and b, the Hill's slope (the steepness of the inflection point (c)).

2.6.2. Cluster Analysis of Temperature and pH Growth Profiles of Yeast Isolates

To predict the distribution of yeast isolates as a function of environmental parameters, a cluster analysis was performed. Cluster analysis (CA) was used to determine the possible correlations among different yeast isolates and to classify objects according to temperature and pH growth profiles of selected yeast isolates. The complete linkage algorithm and city block (Manhattan) distances were used to explain the temperature and pH growth profile of the natural isolates grouped it in different clusters.

2.6.3. Ranking Procedure

Yeast isolates were ranked based on their assimilation/enzyme production capabilities [25], which were tested in the primary step of this study (see Section 2.1). The obtained scores of yeast isolates for utilization of different C and N sources were ranked in the following order: D-glucose, glycerol, calcium, 2-keto-gluconate, L-arabinose, D-xylose, adonitol, xylitol, D-galactose, inositol, D-sorbitol, methyl-α-D-glucopyranoside, N-acetyl-glucosamine, D-cellobiose, D-lactose, D-maltose, D-sucrose, D-trehalose, D-melesitosis, and D-raffinose. With respect to the production of different enzymes, the isolates were ranked in the following order: protease, lipase, and cellulase. A negative reaction was coded as "0", whereas a positive reaction of the yeast isolates was coded as "1". Several quality parameters were calculated within the standard score analysis for each yeast isolate: the sum of utilization scores, the sum of production scores, the total sum of scores, the number of positive reactions, and the number of negative reactions.

2.6.4. Artificial Neural Network (ANN) Optimization

A multilayer perceptron model (MLP) with three layers was used for modelling. This model construction was proven accurate in approximating non-linear functions [26,27]. The experimental database for ANN was randomly divided into training, cross-validation, and testing data (60%, 20%, and 20% of experimental data, respectively). The Broyden–Fletcher–Goldfarb–Shanno (BFGS) algorithm was applied as an iterative method to solve unconstrained non-linear optimization during ANN modelling. The number of repeating steps during ANN construction was 100,000. Coefficients associated with the hidden and output layer (weights and biases) were grouped in matrices W_1 and B_1 and W_2 and B_2, respectively, as represented in Equation (3).

$$Y = f_1(W_2 \cdot f_2(W_1 \cdot X + B_1) + B_2) \tag{3}$$

Weight coefficients were determined during the ANN learning cycle to minimize the error between network results and experimental values [28,29]. The developed models were numerically verified using reduced chi-square (χ^2), root mean square error (RMSE), coefficient of determination (r^2), mean bias error (MBE), and mean percentage error (MPE). These parameters can be calculated using Equations (4)–(7).

$$\chi^2 = \frac{\sum\limits_{i=1}^{N} (x_{\exp,i} - x_{pre,i})^2}{N - n} \tag{4}$$

$$RMSE = \left[\frac{1}{N} \cdot \sum_{i=1}^{N} (x_{pre,i} - x_{\exp,i})^2 \right]^{1/2} \tag{5}$$

$$MBE = \frac{1}{N} \cdot \sum_{i=1}^{N} (x_{pre,i} - x_{\exp,i}) \tag{6}$$

$$MPE = \frac{100}{N} \cdot \sum_{i=1}^{N} \left(\frac{|x_{pre,i} - x_{\exp,i}|}{x_{\exp,i}} \right) \tag{7}$$

where $x_{\exp,i}$ represents the experimental values, and $x_{pre,i}$ represents the predicted values obtained by calculating the model for these measurements. N and n are the number of observations and constants, respectively.

2.7. Experimental Validation of the Universality of the Chosen Carotenoid Extraction Procedure

The chosen extraction procedure was experimentally verification in a microbiology laboratory in order to demonstrate its accuracy. The best experimental procedure was tested for all other yeast strains investigated in this study, and the obtained results were compared with the best-performing yeast isolate (based on the ranking procedure described in Section 2.6.3). All yeasts were grown under the same conditions, as described in Section 2.2.

3. Results and Discussion

Yeasts were isolated from food and food-related samples, and the obtained average number of yeast in samples was between 34 and 117 CFU/g. According to the targeted macromorphology (orange–red-pigmented colonies), the total number of potential carotenoid-producing isolates was five. All five isolates were obtained from the different samples (Jerusalem artichoke, tomato juice, sugar cane, agricultural soil, and yogurt) and designated as top 30, KV1105, CRV, 4/34, and FK3, respectively. Their macromorphology after 72 h of incubation at 30 °C is shown in Figure 1. The obtained colonies have a typical orange color, but size, shape, surface appearance, and texture differ the strain level. Therefore, the isolates were identified with an API identification system. Four isolates represent *Rhodotorula mucilaginosa* strains (top 30, KV1105, CRV, and 4_34), whereas only one is *Rhodotorula glutinis* (strain FK3). The *Rhodotorula* genus represents comprises environmental basidiomycetous yeasts, and species that are commonly isolated from soil, as well as fermented food, fruit juice, milk, and food waste [30].

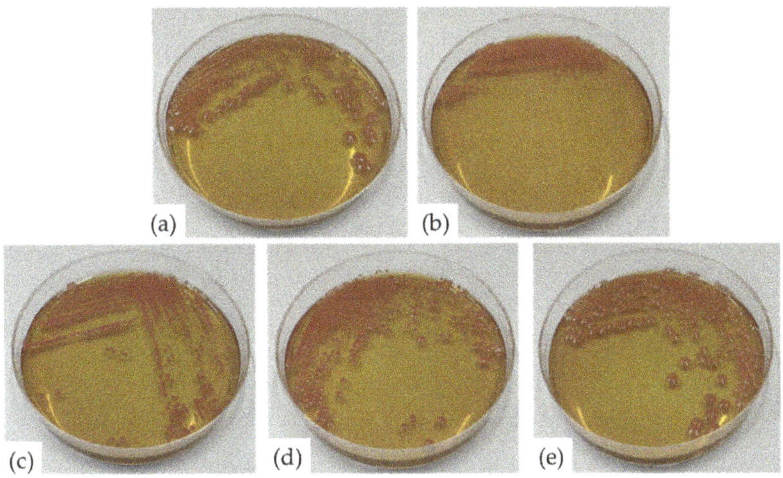

Figure 1. *Rhodotorula* isolates on a solid growth medium after 72 h of incubation: (**a**) top 30; (**b**) KV1105; (**c**) CRV; (**d**) 4_34; (**e**) FK3.

However, the significance of *R. mucilaginosa* and *R. glutinis* lies in their ability to produce carotenoids, which has been described in numerous literature reports. Briefly, representatives of this genus are capable of synthesizing carotenoids, such as β-carotene, astaxanthin, torulene, and torularhodin [30–32]. Their colonies vary from coral red/salmon to slightly orange in color (depending on the nutrient medium and incubation conditions), with mucoid to slightly tough, smooth to wrinkled, or highly glossy to semi-glossy surface texture [33], which is in agreement with the colonies observed in this study (Figure 1). Figure 1a–d depicts the macromorphological details of *R. muciliginosa* strains, whereas Figure 1e correspond to the only *R. glutinis* strain isolated in this study.

As the next experimental step, the growth curve (growth as a function of incubation time) was determined for all selected *Rhodotorula* isolates, as shown in Figure 2.

Dyaa et al. [34] reported that the optimal temperature for carotenoid production in *Rhodotorula* strains is around 30 °C, and the carotenoid rate was reduced at higher temperatures. Therefore, growth was monitored at 30 °C. During biomass determination (Figure 2), the cells multiplied intensively for all isolates, following the sigmoidal shape of the curve.

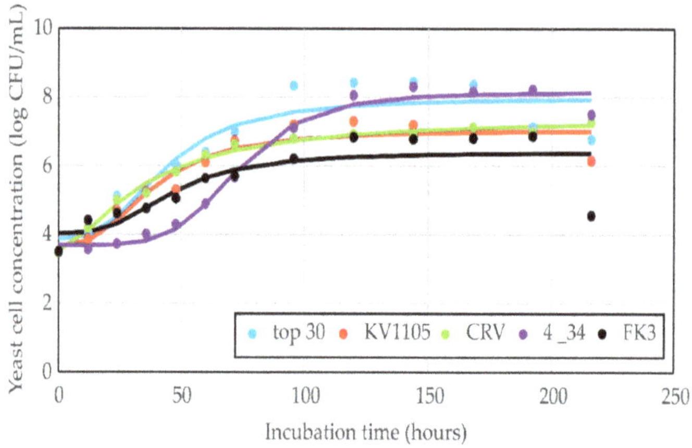

Figure 2. Growth curves for *Rhodotorula* isolates (markers signify the experimental data; lines indicate predictive results).

A growth adjustment phase (lag phase) was observed during the first 12 h. A moment of transition from the exponential phase (log phase) to the stationary growth phase was observed after 4 days (96 h) for all strains, except for *R. glutinis* strain 4_34, for which the starting point of the stationary phase was observed after approximately 120 h. According to Figure 2, the highest cell concentration (approx. 8 log CFU/mL) was achieved for the strains *R. glutinis* 4_34 and *R. mucilaginosa* top 30. In the case of all other examined yeasts, the maximum cell concentration was 1–2 log CFU/mL lower.

To better understand the growth profile of the tested yeasts, a growth kinetic model was developed. In Figure 2, markers indicate the experimentally obtained values of cell concentration at a given point in time, whereas lines indicate predictive results of the developed growth kinetic model. According to the regression coefficients of the obtained models for yeast isolate number during incubation (Table 1) and the goodness of fit between experimentally and model-obtained (predicted) results (Table 2), it can be concluded that the developed models for yeast growth were accurate, with high coefficients of determination (up to 0.989) and proposed models that fit well with experimental data. It can be concluded that the obtained kinetic models for the growth of yeast isolates can be used for the prediction of yeast isolate number during the whole incubation period.

Table 1. Regression coefficients for kinetic models of the growth of yeast isolates.

Regression Coefficient	Yeast Isolate				
	Top 30	**KV1105**	**CRV**	**4_34**	**FK3**
d	7.977	7.059	7.421	8.152	6.393
a	3.872	3.670	3.525	3.698	4.021
c	43.335	38.766	35.446	73.525	46.521
b	2.829	2.654	1.571	5.041	3.016

Table 2. "Goodness of fit" of the kinetic models for the growth of yeast isolates.

Parameter	Yeast Isolate				
	Top 30	KV1105	CRV	4_34	FK3
Reduced chi-square (χ^2)	0.362	0.135	0.017	0.060	0.406
Root mean square error (RMSE)	0.578	0.353	0.124	0.235	0.612
Mean bias error (MBE)	0.000	0.000	0.000	0.000	0.000
Mean percentage error (MPE)	7.551	4.586	1.791	3.099	8.242
Coefficient of determination (r^2)	0.872	0.922	0.989	0.985	0.680
Skewness (Skew)	−0.382	−1.127	−0.863	−1.296	−2.090
Kurtosis (Kurt)	−0.712	1.240	0.514	2.523	5.373
Mean of residuals (Mean)	0.000	0.000	0.000	0.000	0.000
Standard deviation of residuals (StDev)	0.601	0.368	0.129	0.245	0.637
Variance of residuals (Var)	0.362	0.135	0.017	0.060	0.406

The next steps involved a comprehensive study of biokinetic zones with respect to temperature and pH values for the growth of yeasts. The elementary goal of all in vitro bioprocesses is to create an environment that, for each set of parameters, simulates the optimal and specific environment for initiation of production of the desired metabolite. The growth of microorganisms is a function of environmental factors, of which the most significant are the temperature and pH values of the environment. Temperature and pH are the most important environmental factors influencing the growth of microorganisms [35]. Temperature tolerance is an essential parameter for biotechnology processes due to significant changes in this factor [13,36]. Therefore, it is imperative to investigate the behavior of natural isolates at varying incubation temperatures. On the other hand, a similar dependence can be observed for pH values as one of the most influential factors with respect to the growth of yeasts [37]. Yeasts exhibiting pH tolerance are considered potentially beneficial candidates for efficient biotechnology production, whereas most yeasts as bioagents show high metabolic activity in neutral or acidic environments. Therefore, it is necessary to evaluate the effect of pH values on yeast growth in a wide range (between 3 and 9) [38]. To determine the optimal growth rate, biomass production was monitored for 5 days at temperatures of 20, 30, 37, and 44 °C and pH values of 3, 5, 7, and 9. A summary statistical analysis of growth tolerance at different incubation temperatures and pH values is presented in Supplementary Table S1. For most of the selected strains, the optimal growth temperature is between 20 and 30 °C. On the other hand, a temperature of 44 °C is beyond the biokinetic zone for all strains. The majority of *R. mucilaginosa* and *R. glutinis* strains are considered mesophilic [32], in agreement the results obtained in this study. The optimal pH value for the growth of *Rhodotorula* strains is 7, although growth can occur in a more expansive biokinetic zone (pH value range of 3–7) after the total incubation time for all isolates. In terms of the production of carotenoids by *Rhodotorula* representatives, many researchers have indicated that a pH value of 7 is optimal not only for growth but also for carotenoid production [39,40].

We further analyzed the behavioral similarity between natural yeast isolates by cluster analysis of temperature and pH growth. According to results presented in Supplementary Table S1, two dendrograms of temperature and pH growth profiles of yeast isolates are plotted in Figure 3, using complete linkage as an amalgamation rule and a city block (Manhattan) distance as a measure of the proximity between yeast isolates. Two dendrograms based on the growth profile over time showed a proper distinction between yeast isolates. According to the temperature profile, yeast isolates *R. mucilaginosa* top 30 and *R. glutinis* 4_34 showed a similar growth profile, as did yeast isolates *R. mucilaginosa* CRV and *R. mucilaginosa* FK3. Based on the pH profile, yeast isolates *R. mucilaginosa* top 30 was similar to *R. mucilaginosa* FK3, whereas sample *R. glutinis* 4_34 corresponded most closely to the *R. mucilaginosa* CRV isolate. As shown in Figure 3, the *R. mucilaginosa* KV1105 strain is the least similar to the other tested strains in both cases.

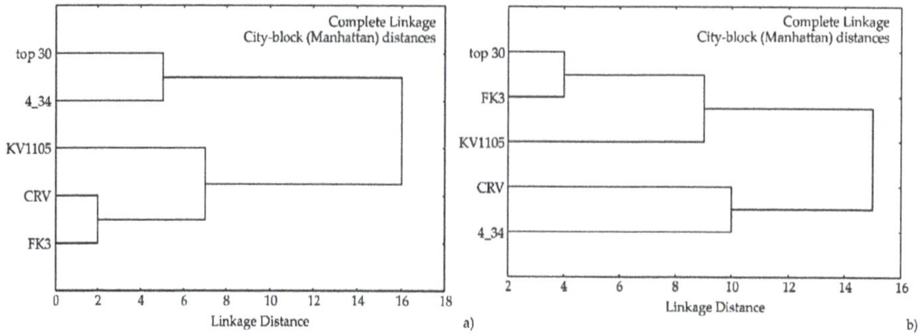

Figure 3. Dendrograms of cluster analysis of (**a**) temperature and (**b**) pH growth profiles of yeast isolates.

For further selection of the best-performing yeast isolate with respect to application in engineered bioprocesses, the assimilation of different C and N sources, as well as enzyme production, was tested (Figure 4). The data are indicated as positive reactions (green circle) or negative reactions (red circle). Interestingly, the tested yeast isolates do not utilize inositol, methyl-α-D-glucopyranoside, N-acetyl-glucosamine, or D-lactose. It has been proven that inositol has a positive effect on fermentation performance, cell growth, and tolerance to negative environmental conditions in the case of *Saccharomyces cerevisiae* [41], but the tested strains in this study remain neutral to this source, as was the case for strains M14, M22, M23, and M24 in a study by Allahkarami et al. [42]. The negative response of the tested strains was also confirmed in the case of methyl-α-D-glucopyranoside. Taking into account that methyl-α-D-glucopyranoside is usually added to the production medium as an inducer of amylases in microorganisms [43], it can be concluded that the yeasts used in the present study do not have the ability to produce amylase. The same negative results were obtained for N-acetyl-glucosamine and D-lactose, indicating that these sources of C and N atoms cannot be utilized by the tested yeasts. On the contrary, according to Figure 4, there are three common sources of C atoms: D-glucose, D-sucrose, and D-raffinose, all of which were found to be fermented by all tested yeast strains.

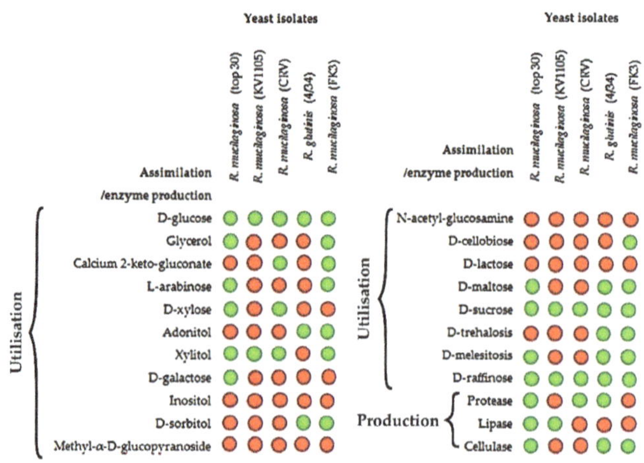

Figure 4. Biochemical and physiological tests (red color indicates a negative result, whereas green color indicates a positive result).

In order to determine the best-performing strain, tested yeasts were statistically ranked based on the mentioned characteristics presented in Figure 4. We evaluated the response of the yeast strains to the assimilation of different C and N sources, as well as enzyme production. In yeast biotechnology, it is believed that wild isolates with high fermentation capabilities in terms of different sources of C and N atoms are a potentially suitable choice for biotechnological processes at the industrial level [44]. The ranking procedure was performed as a method of selection based on the maximum of defined functions (the sum of utilization scores, the sum of production scores, the total sum of scores, and the sum of utilization scores), representing the optimal processing parameters. If the value of the specific function reaches its maximum for a specific yeast isolate, it shows indicates the tendency of the tested parameters to be optimal.

The ranking procedure was based on the discrimination criteria described in Figure 4, and the ranking results are presented in Table 3. According to the ranking procedure, yeast isolate *R. mucilaginosa* top 30 achieved the best performance, showing the highest number of positive reactions (13) and the fewest negative reactions (9). Positive reactions for yeast isolate *R. mucilaginosa* top 30 were observed for D-glucose, glycerol, L-arabinose, D-xylose, xylitol, D-galactose, D-maltose, D-sucrose, D-melesitosis, and D-raffinose assimilation, as well as protease, lipase, and cellulase production. Similar results were obtained for isolate *R. glutinis* 4_34 (which showed 11 positive and 11 negative reactions) and isolate *R. mucilaginosa* FK3 (which showed 12 positive and 10 negative reactions). The scoring of yeast isolates coincides with the cluster analysis (Figure 3).

Table 3. Ranking procedure results.

Yeast Isolate	Sum of Utilization Scores	Sum of Production Scores	Total Sum of Scores	Number of Positive Reactions	Number of Negative Reactions
top 30	10	3	13	13	9
KV1105	4	1	5	5	17
CRV	6	1	7	7	15
4_34	9	2	11	11	11
FK3	11	1	12	12	10

Therefore, *R. mucilaginosa* top 30 was selected as the best-performing yeast for further biotechnological application and used for the experimental design to establish a carotenoid extraction procedure. In this study, 36 cell lysis method–extraction method–solvent combinations were tested to determine the maximum carotenoid yield (Figure 5). Additionally, ANN modelling (Figure 6 and Table 4) was performed to verify the best extraction procedure according to the predicted maximum.

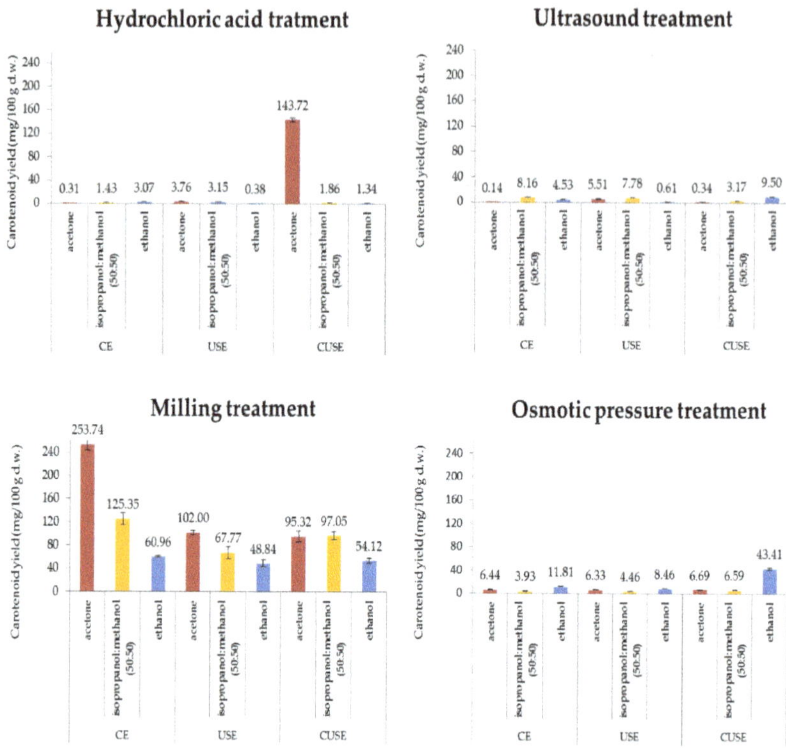

Figure 5. The obtained carotenoid yield (mg/100 g d.w.) using different carotenoid extraction procedures (acetone—red colour; isopropanol:methanol (50:50)—yellow colour; ethanol—blue colour).

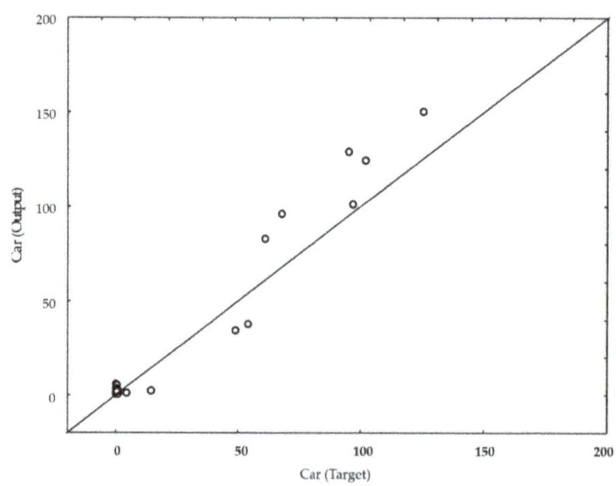

Figure 6. Experimental and predicted values of carotenoid yield for the *R. mucilaginosa* top 30 strain.

Prior to extraction procedure, the obtained biomass was lyophilized. According to Saini and Keum [4], this step enables a high water content in the system (from yeast cells). Because a water environment can be unfavorable for efficient extraction of carotenoids due to their hydrophobic nature, dehydration of samples is an efficient step in yield improvement. Additionally, thermal degradation and isomerization of some dehydration process can be harmful to carotenoids; thus, lyophilization is the most adequate pretreatment for carotenoid extraction.

The experimentally obtained and calculated maximum for carotenoid yield was 253.74 mg/100 g d.w. The optimal output was obtained using milling treatment as a method for cell lysis, conventional extraction as the extraction method, and acetone as the extraction solvent. Cell disruption might be the key step in the production and purification of intracellular compounds from yeast cells, as it exerts important effects on their recovery and quality [45,46], as well as for analytical aims. Owing to its large size and varying cell wall structure, disruption of the ascomycetous yeast cell wall is generally easier than that of other microorganisms, such as bacteria. Various methods, such as mechanical, chemical, or enzymatic approaches, can be used to disrupt cell walls; the optimal method depends on both the yeast characteristics (culture time, specific growth rate, etc.) and the target substance [47–49].

As demonstrated by the results presented in Figure 5, among all chosen methods for cell lysis applied in the present study, the milling treatment showed a significantly higher carotenoid yield than the other methods. This mechanical treatment coupled with conventional extraction with acetone as the extraction solvent led to the highest carotenoid yield, i.e., 253.74 $\pm$ 9.74 mg/100 g d.w. According to Nemer et al. [23], a significant advantage of mechanical cell disruption techniques is that they can generally be scaled-up with relative ease, in addition to being quite effective. This is a crucial step for industrial applications, but special attention needs to be directed to potential heating during the process, as well as carotenoid degradation. Aksu and Eren [50] used a milling process and acetone-dependent extraction (specific procedure data not available) to obtain carotenoids from *Rhodotorula mucilaginosa* cells, achieving a yield of 69.8 mg L^{-1}, which is lower than that achieved in the present study.

In the case of all other tested methods for cell lysis, the carotenoid content was 10–1000 times lower in comparison to the milling treatment. In the present study, the conventional extraction method achieved better results than ultrasonic extraction and the combination of conventional and ultrasound extraction methods, possibly because during the ultrasound treatment, the temperature usually increased, and heat shocks may have a negative influence on carotenoid content [51]. The answer to this hypothesis might be a further crucial moment for optimizing the extraction time and combining classic and modern methods of extraction. Methods for cell lysis for the extraction of pigments in the available literature are divided between ultrasound and milling treatments, emphasizing that the obtained results may vary depending on the culture type, pigment, and experimental parameters [23,49]. Owing to differences in experimental setups and the used yeast cultures, it is impossible to compare results between studies. It is additionally difficult to compare the results reported in the literature owing to the fact that different commercial and alternative substrates with complex compositions are used for yeast cultivation, which significantly affects the yield of carotenoids.

As previously mentioned, ANN modelling was also performed for analysis of carotenoid yield during the tested extraction procedures, with all three categorical variables for ANN contributing to the anticipation of carotenoid content:

1. Cell lysis methods (hydrochloric acid treatment, ultrasound treatment, milling treatment, and osmotic pressure treatment);
2. Extraction methods (conventional extraction, ultrasound extraction, conventional + ultrasound extraction); and
3. Solvent extraction methods (acetone, isopropanol/methanol (50:50), and ethanol extraction).

The ANN technique proved to be a beneficial method for prediction of the yield of different microbiological products [51,52], but it has not been used to date for the selection of a total carotenoid extraction protocol based on the highest carotenoid yield from yeast cells. For the first time, the obtained results of different extraction protocols were involved in this advanced mathematical analysis, and the optimal number of neurons in the hidden layer was four (network MLP 10-4-1) to obtain high values of coefficients of determination (overall r^2 is 0.86) and low values of the sum of squares (SOS). The obtained r^2 values for training, testing, and validation sequences of ANN modeling were 0.94, 0.98, and 1.00, respectively. BHGS 6 was used as a training algorithm, and the error function was SOS (sum of squares). The optimal hidden and output layer activation function was the logistic function. An ANN model was used to predict experimental variables reasonably well for a broad range of process variables (as shown in Figure 5, in which the experimentally measured and ANN-model-predicted values are presented). The accuracy of the ANN model was visually assessed by the dispersion of points in the diagonal line shown in the graphics presented in Figure 6.

The predicted values were very close to the desired values in most cases in terms of the value of the coefficient of determination [29]. The qualities of the model fit were tested in Table 4; the coefficient of determination (r^2) should be close to 1, whereas the values of other tests (χ^2, RMSE, MBE, and MPE) should be lower to achieve a good fit with the experimental values [53,54]. The obtained results indicate the possibility of using the ANN model in the presented experimental setup. This is in agreement with the results reported by Shafi et al. [55], who emphasized that using ANN modelling for analysis of natural sources of bioactive compounds is more flexible and accurate than the response surface methodology.

Table 4. "Goodness-of-fit" kinetics models for the yeast isolate number.

Yeast Isolate	χ^2	RMSE	MBE	MPE	r^2
***R. mucilaginosa* top 30**	3.1×10^4	17.49	−13.27	14.29	0.89
	Skew	**Kurt**	**Mean**	**StDev**	**Var**
	2.22	11.62	−13.27	17.68	3.1×10^4

Legend: Reduced chi-square (χ^2), root mean square error (RMSE), mean bias error (MBE), mean percentage error (MPE), coefficient of determination (r^2), skewness (Skew), kurtosis (Kurt), mean of residuals (Mean), standard deviation of residuals (StDev), variance of residuals (Var).

In order to reduce the time and reagents required for analysis, the universality of the milling treatment coupled with acetone as an extraction solvent using a conventional extraction procedure was validated for all other yeast isolates in terms of comparative achieved carotenoid yield. A simple, rapid, universally applicable, and reproducible procedure is indispensable in order to reduce laboratory time and cost. This is important for studies involving multiple isolates. Furthermore, spectrophotometric analysis of carotenoids was selected as the output. This method is among the cheapest and most useful methods for quantification of carotenoids [56]. Within the field of yeast biotechnology, experimental validation of the universality of some processes or procedures in a laboratory does not rely on the extensive use of expensive resources, but the obtained data confer multiplied benefits for further applications of bioagents in semi-industrial and industrial processing. The results presented in Figure 7 show that for all tested yeast strains, the accomplished carotenoid yield was significant (above 227 mg/100 g d.w.) under the same extraction conditions compared to *R. mucilaginosa* top 30. Slight variations in the carotenoid content were observed between strains, which positively correlated with the diversity in the results obtained in terms of the ranking of yeast isolates (Table 4.). Nevertheless, in this experiment, a similar carotenoid content was recorded across all tested yeast strains when the milling treatment was applied in combination with conventional extraction with acetone as the extraction solvent; therefore, the universality of the best extraction procedure in this study is acknowledged.

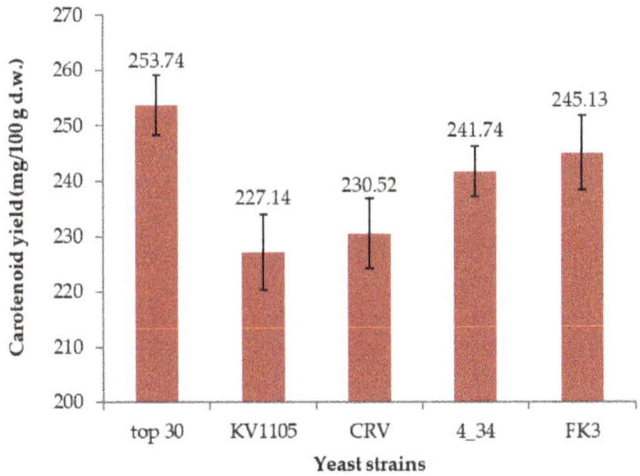

Figure 7. Experimental validation of the universality of the chosen carotenoid extraction procedure for all *Rhotodorula* strains.

Based on the results of the present study, it can be concluded that beneficial effects of carotenoids on human health and their widespread application in various industries have led to increased interest in biotechnological study of these substances. This work can contributes to current research oriented toward the identification, quantification, and optimization of microbial carotenoid production. Various possibilities for the isolation of carotenoid-producing yeasts and multiple sets of extraction procedures should be further explored in order to define the biological activity of effective substances in yeast resources for utilization in the biotechnological field. One such approach involves the use of advanced mathematical tools, as explored in the present study for the selection of yeast isolates by determination of growth characteristics, pH, and temperature profiles to optimize carotenoid production protocols and verify the universality of the selected procedures.

4. Conclusions

Following isolation and identification of five carotenoid-producing yeasts, the best-performing strain, identified as *Rhodotorula mucilaginosa* (strain name: top 30) was selected based on an advanced ranking procedure. Using this strain, we optimized an experimental design for carotenoid extraction. Based on the obtained maximal carotenoid yield (253.74 ± 9.74 mg/100 g d.w), this procedure, which involves milling treatment (for cell lysis) and conventional extraction with acetone, convincingly stood out compared to the other 35 tested protocols. Finally, we validated the universality of the optimized procedure.

Based on a statistical ranking of microorganisms, the universality of an experimental design to obtain an optimized procedure for carotenoid extraction from yeast cells was validated. This method represents an economically feasible calculation approach to an experimental setup involving multiple natural isolates and tested protocols.

Future research should investigate the use of agro-industrial waste as a substrate for biomass and carotenoid production by the best-performing yeast isolate, *R. mucilaginosa* top 30. Coupled with the optimized extraction procedure, this step will contribute to an economically and ecologically acceptable solution for carotenoid production.

Supplementary Materials: The following supporting information can be downloaded at: https://www.mdpi.com/article/10.3390/pr10091699/s1, Table S1: Temperature and pH growth profiles of *Rhodotorula* isolates during incubation.

Author Contributions: Conceptualization, O.Š. and A.S.; methodology, V.Š. and A.T.; validation, B.L., L.P. and V.Š.; formal analysis, A.S. and O.Š.; investigation, A.R., A.T. and D.C.; resources, S.M. and G.Ć.; data curation, B.L. and L.P.; writing—original draft preparation, O.Š. and A.T.; writing—review and editing, A.T.; visualization, B.L. and O.Š.; supervision, J.Č.-B. All authors have read and agreed to the published version of the manuscript.

Funding: This research study was supported by the Provincial Secretariat for Higher Education and Scientific Research of Autonomous Province Vojvodina (Serbia) within the project "From biowaste to pigments—a biotechnological concept of obtaining carotenoids" (grant No. 142-451-2364/2022-01/01) and the Ministry of Education, Science, and Technological Development of the Republic of Serbia (contract no. 451-03-68/2022-14/200134).

Institutional Review Board Statement: Not applicable.

Informed Consent Statement: Not applicable.

Data Availability Statement: Not applicable.

Conflicts of Interest: The authors declare no conflict of interest.

References

1. Cheng, Y.-T.; Yang, C.-F. Using strain *Rhodotorula mucilaginosa* to produce carotenoids using food wastes. *J. Taiwan Inst. Chem. Eng.* **2016**, *61*, 270–275. [CrossRef]
2. Mannazzu, I.; Landolfo, S.; da Silva, T.L.; Buzzini, P. Red yeasts and carotenoid production: Outlining a future for non-conventional yeasts of biotechnological interest. *World J. Microbiol. Biotechnol.* **2015**, *31*, 1665–1673. [CrossRef]
3. Kultys, E.; Kurek, M.A. Green Extraction of Carotenoids from Fruit and Vegetable Byproducts: A Review. *Molecules* **2022**, *27*, 518. [CrossRef] [PubMed]
4. Saini, R.K.; Keum, Y.-S. Carotenoid extraction methods: A review of recent developments. *Food Chem.* **2018**, *240*, 90–103. [CrossRef] [PubMed]
5. Pisoschi, A.M.; Pop, A.; Cimpeanu, C.; Predoi, G. Antioxidant capacity determination in plants and plant-derived products: A review. *Oxidative Med. Cell. Longev.* **2016**, *2016*, 9130976. [CrossRef] [PubMed]
6. Cikoš, A.M.; Jokić, S.; Šubarić, D.; Jerković, I. Overview on the application of modern methods for the extraction of bioactive compounds from marine macroalgae. *Mar. Drugs* **2018**, *16*, 348. [CrossRef]
7. Mussagy, C.U.; Winterburn, J.; Santos-Ebinuma, V.C.; Pereira, J. Production and extraction of carotenoids produced by microorganisms. *Appl. Microbiol. Biotechnol.* **2018**, *103*, 1095–1114. [CrossRef]
8. Zhao, Y.; Guo, L.; Xia, Y.; Zhuang, X.; Chu, W. Isolation, Identification of Carotenoid-Producing *Rhodotorula* sp. from Marine Environment and Optimization for Carotenoid Production. *Mar. Drugs* **2019**, *17*, 161. [CrossRef]
9. Igreja, W.S.; Maia, F.A.; Lopes, A.S.; Chisté, R.C. Biotechnological Production of Carotenoids Using Low Cost-Substrates is Influenced by Cultivation Parameters: A Review. *Int. J. Mol. Sci.* **2021**, *22*, 8819. [CrossRef]
10. Willemsen, M.; Breynaert, J.; Lauwers, S. Comparison of Auxacolor with API 20 C Aux in yeast identification. *Clin. Microbiol. Infect.* **1997**, *3*, 369–375. [CrossRef]
11. Mateo, J.J.; Garcerà, P.; Maicas, S. Unusual Non-Saccharomyces Yeasts Isolated from Unripened Grapes without Antifungal Treatments. *Fermentation* **2020**, *6*, 41. [CrossRef]
12. Arastehfar, A.; Daneshnia, F.; Kord, M.; Roudbary, M.; Zarrinfar, H.; Fang, W.; Hashemi, S.J.; Najafzadeh, M.J.; Khodavaisy, S.; Pan, W.; et al. Comparison of 21-Plex PCR and API 20C AUX, MALDI-TOF MS, and rDNA Sequencing for a Wide Range of Clinically Isolated Yeast Species: Improved Identification by Combining 21-Plex PCR and API 20C AUX as an Alternative Strategy for Developing Countries. *Front. Cell. Infect. Microbiol.* **2019**, *9*, 21. [CrossRef]
13. Šovljanski, O.; Tomić, A.; Pezo, L.; Markov, S. Temperature and pH growth profile prediction of newly isolated bacterial strains from alkaline soils. *J. Sci. Food Agric.* **2019**, *100*, 1155–1163. [CrossRef]
14. González-García, Y.; Rábago-Panduro, L.M.; French, T.; Camacho-Córdova, D.I.; Gutiérrez-González, P.; Córdova, J. High lipids accumulation in *Rhodosporidium toruloides* by applying single and multiple nutrients limitation in a simple chemically defined medium. *Ann. Microbiol.* **2019**, *67*, 519–527. [CrossRef]
15. De Bonadio, M.P.; De Freita, L.A.; Mutton, M.J.R. Carotenoid production in sugarcane juice and synthetic media supplemented with nutrients by *Rhodotorula rubra* l02. *Braz. J. Microbiol.* **2018**, *394*, 7. [CrossRef]
16. Kot, A.M.; Błażejak, S.; Kieliszek, M.; Gientka, I.; Bryś, J. Simultaneous Production of Lipids and Carotenoids by the Red Yeast *Rhodotorula* from Waste Glycerol Fraction and Potato Wastewater. *Appl. Biochem. Biotechnol.* **2019**, *189*, 589–607. [CrossRef]
17. Hien Ly, T.M.; Nga Pham, T.H. Effects of some factors on carotenoid biosynthesis by *Rhodotorula Mucilaginosa*. *Ho Chi Minh City Open Univ. J. Sci.* **2019**, *9*, 62–69.
18. Rovinaru, C.; Pasarin, D.; Capra, L.; Stoica, R. The effect of $ZnSO_4$ in the cultivation medium of *Rhodotorula glutinis* CCY 020-002-033 yeast biomass growth, β-carotene production and zinc accumulation. *J. Microbiol. Biotechnol. Food Sci.* **2019**, *8*, 931–935. [CrossRef]
19. Jeliński, T.; Przybyłek, M.; Cysewski, P. Natural deep eutectic solvents as agents for improving solubility, stability and delivery of curcumin. *Pharm. Res.* **2019**, *36*, 116. [CrossRef]
20. Mata-Gómez, L.C.; Montañez, J.C.; Méndez-Zavala, A.; Aguilar, C.N. Biotechnological production of carotenoids by yeasts: An overview. *Microb. Cell Fact.* **2014**, *13*, 12. [CrossRef]
21. Lopes, N.A.; Remedi, R.D.; Dos Santos Sá, C.; Burkert, C.A.V.; de Medeiros Burkert, J.F. Different cell disruption methods for obtaining carotenoids by *Sporodiobolus pararoseus* and *Rhodothorula mucilaginosa*. *Food Sci. Biotechnol.* **2017**, *26*, 759–766. [CrossRef]

22. Michelon, M.; de Matos de Borba, T.; da Silva Rafael, R.; Burkert, C.A.V.; de Medeiros Burkert, J.F. Extraction of carotenoids from *Phaffia rhodozyma*: A comparison between different techniques of cell disruption. *Food Sci. Biotechnol.* **2012**, *21*, 1–8. [CrossRef]
23. Nemer, G.; Louka, N.; Vorobiev, E.; Salameh, D.; Nicaud, J.-M.; Maroun, R.G.; Koubaa, M. Mechanical Cell Disruption Technologies for the Extraction of Dyes and Pigments from Microorganisms: A Review. *Fermentation* **2021**, *7*, 36. [CrossRef]
24. Brlek, T.; Pezo, L.; Voća, N.; Krička, T.; Vukmirović, Đ.; Čolović, R.; Bodroža-Solarov, M. Chemometric approach for assessing the quality of olive cake pellets. *Fuel Process. Technol.* **2013**, *116*, 250–256. [CrossRef]
25. Momenzadeh, L.; Zomorodian, A.; Mowla, D. Experimental and theoretical investigation of shelled corn drying in a microwave-assisted fluidized bed dryer using Artificial Neural Network. *Food Bioprod. Process.* **2011**, *89*, 15–21. [CrossRef]
26. Trelea, I.C.; Raoult-Wack, A.L.; Trystram, G. Note: Application of neural network modelling for the control of dewatering and impregnation soaking process (osmotic dehydration). *Food Sci. Technol. Int.* **1997**, *3*, 459–465. [CrossRef]
27. Šovljanski, O.; Pezo, L.; Tomić, A.; Ranitović, A.; Cvetković, D.; Markov, S. Contribution of bacterial cells as nucleation centers in microbiologically induced $CaCO_3$ precipitation—A mathematical modeling approach. *J. Basic Microbiol.* **2021**, *61*, 835–848. [CrossRef] [PubMed]
28. Basheer, I.A.; Hajmeer, M. Artificial neural networks: Fundamentals, computing, design, and application. *J. Microbiol. Methods* **2000**, *43*, 3–31. [CrossRef]
29. Otero, D.M.; Bulsing, B.A.; da Huerta, K.M.; Rosa, C.A.; Zambiazi, R.C.; Burkert, C.A.V.; de Burkert, J.F.M. Carotenoid-producing yeasts in the Brazilian biodiversity: Isolation, identification and cultivation in agroindustrial waste. *Braz. J. Chem. Eng.* **2019**, *36*, 117–129. [CrossRef]
30. Bhosale, P.; Gadre, R.V. Optimization of carotenoid production from hyper-producing *Rhodotorula glutinis* mutant 32 by a factorial approach. *Lett. Appl. Microbiol.* **2001**, *33*, 12–16. [CrossRef]
31. Kot, A.M.; Błażejak, S.; Kurcz, A.; Gientka, I.; Kieliszek, M. *Rhodotorula glutinis*—potential source of lipids, carotenoids, and enzymes for use in industries. *Appl. Microbiol. Biotechnol.* **2016**, *100*, 6103–6117. [CrossRef] [PubMed]
32. Keceli, T.M.; Erginkaya, Z.; Turkkan, E.; Kaya, U. Antioxidant and Antibacterial Effects of Carotenoids Extracted from *Rhodotorula glutinis* Strains. *Asian J. Chem.* **2013**, *25*, 42–46. [CrossRef]
33. Wirth, F.; Goldani, L.Z. Epidemiology of *Rhodotorula*: An Emerging Pathogen. *Interdiscip. Perspect. Infect. Dis.* **2012**, *2012*, 465717. [CrossRef]
34. Dyaa, A.; Soliman, H.; Abdelrazak, A.; Samra, B.N.; Khojah, E.; Ahmed, A.F.; El-Esawi, M.A.; Elsayed, A. Optimization of Carotenoids Production from *Rhodotorula* sp. Strain ATL72 for Enhancing its Biotechnological Applications. *J. Fungi* **2022**, *8*, 160. [CrossRef]
35. Rosso, L.; Lobry, J.R.; Bajard, S.; Flandrois, J.P. Convenient Model to Describe the Combined Effects of Temperature and pH on Microbial Growth. *Appl. Environ. Microbiol.* **1995**, *61*, 610–616. [CrossRef]
36. Endeshaw, A.; Birhanu, G.; Zerihun, T.; Misganaw, W. Application of microorganisms in bioremediation-review. *J. Environ. Microbiol.* **2017**, *1*, 2–9.
37. Liu, X.; Jia, B.; Sun, X.; Ai, J.; Wang, L.; Wang, C.; Huang, W. Effect of Initial PH on Growth Characteristics and Fermentation Properties of *Saccharomyces cerevisiae*. *J. Food Sci.* **2015**, *80*, M800–M808. [CrossRef]
38. Wang, Q.; Liu, D.; Yang, Q.; Wang, P. Enhancing carotenoid production in *Rhodotorula mucilaginosa* KC8 by combining mutation and metabolic engineering. *Ann. Microbiol.* **2017**, *67*, 425–431. [CrossRef]
39. Urano, N.; Shirao, A.; Naito, Y.; Okail, M.; Ishida, M.; Takashio, M. Molecular Phylogeny and Phenotypic Characterization of Yeasts with a Broad Range of pH Tolerance Isolated from Natural Aquatic Environments. *Adv. Microbiol.* **2019**, *9*, 56–73. [CrossRef]
40. Tinoi, J.; Rakariyatham, N.; Deming, R.L. Simplex optimization of carotenoid production by *Rhodotorula glutinis* using hydrolyzed mung bean waste flour as substrate. *Process. Biochem.* **2005**, *40*, 2551–2557. [CrossRef]
41. Ishmayana, S.; Kennedy, U.J. Preliminary Evidence of Inositol Supplementation Effect on Cell Growth, Viability and Plasma Membrane Fluidity of the Yeast Saccharomyces cerevisiae. *Procedia Chem.* **2015**, *17*, 162–169. [CrossRef]
42. Allahkarami, S.; Akhavan Sepahi, A. Isolation and identification of carotenoid-producing *Rhodotorula* sp. from Pinaceae forest ecosystems and optimization of in vitro carotenoid production. *Biotechnol. Rep.* **2021**, *32*, e00687. [CrossRef]
43. Moreira, F.G.; Lenartovicz, V. The use of α-methyl-D-glucoside, a synthetic analogue of maltose, as inducer of amylase by *Aspergillus* sp. in soild-state and submerged fermentations. *Braz. J. Microbiol.* **2001**, *32*, 15–19. [CrossRef]
44. Nascimento, V.M.; Antoniolli, G.T.U.; Leite, R.S.R.; Fonseca, G.G. Effects of the carbon source on the physiology and invertase activity of the yeast *Saccharomyces cerevisiae* FT858. *3 Biotech* **2020**, *10*, 348. [CrossRef]
45. Becerra, M.; Belmonte, E.R.; Cerdán, M.E.; Siso, M.I.G. Extraction of intra cell proteins from *Kluyveromyces lactis*. *Food Technol. Biotechnol.* **2001**, *39*, 135–139.
46. Buxadó, J.A.; Heynngnezz, L.E.; Juiz, A.G.; Tamayo, G.; Lima, I.R.; Marshalleck, H.D.; Mola, E.L. Scale-up of processes to isolate the misstargeted rBm86 protein from *Pichia pastoris*. *Afr. J. Biotechnol.* **2004**, *11*, 599–605.
47. Kula, M.; Schütte, H.; Vogels, G.; Frank, A. Cell disintegration for the purification of intracellular proteins. *Food Biotechnol.* **1990**, *4*, 169–183. [CrossRef]
48. Liu, D.; Ding, L.; Sun, J.; Boussetta, N.; Vorobiev, E. Yeast cell disruption strategies for recovery of intracellular bio-active compounds—A review. *Innov. Food Sci. Emerg. Technol.* **2016**, *36*, 181–192. [CrossRef]

49. Desai, K.M.; Survase, S.A.; Saudagar, P.S.; Lele, S.S.; Singhal, R.S. Comparison of artificial neural network (ANN) and response surface methodology (RSM) in fermentation media optimization: Case study of fermentative production of scleroglucan. *Biochem. Eng. J.* **2008**, *41*, 266–273. [CrossRef]
50. Aksu, Z.; Eren, A.T. Carotenoids production by the yeast *Rhodotorula mucilaginosa*: Use of agricultural wastes as a carbon source. *Process Biochem.* **2005**, *40*, 2985–2991. [CrossRef]
51. D'Evoli, L.; Lombardi-Boccia, G. Influence of Heat Treatments on Carotenoid Content of Cherry Tomatoes. *Foods* **2013**, *2*, 352–363. [CrossRef] [PubMed]
52. Joshi, C.; Singhal, R.S. Modelling and optimization of zeaxanthin production by *Paracoccus zeaxanthinifaciens* ATCC 21588 using hybrid genetic algorithm techniques. *Biocatal. Agric. Biotechnol.* **2016**, *8*, 228–235. [CrossRef]
53. Montgomery, D.C. *Design and Analysis of Experiments*; John Wiley and Sons Ltd.: Hoboken, NJ, USA, 1984.
54. Madamba, P.S. The Response Surface Methodology: An Application to Optimize Dehydration Operations of Selected Agricultural Crops. *LWT* **2002**, *35*, 584–592. [CrossRef]
55. Shafi, J.; Sun, Z.; Ji, M.; Gu, Z.; Ahmad, W. ANN and RSM based modelling for optimization of cell dry mass of *Bacillus* sp. strain B67 and its antifungal activity against *Botrytis cinerea*. *Biotechnol. Biotechnol. Equip.* **2017**, *32*, 58–68. [CrossRef]
56. Popescu, M.; Iancu, P.; Pleşu, V.; Sorin, C.; Todasca, C.M. Different spectrophotometric methods for simultaneous quantification of lycopene and β-carotene from a binary mixture. *LWT* **2022**, *160*, 113238. [CrossRef]

processes

MDPI

Article

Adsorption and Desorption Behavior of Ectoine Using Dowex® HCR-S Ion-Exchange Resin

Yu-Chi Wu [1], Yu-Hong Wei [2] and Ho-Shing Wu [1,*]

1 Department of Chemical Engineering and Materials Science, Yuan Ze University, 135 Yuan Tung Road Chung Li, Taoyuan 320315, Taiwan; s1095207@mail.yzu.edu.tw
2 Graduate School of Biotechnology and Bioengineering, Yuan Ze University, 135 Yuan Tung Road Chung Li, Taoyuan 320315, Taiwan; yhwei@saturn.yzu.edu.tw
* Correspondence: cehswu@saturn.yzu.edu.tw; Tel.: +886-3-4638800-2564

Abstract: Dowex® HCR-S ion-exchange resin was used to adsorb ectoine in a batch system under varying operation conditions in terms of contact time, temperature, pH value, initial concentration of ectoine, and type of salt. Six adsorption isotherm models (Langmuir, Freundlich, Temkin, Dubinin–Radushkevich, Sips, and Redlich–Peterson) and three kinetic models (pseudo-first-order, pseudo-second-order, and intraparticle diffusion) were used to investigate the ectoine adsorption mechanism of ion-exchange resin. According to the experimental results, the mechanism of ectoine adsorption using an ion exchanger includes the ion-exchange reaction and physisorption. Both the Langmuir and Freundlich models were found to have a high fitting. For the kinetic analysis, the pseudo-second-order and intraparticle diffusion models were suitable to describe the ectoine adsorption. Dowex® HCR-S resin has an average saturated adsorption capacity of 0.57 g/g and 93.6% of ectoine adsorption at 25~65 °C, with an initial concentration of 125 g/L. By changing the pH of the environment using NaOH solution, the adsorbed ectoine on the ion-exchange resin can be desorbed to 87.7%.

Keywords: ectoine; ion exchange; separation and purification; kinetic and isotherm simulation

Citation: Wu, Y.-C.; Wei, Y.-H.; Wu, H.-S. Adsorption and Desorption Behavior of Ectoine Using Dowex® HCR-S Ion-Exchange Resin. *Processes* **2021**, *9*, 2068. https://doi.org/10.3390/pr9112068

Academic Editors: Nicola Gargiulo and Suresh K. Bhatia

Received: 16 September 2021
Accepted: 17 November 2021
Published: 18 November 2021

1. Introduction

In recent years, because of increasing ultraviolet ray exposure and changes in social awareness, people have gradually paid more attention to skin and health care products. Ectoine ($C_6H_{10}O_2N_2$) [(S)-2-methyl-1,4,5,6-tetrahydropyrimidine-4-carboxylic acid]) is a chemical with a potentially high value and promising biotechnology, agriculture, pharmaceutical, and cosmiceutical applications. With its strong water-binding ability, ectoine can have a long-term moisturizing effect on human skin that is better than that of glycerol and relieve inflammation symptoms in allergic conjunctivitis. Ectoine can also block and protect biomolecules from the exposure damage of UV radiation and oxidative stress. Cosmetics became the first commercial product containing ectoine, giving them super hydrophilic characteristics, anti-wrinkle effects, and delayed skin-aging effects. Ectoine, which is an extremolyte, maintains the osmotic balance to protect biomolecules and cell structures by forming and stabilizing protective water layers around them. It can also protect skin under conditions of extreme dehydration, heating, drying, and freezing [1,2]. Accordingly, ectoine works as a multifunctional agent and demonstrates protective effects for proteins, DNA, food, human cells, and tissues. Ectoine has potential in other fields as well, and further research and innovations are possible [3,4].

Ectoine is one of the most widely present compatible solutes throughout various halophilic and halotolerant microorganisms in a broad range of Gram-negative and Gram-positive bacteria, such as α- and γ-*Proteobacteria*, *Actinobacteria*, and *Firmicutes* [4,5]. Heterotrophic bacteria usually produce it in the genera *Halomonas*, *Halorhodospira*, *Vibrio*, *Chromohalobacter*, and *Pseudomonas* of the γ-*Proteobacteria* class [4,6–8]. Commercially, ectoine is usually synthesized by a bioprocess technique called "bacterial milking" using the halophilic bacterium *H. elongata* [5,9]. At present, in order to meet the increasing demand for commercial-scale production of ectoine, batch culture, fed-batch culture, bacterial milking [3,10], or non-halophilic bacteria (e.g., *E. coli*) are used in the fabrication process. These methods avoid high-salinity concentration strategies that would cause corrosion damage to equipment and reduce growth rates. However, ectoine fabrication still suffers from low production, whether halophilic or non-halophilic bacteria are used, with ectoine titers of 65 g/L and a specific productivity of 1.16 g/L/h (120 mg/g/h) achieved [11]. Hence, the isolation and purification of ectoine is the most critical step in improving downstream processing and reducing manufacturing costs.

Owing to its broad applications, ectoine it is a highly priced chemical, with a current sales price of USD 100/g when provided by Echochemical, Taiwan. Whether ectoine production is performed by bacterial milking, leaky mutants, or recombinant genes on non-halophilic bacteria, the loss of ectoine from the downstream processes, such as during separation and purification, must be reduced. Because ectoine is a highly hydrophilic compound, current separation methods, such as solvent extraction and salt extraction, cannot easily separate it from the fermentation broth. It is challenging to separate ectoine with high purity and low loss using mass separation methods like extraction or energy separation methods like distillation. Kunte et al. reported the production stages of industrial production of the cell protectant ectoine to include fermentation, microfiltration, desalting, capture, and refining [3]. The capture stage includes cation exchange and crystallization. Hence, many researchers use ion-exchange technology to separate it [2,12]. Sauer et al. described the use of Dowex 50 WX8 cation-exchange resin (Na^+ form) packed in a 2.8 × 18 cm cation-exchange column to elute ectoine [10]. Fülberth et al. reported that the ion-exchange resins used in ectoine separation include Dowex Marathon C, Dowex HCR-S, Dowex50WX8, Amberlite IR 122, and Fractogel EMD SE Hicap (M) [13].

Ion exchange is an interface-based process that is particularly efficient for low-concentration systems, and it involves mass transfer without the creation of byproducts. There is currently little in the literature in terms of detailed research on ectoine kinetic and isotherm parameters using ion exchangers, nor their adsorption/desorption behavior. Therefore, this study aimed to find the optimal operating conditions for ectoine purification from an aqueous solution using an ion-exchange strategy and reveal the underlying adsorption/desorption mechanism. Adsorption experiments were designed in a batch system with alternations in reaction time, pH, type of salt, temperature, and initial ectoine concentration. The adsorption isotherm and kinetic study were used to understand the nature of the sorption process, depending on the adsorbent characteristics, and inform the design of industrial sorption columns. The desorption process was performed with changes in pH, temperature, and NaOH concentration, and the desorption mechanism, parameters, and cost evaluation are also discussed.

2. Materials and Methods

2.1. Materials

Ectoine ((S)-2-methyl-1, 4, 5, 6-tetra-hydropyrimidine-4-carboxylic acid, CAS 96702-03-3, >95%) was purchased from Sigma-Aldrich, Munich, Germany; it was produced by the natural Gram-negative bacteria *Halomonas elongate*. Dowex® HCR-S is a strong acid, styrene-divinylbenzene cation-exchange resin (CAS 64082-73-1, mesh size: 20–50 mesh, ionic form: H, moisture: 50–56%, capacity: 4.8 meq/g, Sigma-Aldrich, Munich, Germany). Dowex®50WX8 is a strong acid, styrene-divinylbenzene cation-exchange resin (CAS 69011-22-9, mesh size: 100–200 mesh, ionic form: Na, moisture: 54%, capacity: 4.8 meq/g, Sigma-Aldrich, Munich, Germany). HCl (35%) and NaOH (97%) were purchased from SHOWA (Tokyo, Japan). NaCl (99.8%) was purchased from Sigma-Aldrich (Munich, Germany). The water was deionized. All reagents were used as received, except for the resin.

Pretreatment of Resin

The resin (10 g) was washed using deionized water (20 mL) at least three times until the residual solution became colorless. The resin was regenerated using 15% HCl (120 mL) with a flow rate of 120 mL/h, then washed using deionized water (240 mL) with a flow rate of 240 mL/h. Next, we immersed the resin in acetone for 30 min to remove compounds such as alkylbenzenes, styrene, and uncrosslinked polymers, then dried the resin at 45 °C in an oven.

2.2. Batch Adsorption of Ectoine

Batch adsorption experiments were conducted by mixing the ion-exchange resin (1 g) and ectoine solution (10 mL) in a 30 mL flask. The desired concentration of ectoine solution (1000 mg/L) was prepared by diluting from the stock solution in a 30 mL sample bottle. The concentration of the stock solution was 10,000 mg/L. The solution was agitated at 100 rpm at a selected temperature from 25 °C to 65 °C in a reciprocating water bath shaker (DK W-20, DENG YNG, Taipei, Taiwan). The mixture was shaken at a selected contact time, varying from 0.5 to 6 h for the kinetic experiment and 8 h for the equilibrium isotherm experiment. After adsorption, the samples were filtered through a polyvinylidene difluoride (PVDF) filter with a pore size of 0.45 μm. The residual concentration of ectoine in the filtrate was determined using HPLC with a wavelength of 210 nm.

Since ectoines are biomolecules and act as energy resources for microorganisms, they are easily decomposed when dissolved in water. Hence, capture or release of ectoine was performed in as short a time as possible.

The ectoine ion-exchange reaction is:

$$\overline{(R-SO_3^-)H^+} + \text{Ectoine ion}^+{}_{aq} \leftrightarrow \overline{(R-SO_3^-)\text{Ectoine ion}^+} + H^+_{(aq)} \quad (1)$$

The adsorption percentage of ectoine ions was calculated using Equation (2):

$$\text{Ectoine adsorption } (\%) = \left(\frac{C_0 - C_t}{C_0}\right) \times 100 \quad (2)$$

where C_0 is the initial ectoine concentration (mg/L) and C_t is the ectoine concentration at time t (mg/L).

Equilibrium isotherm experiments were performed with various ectoine concentrations ranging from 1000 mg/L to 9000 mg/L at selected times. The amount of ectoine adsorbed per unit dried mass of adsorbent q_e (mg/g) was calculated using Equation (3):

$$q_e = \frac{(C_o - C_e)V}{m} \quad (3)$$

where V is the volume of the solution (L); C_0 and C_e are the initial and equilibrium concentrations of ectoine ion residuals in the aqueous solution (mg/L), respectively; and m is the dry weight of the adsorbent (g). SigmaPlot software (Systat Software, Inc., San Jose, CA, USA) was used to fit the nonlinear analysis of the isotherm models.

2.2.1. Kinetic Study of Ectoine Adsorption

Resin (1 g) was added to a 10 mL ectoine solution (1000 mg/L) and agitated at 100 rpm at 35 °C in a reciprocating water bath shaker. The adsorbed rate of ectoine was investigated by measuring the ectoine ion concentration at predetermined time intervals using the HPLC method. The amount of adsorbed ectoine ions q_t (mg/g) at time t (min) was calculated using Equation (4):

$$q_t = \frac{(C_0 - C_t)V}{m} \quad (4)$$

where q_t is the adsorption capacity of ectoine in the resin phase at time t (mg/g); and C_0 and C_t are the initial concentration and concentration of ectoine ion at time t (mg/L), respectively.

2.2.2. Preparation of Saturation Concentration of Ectoine in Ion-Exchange Resin

The saturation experiments were carried out in a 2 mL solution with 0.5 g dried resin (solution/resin = 4:1) under a selected concentration, referenced from the isotherm simulations. The saturation of ectoine in the resin was obtained, and subsequently, the desorption experiments proceeded for the next separation/purification procedure. The saturation capacity, q_{sat} (g/g), of ectoine was calculated using Equation (5):

$$q_{sat} = \frac{(C_0 - C_e)V}{m \times 1000} \quad (5)$$

where q_{sat} is the saturation capacity of ectoine in the resin phase at equilibrium (g/g).

2.2.3. Desorption of Ectoine

The desorption experiments were performed identically to the adsorption tests. Saturated resins were added to 10 mL of the desired desorbed agents, such as pure water, NaOH, or NaCl solution. The mixture of saturated resin and solution with a ratio of 0.1 g/mL was agitated at 100 rpm and at a selected temperature. The ectoine concentration of the solution was determined using HPLC. After the experiment, resins were collected for recycling in a new sorption-desorption cycle. The solution's pH value was adjusted with HCl (or NaOH). The pH was measured by using a digital pH meter (SUNTEX, New Taipei, Taiwan). Because ectoine possesses an isoelectric point with amino acids, the pH value was manipulated to transform the ectoine into the desired positively or negatively charged molecules. The desorption percentage of ectoine was:

$$\text{Ectoine desorption } (\%) = \left(\frac{C_e \cdot V}{q_{sat} \times 1000} \right) \times 100 \quad (6)$$

where C_e is the concentration after reaching desorption equilibrium in the solution phase (mg/L).

2.3. *Swelling Index of the Ion Exchanger*

The dry resin was mixed with pure water at 25 °C for 2 h, and we then removed the surface water of the resin using filtration. The weight of the wet resin was measured. The swelling index, X, was calculated using Equation (7).

$$X = \frac{\text{Weight of wet resin} - \text{Weight of dry resin}}{\text{Weight of dry resin}} \quad (7)$$

2.4. *Determination of the Zeta Potential of Ectoine in the Solution*

Ectoine solution (10,000 mg/L) was diluted to the desired concentration using deionized water, and the pH value was adjusted using diluted HCl and NaOH solution. After this, we carefully washed it with deionized water and assembled the flow cell, where the solution contained analyte injects. Scanning of cells containing analytes at a high enough intensity or concentration was conducted to measure the zeta potential using NanoPlus (Norcross, GA, USA), with the software Particulate Systems (Norcross, GA, USA).

3. Models of Adsorption and Desorption

3.1. *Adsorption Kinetic Models*

Ion exchange and adsorption share various standard features regarding application of the batch and fixed-bed processes, which can be grouped as sorption for a unified treatment. These processes involve the transfer and equilibrium distribution of one or more solutes between a fluid phase and particles [14]. Adsorption kinetics are based on the relationship between the adsorbate–adsorbent and the reactor's operation in the adsorption reaction, which plays a vital role in designing the plant column. Common adsorption kinetic equations are the pseudo-first-order model, pseudo-second-order model, and intraparticle diffusion model.

(i) Pseudo-first-order model

In 1898, Lagergren proposed a pseudo-first-order model describing a liquid–solid phase adsorption system, which showed that the adsorption rate is constructed based on the adsorption capacity [15]. The nonlinear form can be expressed as:

$$\frac{dq_t}{dt} = k_{pf}(q_e - q_t) \tag{8}$$

where k_{pf} is the adsorption-rate constant (min^{-1}), q_e is the equilibrium adsorption capacity ($mg\ g^{-1}$), and q_t is the adsorption capacity ($mg\ g^{-1}$) at time t (min). We can plot the log rate (q_e–q_t) vs. t to calculate the adsorption rate constant [16].

(ii) Pseudo-second-order model

In 1995, Ho established a pseudo-second-order model in a batch reactor system [17]. The differential form is:

$$\frac{dq_t}{dt} = k_{ps}(q_e - q_t)^2 \tag{9}$$

where k_{ps} is the adsorption-rate constant ($g\ mg^{-1}\ min^{-1}$).

(iii) Intraparticle diffusion model

The intraparticle diffusion kinetic model focuses on the diffusion process and was proposed by Weber and Morris in 1962 [18]. During the solid–liquid adsorption reaction, the solute moves from the solution to the pores of the solid-phase adsorbent. This process is usually the rate-determining step in the entire adsorption reaction [16,19]:

$$q_t = k_{ip}t^{\frac{1}{2}} + C \tag{10}$$

where k_{ip} is the intraparticle–diffusion constant ($mg\ g^{-1}\ min^{-1/2}$) and C is the intercept constant ($mg\ g^{-1}$) of the model, which is proportional to the thickness of its boundary layer. The larger C is, the more significant the adsorption boundary layer's effect [19,20].

(iv) Elovich model

Elovich kinetics are widely applied to chemisorption (chemical reaction) to describe its mechanism and are suitable for heterogeneous adsorption systems. They were first formulated to describe the kinetics of the oxidation process and later developed for adsorption kinetics by Elovich and his collaborators from 1934 to 1939 [21]. Equation (11) covers an extensive range of adsorption systems with a mildly rising tendency.

$$\frac{dq_t}{dt} = \alpha e^{-\beta q_t} \tag{11}$$

where t is the adsorption time (min), α can be considered the initial adsorption rate ($mg\ g^{-1}\ min^{-1}$) because (dq_t/dt) approaches α when q_t approaches 0 [22], and β is the Elovich constant ($g\ mg^{-1}$).

In the kinetic system, the differential equation is integrated first. We can plot the integral form of the adsorption capacity, qt, with adsorption time. SigmaPlot software (Systat Software, Inc., San Jose, CA, USA) was used to fit the nonlinear analysis to obtain the parameters.

(v) Arrhenius activation energy

The adsorption activation energy is calculated from the Arrhenius equation:

$$\ln k = \ln A - \frac{E_a}{RT} \quad (12)$$

where k is the adsorption rate constant; A is the frequency factor and is independent of the temperature variable; E_a is the adsorption activation energy (J mol^{-1}); R is the universal gas constant (8.314 J mol^{-1} K^{-1}); and T is the solution temperature (K).

3.2. Adsorption Isotherm Models

(i) Langmuir isotherm

In 1916, Langmuir deduced the isothermal adsorption model based on the equilibrium between adsorption and desorption and concluded the relationship between the equilibrium adsorption, q_e, and the equilibrium concentration, C_e, of the liquid phase as a nonlinear form:

$$q_e = \frac{q_m K_L C_e}{1 + K_L C_e} \quad (13)$$

where q_m is the maximum adsorption capacity of the monolayer (mg g^{-1}); C_e is the equilibrium concentration in the solution (mg L^{-1}); and K_L is the Langmuir equilibrium parameter (L mg^{-1}), defining the affinity of binding sites and the energy of sorption.

Furthermore, the constant R_L can determine whether the adsorption process is spontaneous [23,24]. The R_L equation is defined as:

$$R_L = \frac{1}{1 + K_L \times C_0} \quad (14)$$

(1) When $0 < R_L < 1$, adsorption proceeds in a favorable direction. (2) When $R_L > 1$, the adsorption process proceeds in an unfavorable direction. (3) When $R_L = 1$, the adsorption process is linear. (4) When $R_L = 0$, the adsorption process is irreversible [25–27]:

(ii) Freundlich isotherm

In 1906, Freundlich deduced an empirical formula based on isothermal equilibrium experiments which that widely employed to describe solid–liquid adsorption [28]. Freundlich isotherm theory supports reversible and nonideal adsorption on uneven surfaces. The Freundlich isotherm equation can be described as:

$$q_e = K_F \times C_e^{\frac{1}{n}} \quad (15)$$

where K_F and n are Freundlich constants and the adsorption intensity, respectively.

(iii) Temkin isotherm

The Temkin isotherm is generally applied to describe monolayer adsorption behavior in a nonideal sorption system. By ignoring the extremely low and large value of concentrations, this isotherm assumes that the adsorption heat of all molecules in the layer would decrease, rather than logarithmically, with the increased coverage of the adsorbent surface. The Temkin isotherm model is:

$$q_e = \frac{RT}{b_T} \ln(A_T C_e) \quad (16)$$

where b_T is the Temkin constant related to the sorption heat (J mol^{-1}) and A_T is the Temkin isotherm constant related to equilibrium binding (L mg^{-1}).

(iv) Dubinin–Radushkevich isotherm

Dubinin and Radushkevich proposed the D–R isotherm adsorption model in 1947, which is a semi-empirical model. The D–R model assumes multilayer adsorption, which relates to Van der Waal's forces, and thus, it can be utilized in the simulation of the physisorption process [27]. The D–R isotherm equation is:

$$q_e = q_m\left[\exp\left(-B\varepsilon_d^2\right)\right] \quad (17)$$

where q_m is the maximum adsorption capacity (mg g^{-1}); B is the energy constant (mol^2 kJ^{-2}); and ε_d is the D-R constant (mol^2 kJ^{-2}). The D–R model is usually applied to distinguish adsorption types among chemisorption, physisorption, and ion exchange.

(v) Redlich–Peterson isotherm

Redlich–Peterson (R–P) is a hybrid isotherm with three parameters and is characterized by the combination of the Freundlich and Langmuir isotherm models. It can be described using Equation (18):

$$q_e = \frac{K_R C_e}{1 + \alpha_R C_e^g} \quad (18)$$

where K_R is the R–P isotherm constant (L g^{-1}); α_R is the R-P isotherm constant (L mg^{-1}); and g is the R–P isotherm exponent.

(vi) Sips isotherm

Sips isotherm also features the heterogeneity factor β_S. The general expression of the Sips isotherm equation is:

$$q_e = \frac{K_S C_e^{\beta_s}}{1 + \alpha_S C_e^{\beta_s}} \quad (19)$$

where K_S is the Sips isotherm constant (L g^{-1}); α_S is the Sips isotherm constant (L mg^{-1}); and β_S is the Sips isotherm exponent. A summary of the adsorption isotherm and kinetic models is listed in Table 1.

Table 1. Summary of the adsorption isotherms and kinetic models.

Kinetic Model		**Ref.**
Elovich	$\frac{dq_t}{dt} = \alpha \exp(-\beta q_t)$	[21]
Intraparticle diffusion	$q_t = k_{ip}t^{1/2} + C$	[18]
Pseudo-first-order	$\frac{dq_t}{dt} = k_{pf}(q_e - q_t)$	[15]
Pseudo-second-order	$\frac{dq_t}{dt} = k_{ps}(q_e - q_t)^2$	[29]
Isothermal Model		**Ref.**
Dubinin–Radushkevich	$q_e = q_m e^{-B\varepsilon^2}$	[30]
Freundlich	$q_e = K_F C_e{}^{1/n}$	[28]
Langmuir	$q_e = \frac{q_m K_L C_e}{1+K_L C_e}$	[31]
Redlich–Peterson	$q_e = \frac{K_R C_e}{1+\alpha_R C_e^g}$	[32]
Sips	$q_e = \frac{K_S C_e^{\beta_s}}{1+\alpha_S C_e^{\beta_s}}$	[33]
Temkin	$q_e = \frac{RT}{b_T}\ln(A_T C_e)$	[34]

3.3. Adsorption Thermodynamics

In order to describe the thermodynamic behavior of ectoine ion sorption on cation-exchange resin, the thermodynamic parameters were calculated, including the Gibbs free energy (ΔG), entropy change (ΔS), and heat of adsorption (ΔH), using equilibrium constants at different temperatures (298–338 K). They can be obtained by using the following equations:

$$\Delta G = -\Delta H - T\Delta S \quad (20)$$

$$\Delta G = -RT \ln K_D \quad (21)$$

$$K_D = \frac{C_s}{C_e} \tag{22}$$

$$\ln K_D = \frac{-\Delta H}{RT} + \frac{\Delta S}{R} \tag{23}$$

where K_D is the equilibrium distribution constant and C_s (mg L^{-1}) and C_e (mg L^{-1}) are the ectoine equilibrium concentration in the resin and aqueous phase, respectively. ΔH (kJ mol^{-1}), ΔG (kJ mol^{-1}), and ΔS (kJ mol^{-1}) can be calculated from a plot of ln K_D versus 1/T. Inglezakis and Zorpas reported that ion exchange could be endothermic or exothermic with the involvement of heat of adsorption in the range of −24 to 38 kJ/mol, while the adsorption energy limits are found to be in the range of 0.6 kJ/ to 25 kJ/mol, and the activation energy is higher than 40 kJ/mol [35].

3.4. Desorption Kinetic Models

The external diffusion model (EDM) describes the desorption of ectoine from ion-exchange resin.

The EDM model assumes the ion-exchange reaction of ectoine molecules through the external film by a liquid-phase diffusion mechanism. The mass transfer is expressed as:

$$\frac{d}{dt} q_t = \frac{k_F}{\rho_P}(C_t - C_e) \tag{24}$$

where q_t is the ectoine concentration on the resin (mg g^{-1}) at time t; k_F is the volumetric mass-transfer coefficient in the external liquid film (min^{-1}); ρ_P is the resin particle density (g/mL); and C_t and C_e are the ectoine concentrations, desorbed in the aqueous phase at time t [36].

4. Results and Discussion

4.1. Adsorption of Ectoine Using Ion-Exchange Resin

4.1.1. Effect of Types of Ion-Exchange Resin on the Adsorption of Ectoine

This study used two kinds of strong-acid cation-exchange resins (Dowex® 50WX8 and Dowex® HCR-S) to conduct adsorption experiments on ectoine in an aqueous solution. Figure 1 shows the effect of adsorption time on the strong-acid cation-exchange resin adsorption in a reciprocating water bath shaker. Dowex® 50WX8 showed a poor equilibrium adsorption efficiency in ectoine adsorption. The adsorption rate of ectoine using Dowex® 50WX8 was greater than that using Dowex® HCR-S. The swelling index (X) was calculated using Equation (7). The swelling index of Dowex® 50WX8 (X = 0.68) was lower than that of Dowex® HCR-S (X = 0.88). This finding explains why the adsorption capacity was low for Dowex® 50WX8, while the equilibrium time for Dowex® HCR-S was 120 min. Hence, Dowex® HCR-S was chosen as the candidate for adsorption, and the adsorption time was set at 8 h to reach the equilibrium state.

4.1.2. Effect of pH on Ectoine Adsorption

The solution's pH value is crucial to adsorption, especially for ion exchangers. Ectoine is a Zwitterion compound and can transform between cationic and anionic forms based on the pH, as shown in Figure 2. A zwitterion, also called an inner salt, contains an equal number of positively and negatively charged functional groups. Therefore, zwitterions are mostly electrically neutral; the whole molecule's net charge is zero. The isoelectric point is derived from its inner-salt structure. Hence, the pH can affect the adsorption capacity by changing the property of adsorbent and solution analytes. In this work, the initial pH of the solution was adjusted using HCl and NaOH, with and without the ion-exchange resin. Table 2 shows that the initial pH of the solution without ion-exchange resin does not affect the equilibrium adsorption capacity (q_e) of ectoine (mean = 9.83 mg/g). This demonstrates that the surface charge of both the exchanger and analyte is not affected by the surrounding pH, but the concentration gradient of H^+ inside the resin overwhelmed the solution's pH. The acidity of Dowex® HCR-S itself was pH 3.3 when 1 g of dry resin was immersed in 10 mL of neutral water. This demonstrates no effect on ectoine adsorption from adjusting the initial pH in the solution without resin. Therefore, we determined we should adjust the pH in the solution with resin before the experiment.

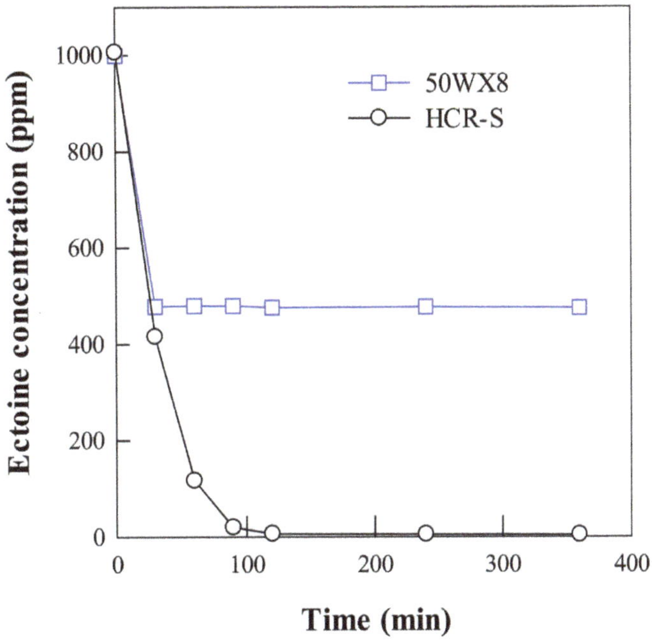

Figure 1. Effect of adsorption time on ectoine adsorption using ion-exchange resin (ectoine = 1000 mg/L, solution/resin = 10 mL/g, reaction time = 6 h, stirring rate = 100 rpm, temperature = 35 °C).

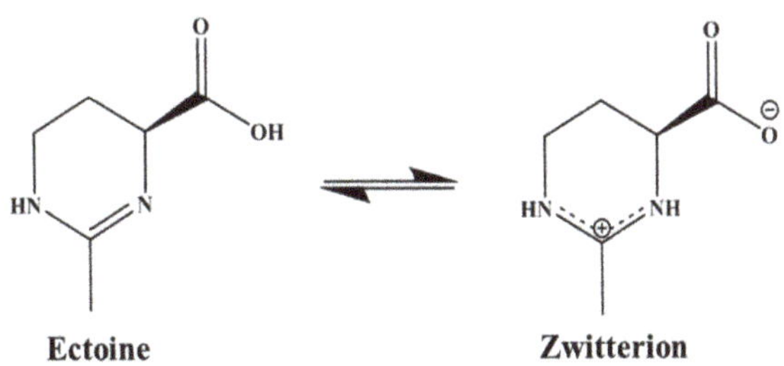

Figure 2. Zwitterion form of ectoine containing both acid and base centers and its isomer.

The pH of the solution with resin was adjusted using NaOH solution (3 M) with a long enough time for equilibrium to be reached. Table 2 indicates that the equilibrium adsorption capacity of ectoine (q_e) is affected by pH values. q_e remained stable before the pH rose to over 3, slightly decreased at pH 4, and dramatically dropped at pH values over 4. After adsorption, the pH was reduced when it was lower than 4; in contrast, it increased when it was higher than 4. Therefore, the maximum adsorption was a 9.8 mg/g (0.07 meq/g) adsorption capacity of ectoine under pH < 3 when using an initial ectoine concentration of 1000 mg/L.

Table 2. Effect of pH in the solution with and without Dowex® HCR-S on ectoine adsorption.

Initial pH without Resin	pH after Adsorption	C_e (mg/L)	q_e (mg/g)	Initial pH with Resin	pH after Adsorption	C_e (mg/L)	q_e (mg/g)
2.00	1.98	15.73	9.84	2.00	1.95	19.90	9.80
3.00	2.82	15.94	9.84	3.00	2.90	18.04	9.82
4.00	3.18	14.44	9.86	4.00	3.80	80.23	9.20
5.00	3.29	16.87	9.83	5.00	6.70	740.76	2.59
6.00	3.30	20.79	9.79	6.00	6.89	869.34	1.31
7.00	3.33	15.23	9.85				

Conditions: initial ectoine = 1000 mg/L, reaction time = 8 h, temperature = 35 °C, stirring rate = 100 rpm.

When the counterion H^+ of the resin is replaced with Na^+, OH^- can neutralize the H^+ released from the resin to adjust the pH in the solution. Equilibrium pH can be attained in both the resin and aqueous phases. The results in Table 2 reveal that ectoine exists as a cationic form under pH < 4, known as its isoelectric point. The acidic nature of the resin results in preferential protonation of the ectoine molecules and thus more favorable ion exchange so that ectoine can exchange with hydrogen ions and maintain electroneutrality.

Aside from the surface-charged state of ectoine, the electrostatic affinity between the ion-exchanger (R-SO_3^-) and ectoine is assumed to be good. The pH in the solution will modify the adsorbent to become negatively charged to enhance the attraction to the desired ions, especially metal ions [37–39]. Dowex® HCR-S ion-exchange resin binding with the sulfonic group (SO_3^-) possesses a positive charge on the surface. This means that the surface charges of ectoine and the ion-exchanger have an opposite trend under pH < 4 in the aqueous medium.

4.1.3. Zeta Potential in Ectoine Solution

The Zeta potential is defined as the magnitude of charge at the slipping plane, separating the mobile bulk fluid from the ionic layer attached to the particle surface. Since the surface charge is the primary key for using ion-exchange as the separation strategy of ectoine from bacterial broth, zeta potential analysis provides the results for the differentiation-charged situation on the particle surface under different pH conditions. Figure 3 shows that the pH of samples containing ectoine with a concentration of 2000 mg/L was titrated to pH = 2, 3, 4, 5, and 6 to determine the zeta potential. The isoelectric point of ectoine is located in the pH range between 3 and 4, close to 3.2.

According to the structure of ectoine in Figure 2, ectoine carries a positive charge when the pH value is below its isoelectric point (pI) and carries a negative charge when the pH value is over its pI. The pI of ectoine is 3.2 when the zeta potential equals 0, as shown in Figure 3. Therefore, the adsorption conditions should be set at a pH value below 3.2 to apply the use of a cation-exchanger as the ectoine separation strategy.

The decline of adsorption capacity with increasing pH resulted from the change of ectoine's surface charge from positive to negative. A considerable electrostatic repulsion between ectoine (Ect^-) and the ion-exchanger surface (R-SO_3^-) under a high pH (>4) confirms the contribution of electrostatic interactions to the adsorption phenomenon. On the other hand, for pH values below 4, ectoine ions take on a cationic form, leading to an electrostatic attraction towards the ion-exchanger matrix, and an ion-exchange reaction occurs between hydrogen ions and ectoine ions. The result of the zeta potential of ectoine corresponds to that in Table 2. Moreover, we proved that the separation mechanism of ectoine using Dowex® HCR-S involves ion exchange and electrostatic force.

4.1.4. Effect of Initial Ectoine Concentration

Figure 4a shows the isotherms of ectoine on Dowex® HCR-S at 25 °C and 65 °C with an initial ectoine concentration from 1000 mg/L to 9000 mg/L. The sequence of adsorption capacity for the effect of temperature was 65 °C > 25 °C > 55 °C > 35 °C > 45 °C. The ion-exchange process of ectoine onto Dowex® HCR-S was only slightly dependent on the operating temperatures because the differences in adsorption capacities among the different temperatures were small (<3%), as shown in Figure 4b. Consequently, the ion-exchange capacity of ectoine might depend on temperature, with a greater adsorptive uptake of ectoine at a higher temperature. The adsorption at 65 °C exhibited the best result in the range of 25 °C to 65 °C.

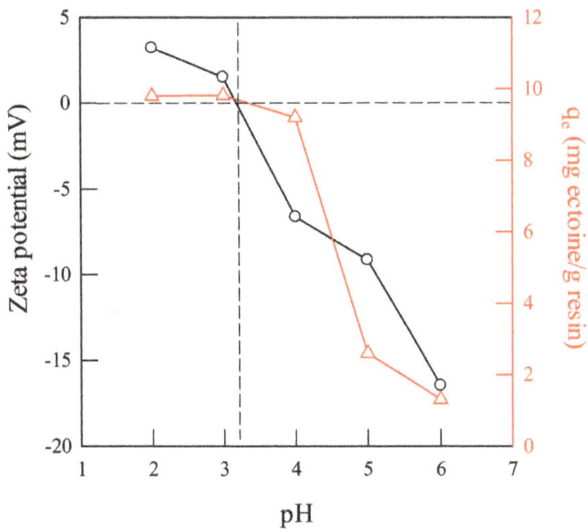

Figure 3. Plot of ectoine's zeta potential and adsorption capacity of ectoine on pH (Dowex® HCR-S resin/aqueous = 0.1 g/mL).

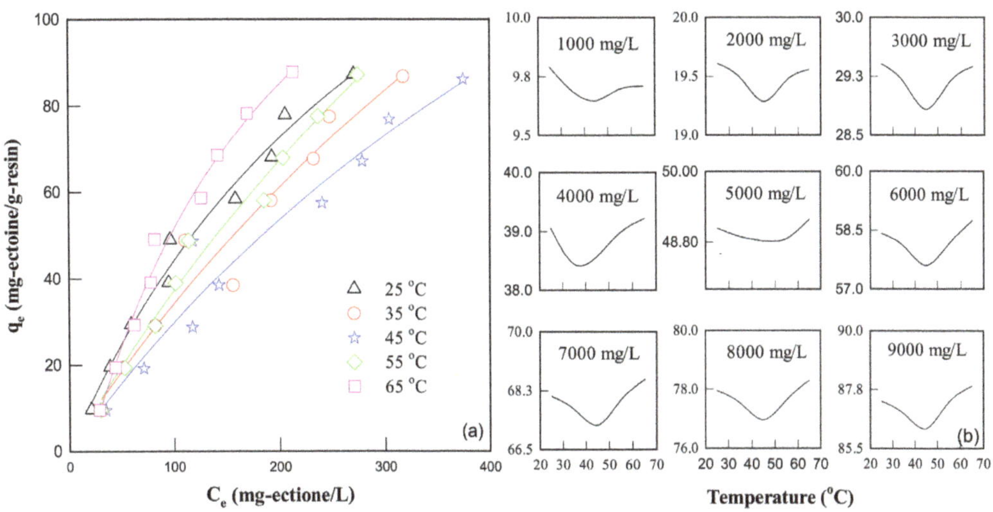

Figure 4. (**a**) Adsorption isotherm of ectoine using Dowex® HCR-S and (**b**) effect of adsorption capacity of ectoine on temperature at different initial ectoine concentrations (initial ectoine = 1000~9000 mg/L, dry resin/aqueous = 0.1 g/mL, agitated rate = 100 rpm, time = 8 h, temperature = 25~65 °C).

The adsorption that occurred can be divided into two temperature ranges: group I (25~45 °C) and group II (45~65 °C), as shown in Figure 4b. The adsorption performance decreased with rising temperatures in group I, while an enhancement of adsorption occurred with increasing temperatures in group II. For group I, the poor adsorption efficiency with increasing temperature can be explained by physisorption. The reduction in equilibrium adsorption capacity (q_e) with increasing temperatures suggests a weak adsorption interaction between the ion-exchanger surface and the ectoine ion, supporting the physisorption assumption. Due to the higher mobility of ectoine ions, there was an increasing tendency for ectoine ions to escape from the ion-exchanger surface in the bulk solution. Hence, decreasing adsorption performance was observed as the temperature increased. Moreover, the thickness of the boundary layer decreases with higher temperatures, leading to poor adsorption efficiency [40].

For group II, the adsorption capacity strengthened with increasing temperature. This phenomenon indicates that it is an endothermic reaction that is enhanced at higher temperatures. Additionally, it shows favorable intermolecular forces between ectoine ions and ion-exchange resin, making the adsorbate easier to adsorb. Chemisorption supports the adsorption in group II. Therefore, the adsorption of ectoine onto ion-exchange resin could be considered on the basis of two groups with the effect of temperature. Group I exhibited the characteristic of physisorption, but group II tended to show chemisorption.

4.1.5. Isotherm Studies of Ectoine

This research used six models to describe the ion-exchange reaction: the Langmuir, Freundlich, Temkin, Dubinin–Radushkevich (D–R), Sips, and Redlich–Peterson (R–P) models. Nonlinear regression is a more general method that can be used to estimate model parameters [38]. For the three-parameter isotherm Sips and Redlich–Peterson models, nonlinear regression is easier than linearizing to find the optimal conditions. This study aimed to utilize a three-parameter isotherm because ectoine adsorption fits both the Langmuir and Freundlich models, while Sips and R–P isotherms, which are known Langmuir–Freundlich isotherms, combine Langmuir with Freundlich and thus exhibit both characteristics. Table 3 lists the parameters using all six isotherms for nonlinear regression. The R–P and Sips isotherms exhibited good fitting, like Langmuir and Freundlich, with the highest R^2 of 0.98 for R–P and 0.99 for the Sips model. Based on the parameters of the R–P isotherm, g was not equal to 1 but was close to it under some temperatures, which means that ectoine adsorption does not follow the Langmuir isotherm fully. In other words, it is not a pure ion-exchange reaction.

Table 3. Isotherm parameters using nonlinear simulation for the adsorption of ectoine onto Dowex® HCR-S using the Freundlich, Langmuir, Dubinin–Radushkevich, Temkin, Redlich–Peterson, and Sips models.

Temperature (°C)	Freundlich Model			Langmuir Model				Dubinin–Radushkevich Model			Temkin Model		
	n	K_F (mg L^{-1})	R^2	q_m (mg g^{-1})	K_L (L mg^{-1})	R^2	R_L	q_m (mg g^{-1})	E [a] (kJ mol^{-1})	R^2	b_T (J mol^{-1})	A_T (L g^{-1})	R^2
25	1.35	1.43	0.98	201	0.0028	0.98		79.7	0.011	0.87	81.0	0.051	0.95
35	1.22	0.80	0.95	299	0.0013	0.95		78.3	0.009	0.82	79.1	0.035	0.91
45	1.24	0.74	0.95	259	0.0013	0.95	0.038–0.454	81.5	0.007	0.87	81.1	0.029	0.91
55	1.17	0.71	0.98	348	0.0012	0.98		84.3	0.008	0.90	77.6	0.034	0.95
65	1.15	0.88	0.97	357	0.0016	0.97		87.7	0.010	0.94	67.3	0.038	0.98

Temperature (°C)	Redlich–Peterson Model				Sips Model			
	K_R (L g^{-1})	α_R (L mg^{-1})	g	R^2	K_S (L g^{-1})	α_S (L mg^{-1})	β_S	R^2
25	0.58	0.0045	0.92	0.98	0.55	0.0028	1.00	0.98
35	0.52	0.0530	0.49	0.95	0.62	0.0009	0.88	0.98
45	0.37	0.0076	0.74	0.95	0.37	0.0013	0.98	0.95
55	0.41	0.0009	1.05	0.98	0.27	0.0012	1.11	0.98
65	0.51	9E-09	3.17	0.98	0.05	0.0004	1.63	0.99

dry resin = 1 g, initial ectoine = 1000~9000 mg/L, equilibrium time = 8 h, [a]: adsorption energy.

The Langmuir equation was found to fit well to the experimental data, with a coefficient of determination (R^2) ranging from 0.94 to 0.99. The Langmuir constants (R_L) calculated by Equation (14) using the Langmuir equation, ranging from 0.038 to 0.454, were between 0 and 1, which indicates that the ion-exchange process proceeded in a favorable direction. The adsorption of ectoine increased concomitantly with the initial concentration.

Figure 5 shows the direct prediction using the q_e vs. C_e diagram fitted with the Langmuir, Freundlich, Temkin, D–R, Sips, and R-P isotherms. One of the benefits of using nonlinear isotherm regression is direct simulation, showing the actual trend. A conclusion could be made that nonlinear regression is more general than linear fitting. Still, under some circumstances with restricted conditions, such as non-saturated curves, linearization provides easier and fast prediction with raw data.

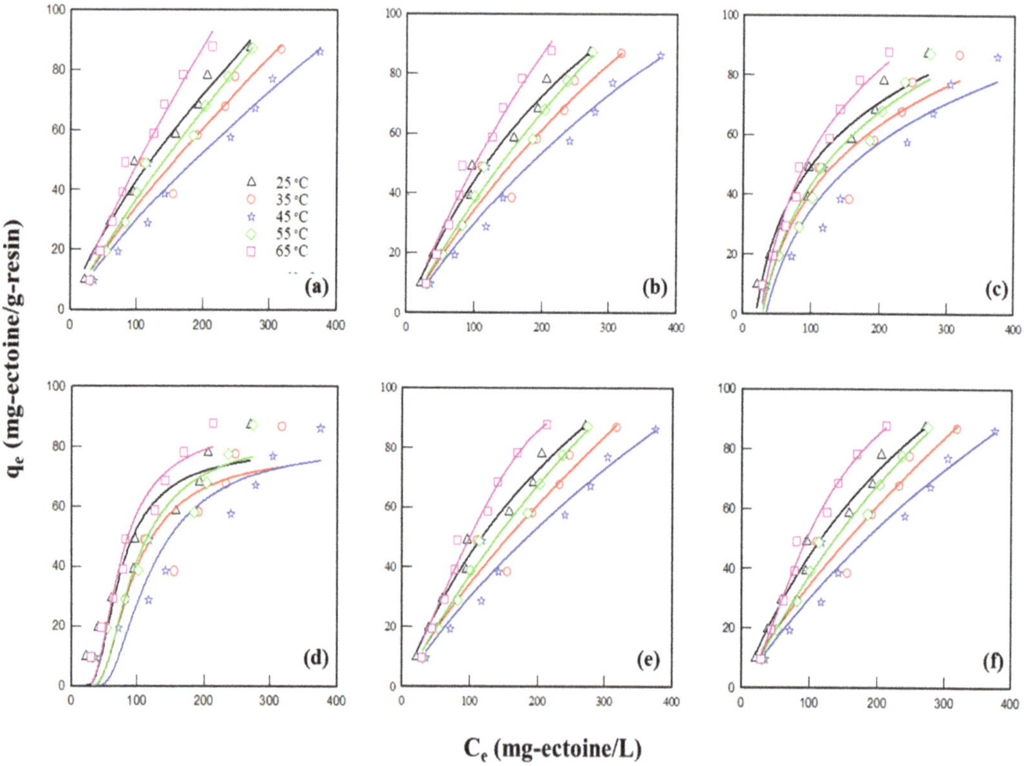

Figure 5. Nonlinear regression fitting curve using the isotherm models (**a**) Freundlich, (**b**) Langmuir, (**c**) Temkin, (**d**) D–R, (**e**) R–P, and (**f**) Sips for the simulation of ectoine adsorption.

Since nonlinear regression can reflect a more accurate and real system than linear regression, the nonlinear regression results exhibited a smaller and more precise range of data. In conclusion, the mechanism of ectoine capture using Dowex® HCR-S resin is not a pure ion-exchange reaction but a mixture of physisorption (van der Waals force) and ion exchange (Coulomb's force). This explains the result from Section 4.1.4 for the effect of temperature on ectoine adsorption, which could be divided into groups I (25~45 °C) and II (45~65 °C). Physisorption based on the van der Waals force was dominant when the temperature was below 45 °C, but chemisorption-like ion exchange based on the electrostatic force dominated when the temperature was above 45 °C.

Based on the isotherm results, the adsorption of ectoine is suggested to include physisorption and ion exchange. The strong hydrophilic force that emanates from the hydrogen bonds that exist between ectoine molecules explains physisorption. In terms of the ion exchange, ectoine as a cation had a higher affinity for exchanging sites than hydrogen ions. The following effect can influence the adsorption of ectoine by using ion-exchange resin: (i) hydrogen bonds existing between ectoine ions; (ii) the mass transfer of ectoine diffuses from the bulk solution to the boundary layer on the surface of the ion-exchange resin; (iii) the pore size of the ion-exchanger. As the temperature increased, physisorption based on hydrogen bonds became weaker due to the higher mobility and the increasing tendency for ectoine ions to escape from the surface of the ion exchanger. In contrast, chemisorption-like ion exchange became stronger when the temperature increased due to the higher mass transfer rate of intraparticle diffusion. Further, Dowex resin is a gel-type resin that is a microporous exchanger. Therefore, the rate-determining phase is either counterion interdiffusion within the ion exchanger (particle diffusion), or counterion interdiffusion in the adhering film (film diffusion).

4.1.6. Effect of Adsorption Time

Adsorption kinetics involve the adsorption uptake concerning the time for measurements of the adsorbate diffusion in the pores of the adsorbent. This study conducted kinetic adsorption experiments in a batch system at different temperatures and adsorption times. The effect of contact time on ectoine adsorption was studied at pH 3.3 and agitating times of 30, 60, 90, 120, 240, and 360 min (Figure 6). The adsorption percentage of ectoine (%) was calculated by Equation (2) and was proportional to adsorption time until equilibrium was reached. Fast adsorption was spotted through the first 90 min of the process, reaching 91.7%, 97.5%, and 91.2% at 35 °C, 45 °C, and 55 °C, respectively. Fast adsorption might reflect the high accessibility of ectoine ions at the ion-exchange sites when the resin was immersed with ectoine molecules and swelled in the aqueous medium. Simultaneously, the stronger driving force resulting from the concentration gradient moved fast ectoine ions to the exchanging sites on the surface of the Dowex® HCR-S resin. On the other hand, with increasing time, the adsorption rate decreased and gradually reached equilibrium due to the crosslinking limit from the ion-exchange resin and the lower availability of the remaining exchange sites.

When the operating temperature was 45 °C, the adsorption rate was the highest; the adsorption percentage reached equilibrium faster than at 35 °C and 55 °C. However, there was no significant effect of temperature on the equilibrium adsorption percentage, which is close to 100% removal. Therefore, the effect of temperature is demonstrated in Figure 6: (1) the adsorption rate was influenced by different temperatures with the trend of 45 °C > 35 °C ≈ 55 °C; (2) there was no obvious effect of temperature on the equilibrium adsorption, which was almost 100% adsorption, resulting from the unsaturated adsorption of ectoine on the ion-exchange resin because of insufficient loading of the initial ectoine concentration.

Therefore, the adsorption of ectoine on the cation-exchange resin is affected by reaction time and temperature. A reaction time of 120 min is required to reach equilibrium. The adsorption rate was influenced by different temperatures, while the equilibrium adsorption did not significantly differ at different temperatures.

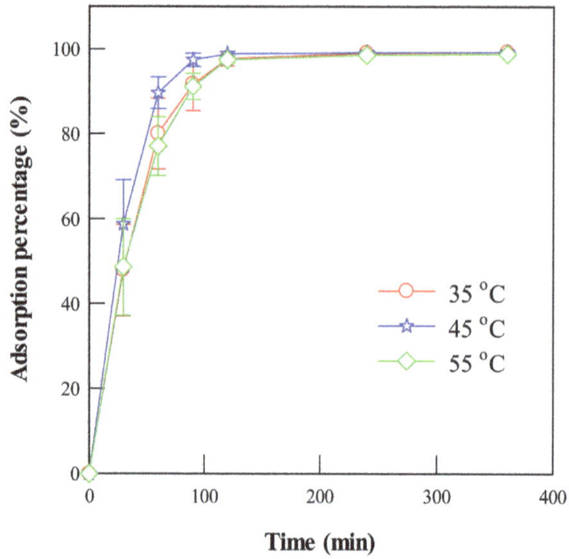

Figure 6. Effect of adsorption time on ectoine adsorption percentage at different temperatures (Dowex® HCR-S = 1 g, initial ectoine = 1000 mg/L, aqueous solution = 10 mL).

4.1.7. Adsorption Kinetics of Ectoine onto the Ion-Exchanger

Adsorption kinetics are a crucial factor in determining the mass-transfer constant and kinetics parameters. This information is essential for the design of an adsorbent-packed column in actual industrial applications. The adsorption kinetics simulation depicted the adsorbate's retention rate or release from the bulk solution to the solid-phase surface of adsorbents at a given adsorbent dose, temperature, flow rate, and pH [41]. When designing an adsorption system, the adsorption rate, adsorbent surface complexity, concentration of adsorbate, and flow rate can affect the adsorption kinetics.

The simulation results obtained by the pseudo-first-order model, pseudo-second-order model, Elovich equation, and intraparticle diffusion model are shown in Figure 7. The activation energy of the physisorption process was lower than 40 kJ/mol. The activation energy was between 24 and 40 kJ/mol for the ion-exchange reaction [35]. According to Table 4, the activation energies of the pseudo-second-order and intraparticle diffusion models were 36.5 kJ/mol and 35.1 kJ/mol, respectively. The pseudo-second-order model can describe the electrostatic interaction between ectoine and sulfonic groups (SO_3^-) covalently bonded in the ion-exchange matrix because it is related not only to ectoine ions or exchange sites but also to the attractions and repulsions between them.

Figure 7d shows two-step adsorption by plotting q_t against $t^{1/2}$. The first half of the curve belongs to the diffusion during the solid-liquid adsorption process. The second half of the straight line is the saturation condition. The adsorption reaction mechanism usually consists of three consecutive steps, starting from the adsorption on the outer surface of the solid phase, moving the solute in the liquid phase to the solid surface layer, and then performing intraparticle diffusion, transferring from the solid surface layer to the pores. The solute slowly moves from the large pores to the small pores to reach an equilibrium state among the pores. The intercept is usually a positive value, indicating a rapid adsorption reaction in a short period of time.

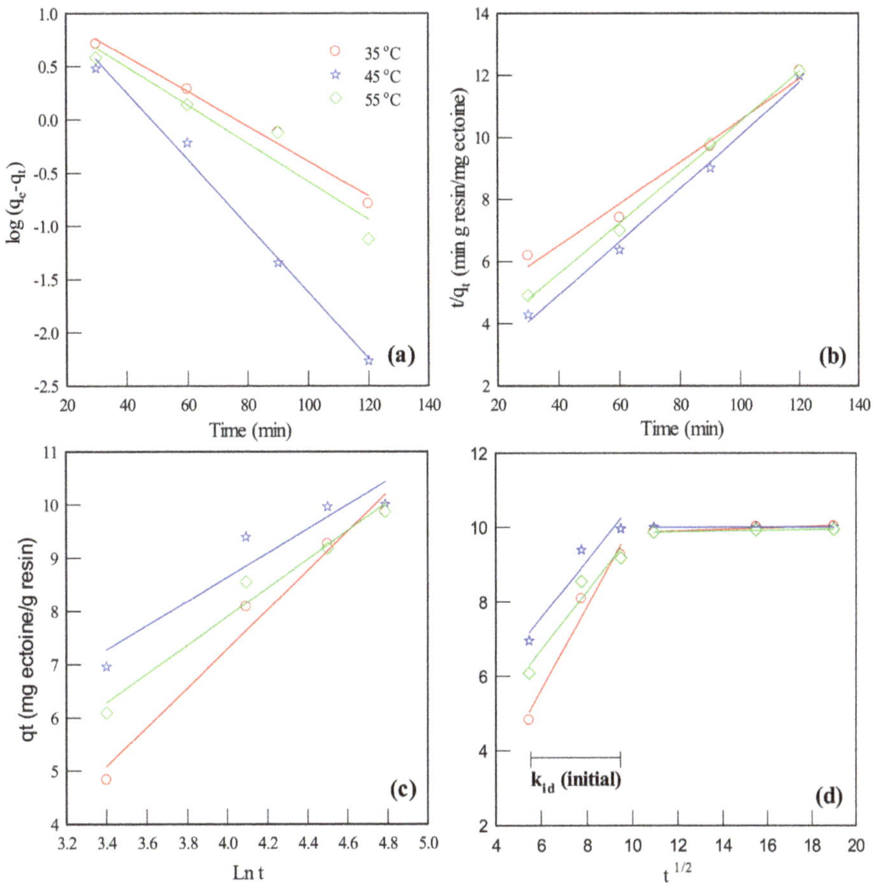

Figure 7. Regression analysis of the adsorption of ectoine by the (**a**) pseudo-first-order equation, (**b**) pseudo-second-order equation, (**c**) Elovich equation, and (**d**) intraparticle diffusion model.

Table 4. Kinetic parameters for ectoine ions using Dowex® HCR-S as the adsorbent.

	Pseudo-First-Order					Pseudo-Second-Order			
T (°C)	q_e [a] (mg g^{-1})	k_{pf} (Min^{-1})	q_e [b] (mg g^{-1})	R^2	E_a (kJ Mol^{-1})	k_{ps} (g mg^{-1} Min^{-1})	q_e [b] (mg g^{-1})	R^2	E_a (kJ Mol^{-1})
35	10.03	0.04	18.00	0.98		0.0012	14.90	0.97	
45	10.01	0.07	32.34	0.99	3.72	0.0049	11.67	0.99	36.5
55	9.95	0.04	16.50	0.92		0.0028	12.25	1.00	

	Elovich				Intraparticle diffusion		
T (°C)	α (mg g^{-1} min^{-1})	β (mg g^{-1})	R^2	E_a (kJ mol^{-1})	k_{id} (initial) (mg g^{-1} min$^{1/2}$)	R^2 (initial)	E (kJ mol^{-1})
					1.12	0.97	
35	0.49	0.27	0.97		0.77	0.93	35.1
					0.79	0.94	
45	1.86	0.44	0.90	23.1			
55	0.83	0.35	0.97				

k_{id} (initial): rate constant before equilibrium using the intraparticle diffusion model; a: experimental, b = simulated. Resin = 1 g, ectoine = 1000 mg/L.

4.1.8. Effect of Salt Concentration

Ectoine is produced using fermentation from halophilic and halotolerant bacteria; thus, salts, such as NaCl, occur in the ocean. While the liquid was maintained at a neutral pH, salt was dissolved in the ectoine solution at different concentrations to study ectoine adsorption with ion-exchange resin. Furthermore, the desorption was conducted after capturing ectoine with an ion-exchange resin. Figure 8 shows that the ectoine residual concentration increased with salt in the liquid phase. This shows that ectoine ions compete with sodium ions at the ion-exchange sites.

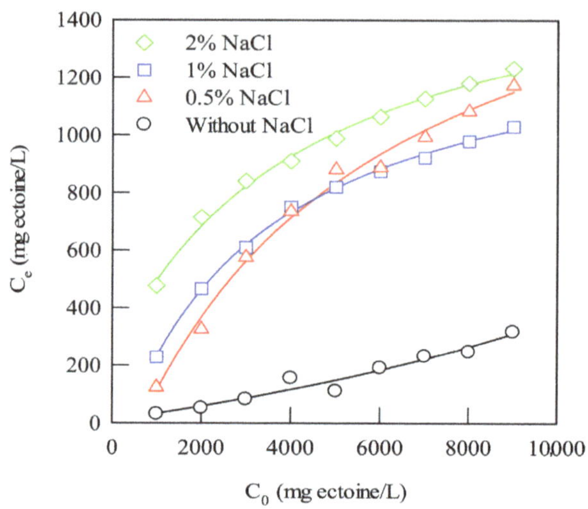

Figure 8. Effect of salt on adsorption of ectoine using a cation-exchange resin (Dowex® HCR-S = 1 g, initial ectoine = 1000–9000 mg/L, aqueous solution = 10 mL).

4.1.9. Adsorption Thermodynamics

Equilibrium adsorption was used to investigate the thermodynamic properties of ectoine adsorption, and the thermodynamic parameters, including the equilibrium distribution coefficient (K_D), Gibbs free energy ($\triangle G$), enthalpy ($\triangle H$), and entropy ($\triangle S$), were calculated. According to the results in Table 5, the negative ΔH values demonstrated that ectoine adsorption is an exothermic process at temperatures between 25 and 45 °C, whereas positive values revealed an endothermic reaction at temperatures between 45 and 65 °C. This supports the idea that ectoine adsorption in group I could be more favorable at lower temperatures and is a physisorption and exothermic process. In contrast, chemisorption in group II exhibits endothermic characteristics. Similarly, in group I, the values of ΔS were negative, indicating that the system's randomness at the boundary layer of a solid solution decreased during the adsorption process. In contrast, positive values of ΔS indicate that the randomness at the resin-solution interface increased. At all experiment temperatures, the values of ΔG were negative, showing that the adsorption process was spontaneous. The ΔG for physical adsorption is usually between −20 and 0 kJ mol^{-1}, while it is between −400 and −80 kJ mol^{-1} for chemical adsorption. Ectoine adsorption had a ΔG of -14.6 kJ mol^{-1} on average, indicating it was mostly physisorption. The heats of adsorption ($\triangle H$) were −256 and 351 J/ mol, which corresponds to the heat range of adsorption of −24–38 kJ/mol [35].

Table 5. Thermodynamic parameters of ectoine adsorption at different temperatures (group I = 25~45 °C; group II = 45~65 °C).

C_0 (g/L)	ΔH (J/mol)		ΔS (J/mol)		ΔG (kJ/mol)				
	I	II	I	II	25 °C	35 °C	45 °C	55 °C	65 °C
1	−302	129	−19.22	74.8	−15.2	−14.2	−13.9	−14.3	−14.4
2	−352	319	−29.74	116.0	−15.4	−14.7	−13.9	−14.6	−15.1
3	−401	422	−41.39	137.5	−15.4	−14.5	−13.6	−14.6	−15.3
4	−243	401	−7.08	133.9	−14.9	−13.7	−13.9	−14.8	−15.4
5	−118	240	24.31	101.8	−15.4	−15.1	−14.9	−15.0	−15.9
6	−248	430	−8.16	138.7	−14.7	−14.2	−13.6	−14.3	−15.2
7	−217	450	−1.55	143.4	−14.5	−14.1	−13.6	−14.4	−15.3
8	−231	390	−4.19	130.5	−14.7	−14.2	−13.7	−14.4	−15.2
9	−193	379	3.35	127.6	−14.3	−13.9	−13.5	−14.3	−14.9
Average	−256	351	−9.30	122.7			−14.6		

4.2. Saturation of Ectoine in the Ion-Exchange Resin

Saturation experiments were carried out to confirm the maximum ectoine loaded in the cation-exchange resin and for further the desorption investigation. According to the isotherm results in Section 4.1.5, the predicted saturation capacities of ectoine on Dowex® HCR-S were 0.22–0.67 and 0.20–0.36 (g/g) based on the nonlinear Langmuir maximum adsorption capacity in Table 3, respectively. Therefore, the experiments were performed by mixing 0.5 g resin with 2 mL solution of ectoine, with the concentrations set as 0.12, 0.125, 0.135, 0.15, 0.17, and 0.20 g/mL. As shown in Figure 9a, there was no significant effect of initial concentration on ectoine saturation at 25 °C, and the adsorption capacity ranged from 0.55 to 0.60 (g/g). Therefore, the adsorption capacity can be confirmed as the saturation capacity (q_{sat}) since no more ectoine could be loaded with the amplification of the concentration gradient. Based on the results in Section 4.1.5, the mechanism of ectoine adsorption is not a pure ion exchange but a mixture of ion exchange, electrostatic force, and physisorption (van der Waals force). The Langmuir isotherm assumes adsorption occurs under a homogeneous system with monolayer adsorption, and thus, the predicted maximum adsorption lacks multilayer adsorption based on the Freundlich theory. Multilayer adsorption can be attributed to physisorption (van der Waals), with ectoine stacked on the first layer sorption (ion exchange). The conclusion that the Sips isotherm, also known as the Langmuir–Freundlich model, can describe the ectoine adsorption by using Dowex® HCR-S shows that the experimental saturation capacity is larger than the Langmuir adsorption capacity. According to Figure 9b, the effect of temperature on ectoine saturation is not significant either, which means ectoine adsorption is not dependent on temperature.

According to Figure 9c, the resin weight did not affect the saturation capacity of ectoine in the ion-exchange resin significantly, and the ectoine-saturated capacity was around 0.56 (g/g) under the same temperature and initial ectoine concentration, which equaled an average of 3.9 (meq/g). The saturation percentage of the resin was 81.8% (=3.9/4.8) based on Dowex® HCR-S's total exchange capacity of 4.8 (meq/g).

4.3. Desorption of Ectoine from Saturated Ion-Exchange Resin

Adsorption and ion exchange are feasible alternatives for the separation and purification processes. These interface-based methods involve mass transfer and are particularly efficient for low-concentration systems, such as ectoine produced from bacteria with low concentrations. In addition to a separate desired compound from the original stream, the recovery of adsorbate and adsorbent regeneration must be considered. The desorption of ectoine from the acidic cation-exchange resin was studied with different NaOH concentrations.

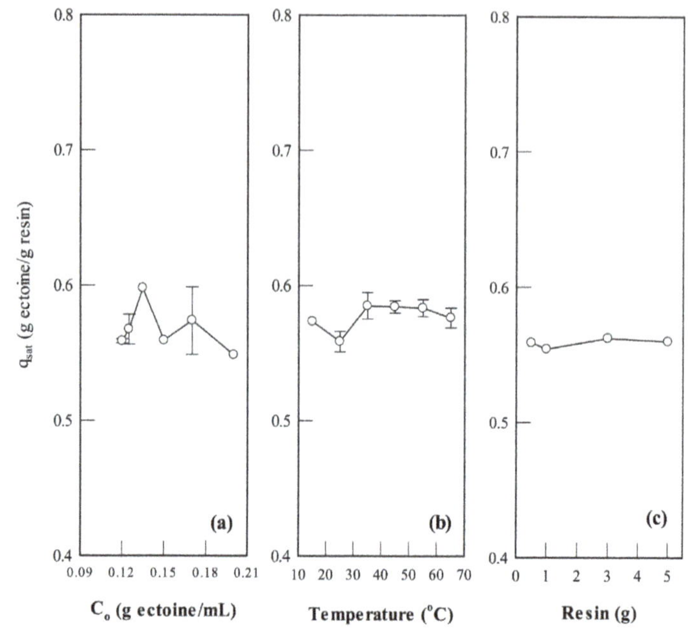

Figure 9. Effect of saturation adsorption capacity of ectoine using Dowex® HCR-S on (**a**) initial ectoine concentration, (**b**) temperature, and (**c**) resin dosage (aqueous/resin = 4 mL/g, agitation = 100 rpm, reaction time = 8 h).

4.3.1. Effect of pH on Ectoine Desorption

Based on the principle of using an ion exchanger to separate and purify amino acids, pH is the critical factor in modifying the analyte's surface charge to become cationic or anionic. To use a cationic-exchange resin in desorption, ectoine is supposed to transform to its anionic form to achieve electrostatic repulsion from the functional group (SO_3^-) fixed in the ion-exchange matrix. Therefore, the pH should be set above the isoelectric point of ectoine to manipulate the charged property on the molecule surface. Based on Figure 3, the pI of ectoine was 3.2 and became the standard for designing the experiments. The experiments were performed under pH = 2, 3, 4, 5, and 6 to investigate the effect on ectoine desorption.

Table 6 shows that not much ectoine was released from the saturated resins with a desorption percentage < 10%. This can be explained by the residual H^+ ions inside the saturated resin. Since the saturation capacity ranges from 0.55 to 0.60 g/g, the resin efficiency ranges between 80.6% and 87.9%, which means that 12.1% to 19.4% of hydrogen ions remain inside the resin. Therefore, the desorption capacity of 0.3~0.4 meq/g can be attributed to the salt desorption (exchanging with Na^+), and a more basic condition is required to allow the charged property of ectoine to change.

Table 6. Effect of pH on ectoine desorption from saturated Dowex® HCR-S resins.

pH	C_e (g/L)	$m_{ect,e}$ (g)	$m_{ect,e}$ (meq)	Desorption (%)
2	5.42	0.06	0.42	9.96
3	4.05	0.04	0.31	7.44
4	3.72	0.04	0.29	6.84
5	3.58	0.04	0.28	6.58
6	3.99	0.04	0.31	7.33

$m_{ect,e}$ = mass of ectoine in solution at equilibrium. Condition: resin/aqueous = 0.1 g/mL; temperature: 20 °C; q_{sat} = 0.60 g/g = 4.2 meq/g; time = 8 h.

4.3.2. Effect of NaOH Concentration

The desorption equilibrium of ectoine released from saturated Dowex® HCR-S resin was found in a batch reactor, as shown in Figure 9. The sequence of temperatures for ectoine desorption was 15 °C ≈ 25 °C > 35 °C > 45 °C > 55 °C. The desorption performance was higher at lower temperatures. The desorption of ectoine was greatest at 15 °C, with 89.7% desorption. Theoretically, with a higher NaOH concentration, better desorption results could be obtained and a plateau reached due to the higher driving force provided with larger OH^- concentrations. The best desorption performance was found at an NaOH concentration of ≈0.5 mol/L, as shown in Figure 10. Furthermore, the ectoine ions were transferred from the solid to the liquid phase when the eluent solution exchanged the ectoine molecules. This phenomenon is a consequence of mass transfer, which was boosted by the high concentration gradient between resin and solution phases and maximized by the ionic strength of the basic eluent. NaOH solution can promote a very high ectoine removal as a desorbed agent since even the lowest concentrations in the resin phase correspond to high concentrations in the solution, resulting in favorable desorption. When the NaOH concentration is larger than 0.5 mol/L, the desorption percentage of ectoine is decreased (Figure 10). This results from ectoine denaturation under extreme alkaline conditions.

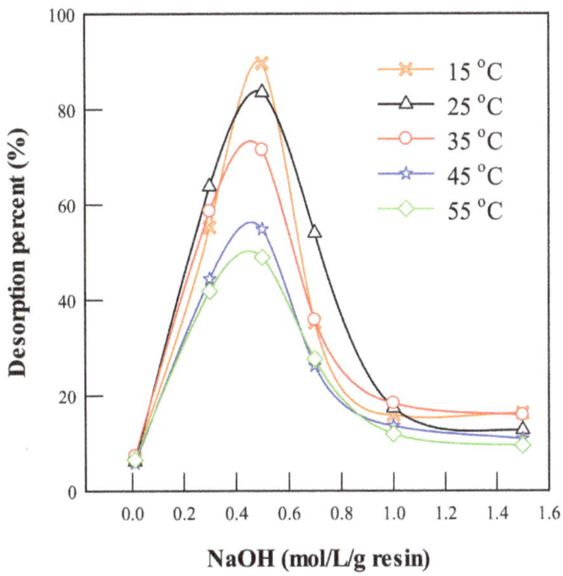

Figure 10. Effect of NaOH on the equilibrium desorption of ectoine using Dowex® HCR-S at different temperatures (saturated resin: 1 g, NaOH solution: 10 mL, time: 8 h).

4.3.3. Effect of Desorption Time and Temperature

The effect of temperature on the equilibrium and the desorption kinetics using NaOH as an adsorbed agent was assessed to study the desorption rate. Figure 11 shows the effect of desorption time on the release of ectoine ions from Dowex® HCR-S resin. Ectoine ions were rapidly released from the acidic cation exchanger before 180 min, followed by slow desorption and equilibrium being attained. Desorption of ectoine ions was found to be proportional to the desorbed time before equilibrium was reached. The sequence of temperatures for ectoine desorption was 25 °C > 15 °C > 45 °C > 55 °C > 35 °C. This shows no trend for the temperature effect. Desorption and ion exchange may simultaneously control the reaction kinetics.

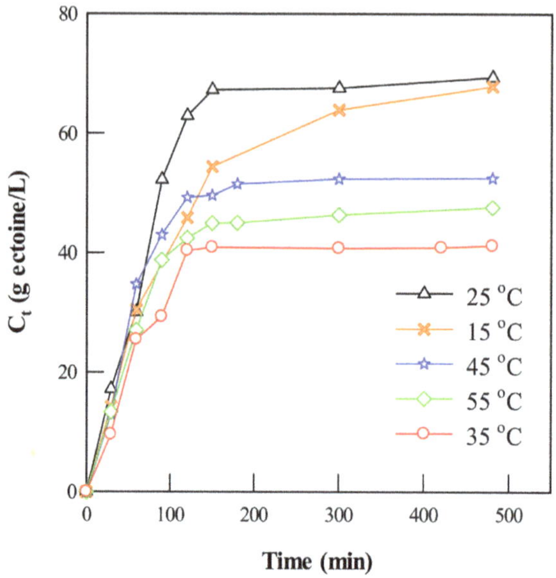

Figure 11. Effect of time on ectoine desorption from Dowex® HCR-S with different temperatures (saturated resin/NaOH solution = 0.1 g/mL, NaOH concentration: 0.5 mol/L).

The desorption of ectoine released from ion exchangers in the first 180 min can be attributed to the strong driving force resulting from the concentration gradient of ectoine between the solid and solution phases and the fact that the resin was saturated first, which means the initial ectoine concentration was null ($C_0 = 0$) in the solution phase. At the same time, the desorbed agent NaOH played an essential role in converting ectoine from the cationic to anionic form in order to lose the electronic attraction and even repel the sulfonic functional group (SO_3^-) covalently fixed in the ion-exchanger matrix. On the other hand, the desorption process decreased as the reaction time increased. This can be explained by the structure of the ion-exchanger synthesized by polystyrene-divinylbenzene since crosslinking can hinder and slow down the immersion and swelling of resin.

4.3.4. Desorption Kinetics of Ectoine from the Ion-Exchange Resin

Mathematical modeling of experimental data from the batch reactor was undertaken to construct a robust and trustworthy mathematical model that could predict the kinetics of desorption in a fixed-bed column [36]. Table 7 shows the desorption results using the external diffusion model (EDM); the simulated parameters were not reliable based on the unstable R^2. Except for the data under 35 °C, the EDM rate constant, k_d, had an average value of 1.13 min^{-1}, and the ectoine desorption was more favorable under lower temperatures. More desorption kinetic models are required for the investigation of ectoine desorption in order to find the rate-limiting step.

Table 7. Parameters of the external solid diffusion model calculated at different temperatures.

Temp. (°C)	k_d (min^{-1})	R^2
15	0.31	0.92
25	1.08	0.82
35	4.90	0.64
45	1.64	0.91
55	1.48	0.98

4.3.5. Mechanism of Ectoine Desorption from Ion-Exchange Resin

Desorption using the ion-exchange method can be classified into two types: pH desorption and salt desorption. Desorption of amino acid depending on pH value is built on the use of the isoelectric point of its nature. When the pH value is over the pI of ectoine, the anionic form is exhibited. In contrast, the cationic form is shown when the pH is below its pI. As for salt desorption, by using the exchange priority depending on their affinity towards ion-exchanging sites, salt ions with a stronger affinity compete with targeting analytes, and thus, analytes can be exchanged out. The effect of pH (Section 4.1.1) and NaOH concentration (Section 4.1.2) on ectoine desorption shows that the desorption strengthened with the increase of the OH^- concentration. Moreover, ectoine desorption relies on the mechanism of pH desorption, with 90% desorption. NaOH solution can promote a strong ectoine removal, resulting in favorable desorption.

Figure 12 shows that the desorption efficiency when using NaCl (54.5%) was not as high as when using NaOH (78.1%) at 35 °C. The desorption of ectoine showed a higher rate in the first 120 min when using NaCl, based on the mechanism of salt desorption, and a higher equilibrium desorption was obtained by using NaOH, which relies on both pH and salt desorption. Therefore, NaOH promotes stronger desorption of ectoine than NaCl since there are two mechanisms that support using NaOH as a desorption agent.

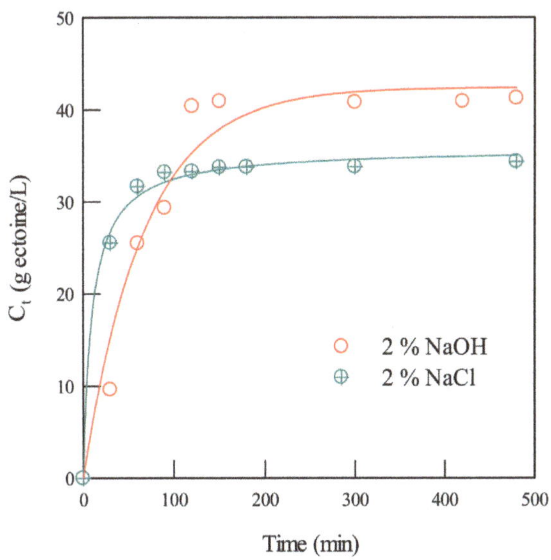

Figure 12. Comparison of NaOH and NaCl in ectoine desorption (temperature: 35 °C).

According to the results of the isotherm analysis, the adsorption of ectoine using Dowex strong-acid cation-exchange resin achieved the best fit to the Sips model, implying that the sorption process is built on both ion exchange and physisorption. Ectoine desorption from the saturated resin using NaOH as a desorbed agent was assumed to have three steps, based on the sorption phenomena and desorption results: concentration-dependent desorption, salt desorption, and pH desorption (Figure 13). The desorption of ectoine under pH values ranging from 2 to 6 was an average of 7.63%, as shown in Table 5. Physisorption resulted in concentration-dependent desorption of ectoine. Secondly, an NaCl solution was applied to the ectoine desorption, represented by salt desorption, obtaining 54.5% desorption under neutral conditions. Finally, using NaOH solution as a pH desorption strategy, the best ectoine desorption reached (89.7%). Van der Waals desorption (concentration desorption) was weaker than electrostatic desorption (salt and pH desorption), and this supports the sorption theory that physisorption is more accessible than ion exchange.

Therefore, ectoine desorption using NaOH solution included ion exchange mechanisms as the significant driving force to repel ectoine from the ion-exchanger phase (pH and salt desorption) and concentration-dependent desorption.

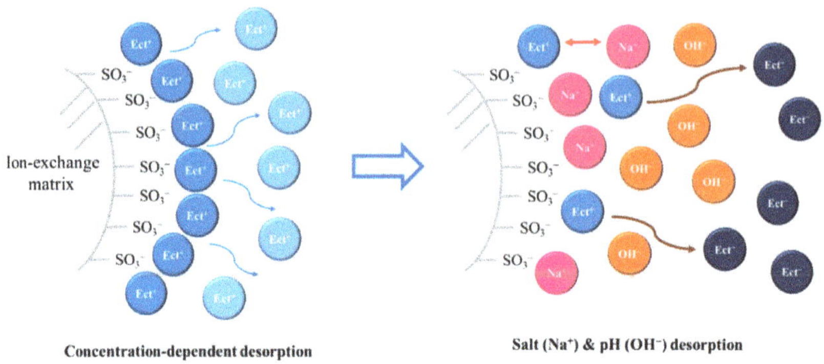

Figure 13. Mechanism of ectoine desorption using Dowex® HCR-S. **: adsorbed ectoine ion, : free ectoine ion, : sodium ion, : hydroxide ion.**

5. Conclusions

In this research, adsorption, saturation, and desorption were utilized to capture ectoine using ion-exchange resin in a batch system in order to investigate ectoine's adsorption/desorption behavior. Regression of isotherms and kinetic models were carried out to elucidate the adsorption behavior and optimum operating conditions. The analysis results showed that the adsorption of ectoine using Dowex® HCR-S resin fits best when using the Sips isotherm. Thus, the ectoine adsorption mechanism is ion exchange, based on electrostatic force (monolayer), and physisorption based on van der Waal's force (multilayer). Moreover, the effect of temperature indicates that the adsorption of ectoine can be divided into two outcomes: group I (25 °C~45 °C) was exothermic, and group II (45 °C~65 °C) was endothermic. In group I, ectoine adsorption is more favorable at lower temperatures, establishing physisorption. In group II, adsorption uptake is enhanced at higher temperatures, supporting ion exchange via the Coulomb force. For the kinetic study, the results indicate that pseudo-second-order models could explain the ectoine adsorption reaction using Dowex® HCR-S resin, and the activation energy was 36.5 kJ mol^{-1}. The intraparticle diffusion model can explain the kinetic mechanism for the multilinear plot and the resin structure (gel-type). The values of ΔG, ΔH, and ΔS were also calculated. The negative values of ΔG at the experimental temperatures indicate that the adsorption was spontaneous, and the negative values of ΔH and ΔS show that the adsorption was exothermic. The randomness decreased during adsorption for group I but increased for group II.

The saturation of ectoine on Dowex® HCR-S resin was independent of both temperature and resin dosage. On average, the saturation capacity was 0.58 g/g, with 93.6% adsorption using 125 g/L of initial ectoine concentration. Desorption of ectoine reached 89.7% of desorption with 0.51 g under an NaOH concentration of 0.5 mol/L and at 15 °C. Comparing the desorbed agent with both NaOH and NaCl, more ectoine ions were desorbed using NaOH, which confirmed that both pH and salt desorption are the mechanisms of ectoine desorption. The thermodynamic and kinetic parameters obtained from this study can be used for adsorption/desorption of the fix-bed column, designed for larger-scale production. The optimal condition of separating the ectoine from the fermentation broth in a fix-bed column can reduce the separation cost of ectoine.

Author Contributions: Conceptualization, Y.-C.W., Y.-H.W. and H.-S.W.; methodology, Y.-C.W.; formal analysis, Y.-C.W.; investigation, Y.-C.W.; writing—original draft preparation, Y.-C.W.; writing—review and editing, H.-S.W.; supervision, H.-S.W. All authors have read and agreed to the published version of the manuscript.

Funding: We thank the Ministry of Science and Technology of Taiwan for financially supporting this study under grant number MOST 104-2632-E-155 -001, 105-2632-E-155 -001, 106-2632-E-155 -001.

Institutional Review Board Statement: Not applicable.

Informed Consent Statement: Not applicable.

Conflicts of Interest: The authors declare no conflict of interest.

References

1. Becker, J.; Wittmann, C. Microbial production of extremolytes—High-value active ingredients for nutrition, health care, and well-being. *Curr. Opin. Biotechnol.* **2020**, *65*, 118–128. [CrossRef] [PubMed]
2. Chen, R.; Zhu, L.; Lv, L.; Yao, S.; Li, B.; Qian, J. Optimization of the extraction and purification of the compatible solute ectoine from Halomonas elongate in the laboratory experiment of a commercial production project. *World J. Microbiol. Biotechnol.* **2017**, *33*, 116. [CrossRef] [PubMed]
3. Kunte, H.J.; Lentzen, G.; Galinski, E.A. Industrial Production of the Cell Protectant Ectoine: Protection Mechanisms, Processes, and Products. *Curr. Biotechnol.* **2014**, *3*, 10–25. [CrossRef]
4. Pastor, J.M.; Salvador, M.; Argandoña, M.; Bernal, V.; Reina-Bueno, M.; Csonka, L.N.; Iborra, J.L.; Vargas, C.; Nieto, J.J.; Cánovas, M. Ectoines in cell stress protection: Uses and biotechnological production. *Biotechnol. Adv.* **2010**, *28*, 782–801. [CrossRef] [PubMed]
5. Behazin, R.; Ebrahimi, A. The physicochemical properties and tyrosinase inhibitory activity of ectoine and its analogues: A theoretical study. *Comput. Theor. Chem.* **2018**, *1130*, 6–14. [CrossRef]
6. Zhao, Q.; Li, S.; Lv, P.; Sun, S.; Ma, C.; Xu, P.; Su, H.; Yang, C. High ectoine production by an engineered Halomonas hydrothermalis Y2 in a reduced salinity medium. *Microb. Cell Factories* **2019**, *18*, 184. [CrossRef]
7. Chen, W.-C.; Hsu, C.-C.; Lan, J.C.-W.; Chang, Y.-K.; Wang, L.-F.; Wei, Y.-H. Production and characterization of ectoine using a moderately halophilic strain Halomonas salina BCRC17875. *J. Biosci. Bioeng.* **2018**, *125*, 578–584. [CrossRef] [PubMed]
8. Chen, W.-C.; Yuan, F.-W.; Wang, L.-F.; Chien, C.-C.; Wei, Y.-H. Ectoine production with indigenous Marinococcus sp. MAR2 isolated from the marine environment. *Prep. Biochem. Biotechnol.* **2020**, *50*, 74–81. [CrossRef]
9. Chen, J.; Liu, P.; Chu, X.; Chen, J.; Zhang, H.; Rowley, D.C.; Wang, H. Metabolic pathway construction and optimization of Escherichia coli for high-level ectoine production. *Curr. Microbiol.* **2020**, *77*, 1412–1418. [CrossRef] [PubMed]
10. Sauer, T.; Galinski, E.A. Bacterial milking: A novel bioprocess for production of compatible solutes. *Biotechnol. Bioeng.* **1998**, *57*, 306–313. [CrossRef]
11. Gießelmann, G.; Dietrich, D.; Jungmann, L.; Kohlstedt, M.; Jeon, E.J.; Yim, S.S.; Sommer, F.; Zimmer, D.; Mühlhaus, T.; Schroda, M. Metabolic engineering of Corynebacterium glutamicum for high-level ectoine production: Design, combinatorial assembly, and implementation of a transcriptionally balanced heterologous ectoine pathway. *Biotechnol. J.* **2019**, *14*, 1800417. [CrossRef] [PubMed]
12. Lang, Y.-j.; Bai, L.; Ren, Y.-n.; Zhang, L.-h.; Nagata, S. Production of ectoine through a combined process that uses both growing and resting cells of Halomonas salina DSM 5928 T. *Extremophiles* **2011**, *15*, 303–310. [CrossRef] [PubMed]
13. Fülberth, K.; Müssig, H.; Wesp, B. Method for the Chromatographic Isolation of Ectoine. European Patent EP1461322A1, 2004.
14. LeVan, M.D.; Carta, G.; Yon, C.M. Adsorption and ion exchange. *Energy* **1997**, *16*, 17.
15. Lagergren, S. Zur theorie der sogenannten adsorption geloster stoffe. *K. Sven. Vetenskapsakademiens. Handl.* **1898**, *24*, 1–39.
16. Vadivelan, V.; Kumar, K.V. Equilibrium, kinetics, mechanism, and process design for the sorption of methylene blue onto rice husk. *J. Colloid Interface Sci.* **2005**, *286*, 90–100. [CrossRef] [PubMed]
17. Ho, Y.S. Removal of copper ions from aqueous solution by tree fern. *Water Res.* **2003**, *37*, 2323–2330. [CrossRef]
18. Weber, W.J.; Morris, J.C. Kinetics of adsorption on carbon from solution. *J. Sanit. Eng. Div.* **1963**, *89*, 31–60. [CrossRef]
19. Wu, F.C.; Tseng, R.L.; Juang, R.S. Initial behavior of intraparticle diffusion model used in the description of adsorption kinetics. *Chem. Eng. J.* **2009**, *153*, 1–8. [CrossRef]
20. McKay, G.; Otterburn, M.; Sweeney, A. The removal of colour from effluent using various adsorbents—III. Silica: Rate processes. *Water Res.* **1980**, *14*, 15–20. [CrossRef]
21. Aharoni, C.; Tompkins, F. Kinetics of adsorption and desorption and the Elovich equation. In *Advances in Catalysis*; Elsevier: Amsterdam, The Netherlands, 1970; Volume 21, pp. 1–49.
22. Wu, F.C.; Tseng, R.L.; Juang, R.S. Characteristics of Elovich equation used for the analysis of adsorption kinetics in dye-chitosan systems. *Chem. Eng. J.* **2009**, *150*, 366–373. [CrossRef]
23. Dada, A.; Olalekan, A.; Olatunya, A.; Dada, O. Langmuir, Freundlich, Temkin and Dubinin–Radushkevich isotherms studies of equilibrium sorption of Zn2+ unto phosphoric acid modified rice husk. *IOSR J. Appl. Chem.* **2012**, *3*, 38–45. [CrossRef]
24. Weber, T.W.; Chakravorti, R.K. Pore and solid diffusion models for fixed-bed adsorbers. *AIChE J.* **1974**, *20*, 228–238. [CrossRef]

25. Prasad, R.K.; Srivastava, S. Sorption of distillery spent wash onto fly ash: Kinetics and mass transfer studies. *Chem. Eng. J.* **2009**, *146*, 90–97. [CrossRef]
26. Osmari, T.A.; Gallon, R.; Schwaab, M.; Barbosa-Coutinho, E.; Severo Jr, J.B.; Pinto, J.C. Statistical analysis of linear and non-linear regression for the estimation of adsorption isotherm parameters. *Adsorpt. Sci. Technol.* **2013**, *31*, 433–458. [CrossRef]
27. Al-Ghouti, M.A.; Da'ana, D.A. Guidelines for the use and interpretation of adsorption isotherm models: A review. *J. Hazard. Mater.* **2020**, 122383. [CrossRef]
28. Freundlich, H.M.F. Over the adsorption in solution. *J. Phys. Chem* **1906**, *57*, 385.
29. Ho, Y.S. *Absorption of Heavy Metals from Waste Streams by Peat*; University of Birmingham: Birmingham, UK, 1995.
30. Dubinin, M. The potential theory of adsorption of gases and vapors for adsorbents with energetically nonuniform surfaces. *Chem. Rev.* **1960**, *60*, 235–241. [CrossRef]
31. Langmuir, I. The constitution and fundamental properties of solids and liquids. Part I. Solids. *J. Am. Chem. Soc.* **1916**, *38*, 2221–2295. [CrossRef]
32. Redlich, O.; Peterson, D.L. A useful adsorption isotherm. *J. Phys. Chem.* **1959**, *63*, 1024. [CrossRef]
33. Sips, R. On the Structure of a Catalyst Surface. *J. Chem. Phys.* **1948**, *16*, 490–495. [CrossRef]
34. Temkin, M.I. Kinetics of ammonia synthesis on promoted iron catalysts. *Acta Physiochim. URSS* **1940**, *12*, 327–356.
35. Inglezakis, V.J.; Zorpas, A.A. Heat of adsorption, adsorption energy and activation energy in adsorption and ion exchange systems. *Desalination Water Treat.* **2012**, *39*, 149–157. [CrossRef]
36. Staudt, J.; Scheufele, F.B.; Ribeiro, C.; Sato, T.Y.; Canevesi, R.; Borba, C.E. Ciprofloxacin desorption from gel type ion exchange resin: Desorption modeling in batch system and fixed bed column. *Sep. Purif. Technol.* **2020**, *230*, 115857. [CrossRef]
37. Ismail, M.; Hanafiah, K.M. Kinetic, isotherm, and possible mechanism of Pb (II) ion adsorption onto xanthated neem (Azadirachta indica) leaf powder. *Sains Malays.* **2020**, *49*, 1585–1596. [CrossRef]
38. Nagy, B.; Mânzatu, C.; Măicăneanu, A.; Indolean, C.; Barbu-Tudoran, L.; Majdik, C. Linear and nonlinear regression analysis for heavy metals removal using Agaricus bisporus macrofungus. *Arab. J. Chem.* **2017**, *10*, S3569–S3579. [CrossRef]
39. Bulut, Y.; Aydın, H. A kinetics and thermodynamics study of methylene blue adsorption on wheat shells. *Desalination* **2006**, *194*, 259–267. [CrossRef]
40. Aksu, Z.; Kutsal, T. A bioseparation process for removing lead (II) ions from waste water by using C. vulgaris. *J. Chem. Technol. Biotechnol.* **1991**, *52*, 109–118. [CrossRef]
41. Kajjumba, G.W.; Emik, S.; Öngen, A.; Özcan, H.K.; Aydın, S. Modelling of adsorption kinetic processes—errors, theory and application. In *Advanced Sorption Process Applications*; IntechOpen: London, UK, 2018.

processes

Article

Adsorption/Desorption Characteristics and Simultaneous Enrichment of Orientin, Isoorientin, Vitexin and Isovitexin from Hydrolyzed Oil Palm Leaf Extract Using Macroporous Resins

Mohamad Shazeli Che Zain [1], Soo Yee Lee [1], Chian Ying Teo [2,*] and Khozirah Shaari [1,*]

1 Natural Medicines and Products Research Laboratory (NaturMeds), Institute of Bioscience, Universiti Putra Malaysia, Serdang 43400, Selangor, Malaysia; shazelizain@gmail.com (M.S.C.Z.); daphne.leesooyee@gmail.com (S.Y.L.)
2 Department of Pharmaceutical Chemistry, School of Pharmacy, International Medical University, Bukit Jalil 57000, Kuala Lumpur, Malaysia
* Correspondence: TeoChianYing@imu.edu.my (C.Y.T.); khozirah@yahoo.com.my (K.S.); Tel.: +60-3-27317742 (C.Y.T.); +60-3-8942148 (K.S.)

Abstract: Oil palm leaves (OPL) containing flavonoid *C*-glycosides are abundantly generated as oil palm byproducts. The performances of three macroporous resins with different physical and chemical properties for the enrichment of isoorientin, orientin, vitexin, and isovitexin from acid-hydrolyzed OPL (OPLAH) extract were screened. The XAD7HP resin exhibited the best sorption capacities for the targeted flavonoid *C*-glycosides and was thus selected for further evaluation. Static adsorption using the XAD7HP resin under optimal conditions (extract adjusted to pH 5, shaken at 298 K for 24 h) gave adsorption kinetics that fit well with a *pseudo*-second-order kinetic model. The adsorption of isoorientin and orientin was well described by Langmuir isotherms, while vitexin and isovitexin fit well with the Freundlich isotherms. Dynamic sorption trials using the column-packed XAD7HP resin produced 55–60-fold enrichment of isovitexin and between 11 and 25-fold enrichment of isoorientin, vitexin, and orientin using aqueous ethanol. The total flavonoid *C*-glycoside-enriched fractions (enriched OPLAH) with isoorientin (247.28–284.18 μg/mg), orientin (104.88–136.19 μg/mg), vitexin (1197.61–1726.11 μg/mg), and isovitexin (13.03–14.61 μg/mg) showed excellent antioxidant free radical scavenging activities compared with their crude extracts, with IC_{50} values of 6.90–70.63 μg/mL and 44.58–200.00 μg/mL, respectively. Hence, this rapid and efficient procedure for the preliminary enrichment of flavonoid *C*-glycosides by using macroporous resin may have practical value in OPL biomass waste utilization programs to produce high value-added products, particularly in the nutraceuticals, cosmeceuticals, pharmaceuticals, and fine chemicals industries.

Keywords: oil palm leaves; total flavonoid *C*-glycosides; acid hydrolysis; macroporous resin; enrichment; antioxidant free radical scavenging activity

Citation: Che Zain, M.S.; Lee, S.Y.; Teo, C.Y.; Shaari, K. Adsorption/Desorption Characteristics and Simultaneous Enrichment of Orientin, Isoorientin, Vitexin and Isovitexin from Hydrolyzed Oil Palm Leaf Extract Using Macroporous Resins. *Processes* **2021**, *9*, 659. https://doi.org/10.3390/pr9040659

Academic Editor: Zongbi Bao

Received: 12 March 2021
Accepted: 7 April 2021
Published: 9 April 2021

1. Introduction

The oil palm (*Elaeis guineensis* Jacq.) tree was introduced into Malaysia in 1875, with the first oil palm tree plantation established at Tennamaran Estate in Kuala Selangor [1]. Fueled by full support from Malaysian government agricultural diversification initiatives, palm oil plantations expanded tremendously and now cover a large acreage of agricultural land areas [2]. Currently, Malaysia is the second-largest oil palm producer in the world after Indonesia [3]. In fact, the latest statistics show that Malaysia was reaching 20 million tons of crude palm oil production in 2020 [4]. However, in the wake of this massive cultivation, a huge amount of oil palm biomass is generated as agricultural waste. Apart from mesocarp fibers (MF), empty fruit bunches (EFB), and palm kernel shells (PKS) from downstream processing in oil palm mills, in parallel, oil palm trunks (OPT), oil palm fronds (OPF), and oil palm leaves (OPL) were also generated, presenting a huge environmental problem if left unutilized [5].

Like many other species in the plant kingdom, OPL byproduct is an excellent source of phytochemicals which could be used for some applications. OPL have in fact been reported to contain bioactive compounds that are responsible for various medicinal properties, such as treating kidney diseases, cancer, cardiovascular diseases, and wounds [6]. A previous study on OPL revealed the presence of both flavonoid *O*- and *C*-glycosides [7]. In general, flavonoid *C*-glycosides are not widely present in plants, and due to this, they have received less attention in comparison with their *O*-glycosyl counterparts. Nevertheless, several recent biological and pharmacological studies have shown that flavonoid *C*-glycosides also possess a wide spectrum of biological properties, which include anticancer, hepatoprotective, antioxidant, and antidiabetic properties [8]. Flavonoid *C*-glycosides differ from flavonoid *O*-glycosides in that they are more resistant to hydrolysis, since the aglycone is linked to the anomeric carbon of the sugar moiety via an acid-resistant C-C bond. Figure 1 shows the structures of four flavonoid *C*-glycosides of OPL, which include orientin, isoorientin, vitexin, and isovitexin. These flavonoid *C*-glycosides were present in considerable amounts in comparison with other luteolin and apigenin derivatives in OPL [7,9–12]. It is worth noting that the global demand for flavonoids, including flavonoid *C*-glycosides, is forecasted to reach USD 1.2 billion by 2024 [13]. Therefore, their presence in a widely available and abundant biomass material warrants further investigation into the development of efficient methods for the preparative purification for downstream purposes and applications.

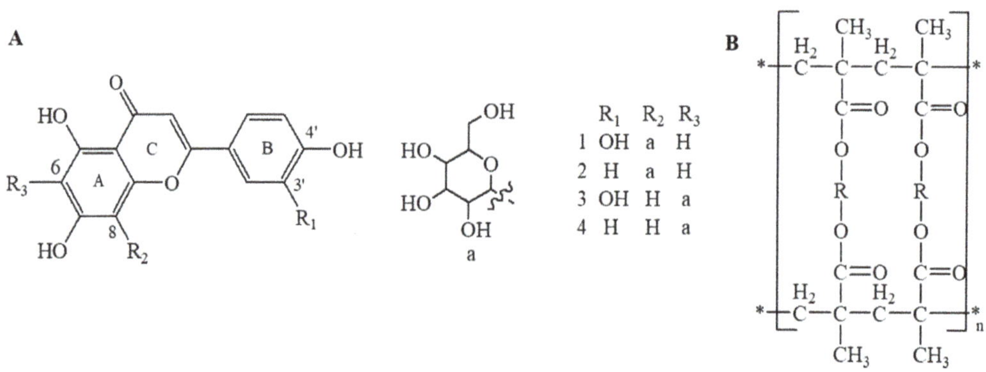

Figure 1. (**A**) Structures of *C*-glycosyl flavonoids identified in oil palm leaves (OPL): orientin (1), vitexin (2), isoorientin (3), isovitexin (4). (**B**) Structure of the XAD7HP resin. R = polyfunctional aliphatic residue.

Utilization of macroporous resins (MARs) in separating and purifying the flavonoid *C*-glycosides present in plant extracts has been practiced in recent years. It offers an alternative to conventional methods, which often start with solid-liquid extraction, followed by liquid-liquid extraction and eventually column chromatography [14]. These conventional approaches are not only time-consuming and inefficient, but they also require high consumption of solvents and energy [15,16]. The chemical nature of MARs allows them to selectively adsorb through hydrogen bonding and Van der Waals interactions with benzene rings and hydrogen groups present in the molecular structure of the targeted flavonoids [17]. The entrapment of these flavonoids on MARs is due to similarity in their physical and chemical characteristics, such as the appropriate surface area, average pore diameter, and polarity of both the MARs and the targeted flavonoids [18–20]. The high flavonoid sorption capacities made MARs a useful and practical adsorbent to enrich and purify flavonoid *C*-glycosides from various plants such as *Cajanus cajan* (L.) Millsp.

(pegionpea) [19], *Ficus microcarpa* L.f leaves [20], *Abrus mollis* (Jigucao) [18], and many others. Additionally, compared with silica gel, which is commonly used as the adsorbent of choice for compound separation, MARs are reusable, fast, simple, and efficient in trapping compounds of interest, particularly flavonoids from various plant materials [18–20].

Previously, we reported the adsorption behavior of the total flavonoids of OPL extract on different macroporous resins [9]. As an extension of this study, we further examine the adsorption and desorption properties of flavonoid C-glycosides, specifically (1) orientin, (2) vitexin, (3) isoorientin, and (4) isovitexin on a selection of MARs. The MARs with the best sorption properties for the target compounds were then used to develop a rapid and efficient method for the enrichment and purification of C-glycosyl flavonoids from acid-hydrolyzed OPL (OPLAH) extract. Factors affecting the sorption properties of the individual flavonoids C-glycoside were optimized, and their kinetics and isotherms were simultaneously evaluated. The method developed in this study presents an improved process for converting OPL biomass into fine chemicals at a high purity for potential applications.

2. Materials and Methods

2.1. Chemicals and Reagents

For the selected macroporous resins (MARs), n-(1-naphthyl) ethylenediamine dihydrochloride; quercetin; sodium acetate; sulphanilamide; 1,1-diphenyl-2-picrylhydrazyl (DPPH); and phosphoric acid were purchased from Sigma-Aldrich (St. Louis, MO, USA). Hydrochloric acid (HCl), dimethyl sulfoxide (DMSO), sodium hydroxide (NaOH), formic acid, acetonitrile, and ammonium formate were acquired from Merck (Darmstadt, Germany). Sodium nitroprusside was obtained from Bendosen Laboratory Chemicals (Bendosen, Norway), methanol (MeOH) and ethanol (EtOH) were acquired from R&R Chemicals (Essex, UK), and aluminum chloride was procured from HmbG Chemicals (Hamburg, Germany). With purities greater than 98.0%, isoorientin, orientin, vitexin, and isovitexin were obtained from Wuhan ChemFaces Biochemical Co., Ltd. (Wuhan, China). Milli-Q ultrapure water (Millipore Lab, Bedford, MA, USA) was used in all experiments.

2.2. Pretreatment of MARs

The chemical and physical properties of the selected MARs (XAD7HP, DAX-8, and XAD4) are summarized in Table 1. All MARs were pretreated prior to use to remove residual monomers and porogenic agents, which could be trapped in the pores of the resins during manufacturing. The MARs were soaked in 95% EtOH at a 1:20 ratio and washed with deionized water. The resins were then immersed in 1 mol/L NaOH and washed several times with deionized water to remove the base. Subsequently, the resins were subjected to a second immersion in 1 mol/L HCl and then washed thoroughly with deionized water to remove the acid. For each stage of the pretreatment, the resins were allowed to soak for 24 h before washing. The washed resins were then dried in a drying oven (model 100–800, Memmert, Schwabach, Germany) at 60 °C until reaching a constant weight.

Table 1. Chemical and physical properties of different macroporous resins (MARs).

	XAD7HP	DAX-8	XAD4
Functional group	Acrylic	Acrylic ester	Styrene-divinylbenzene
Particle diameter (mm)	0.250–0.841	0.250–0.420	0.250–0.841
Surface area (m^2/g)	380	140	750
Pore size (Å)	300–400	225	100
Polarity	Moderate	Moderate	Polar

2.3. Preparation of Crude and Acid-Hydrolyzed Extracts

Mature OPL were harvested from oil palm trees growing in the University Agricultural Park at the Universiti Putra Malaysia (UPM). A voucher specimen (SK 3332/18) was placed in the mini herbarium of the Institute of Bioscience (IBS) at UPM after the species was authenticated by an appointed botanist. The optimized procedures of preparing crude and acid-hydrolyzed OPL extract were described in our recent publication [10]. Briefly, the powdered OPL was mixed with aqueous MeOH (4:1 MeOH:water, v:v) and vortex-mixed for 0.5 min at 3000 g/min. The mixture was ultrasonicated at a frequency of 40 Hz for 30 min at 25 °C.

Subsequently, the crude OPL extract was mixed with distilled water and 6 mol/L HCl in a ratio of 1:10:10 (w:v:v). The mixture was incubated for 45 min at 95 °C. At 25 °C, 40 mL MeOH was added. After centrifuging at 4000 g/min for 15 min, the supernatant was separated, vacuum-evaporated to dryness, and then freeze-dried at 0.064 mbar and −50 °C using a Labconco® FreeZone Freeze Drier System (Kansas, MO, USA) to yield acid-hydrolyzed OPL extract (OPLAH).

2.4. UHPLC Analysis of Orientin, Isoorientin, Vitexin, and Isovitexin

Chromatographic separation was carried out with an ultra-high performance liquid chromatography (UHPLC) system consisting of an Ultimate 3000 LC system (Thermo Scientific™ Dionex™ (Sunnyvale, CA, USA) equipped with a photodiode array detector (PDA-3000) scanning from 200 nm to 600 nm, a thermostatted column compartment, and an autosampler (Exactive™, Thermo Fisher Scientific, Waltham, MA, USA). An Acquity UPLC® BEH C_{18} column (2.1 × 100 mm, 1.7 μm) (Waters, Manchester, UK) was used for analysis. The mobile phase was comprised of solvent A (water containing 0.1% formic acid and 0.063% ammonium formate) and solvent B (acetonitrile containing 0.1% formic acid) flowing at 0.30 mL/min.

The gradient program employed was performed according to the previously reported method [9,12]. Briefly, the program was started with 10% solvent B for 0.6 min, gradually increased to 11.3% until 1.5 min, maintained isocratically until 5.5 min, and slightly increased to 11.4% until 8.0 min and 11.8% until 8.2 min. Solvent B was further increased to 12% until 12.0 min and then decreased to 10% for 1.0 min and maintained until 25 min. The column temperature was maintained at 25 °C, and the UV detector was set to a wavelength of 340 nm. Peak identification was based on the retention time and comparison of UV spectra with the respective reference standards. For sample analysis, 5 mg/mL of each sample solution was prepared and filtered through a 0.22 μm membrane filter. A 2 μL sample injection volume was used for all sample analysis.

The quantification method was developed and validated based on the following characteristics: specificity, linearity, limit of detection (LOD) and quantification (LOQ), accuracy, repeatability, intermediate precision, and robustness, according to the International Conference on Harmonization (ICH) guidelines [21]. The full information with regard to method validation has been recently published [22]. Briefly, the developed method displayed good calibration curves with linearity (R^2 = 0.999) in the ranges of 16–500 μg/mL for isoorientin, 31–800 μg/mL for orientin, 47–1500 μg/mL for vitexin, and 16–500 μg/mL for isovitexin. In addition, the LODs for isoorientin, orientin, vitexin and isovitexin were 17.99, 30.22, 80.63, and 17.69 μg/mL, respectively while the LOQs for these compounds were 54.52, 91.58, 244.35, and 53.61 μg/mL, respectively. The recovery percentages were between 95% and 105% for all tested compounds, while for the inter- and intraday precisions, the relative standard deviation (RSD) values were found to be below 5%. For robustness, the chromatographic conditions, such as the detected changes in wavelength, column temperature, and sample stability showed insignificant changes, as indicated by *t*-test results ($p > 0.05$).

2.5. Preliminary Selection Macroporous Resin as an Effective Adsorbent

The static adsorption capacities of the resins were first screened to select the best resin for flavonoid enrichment. An accurately weighed amount (0.1 g) of each of the pretreated resins was transferred into 15 mL centrifuge tubes. A 5 mL aliquot of OPLAH was then added into the tubes. These centrifuge tubes were capped, placed horizontally, and taped tightly in an orbital shaker (Wisube WIG-10RL Precise Shaking Incubator, Wisd Laboratory Instruments, Wertheim, Germany). The mixture was shaken for 24 h at 298 K with an agitation speed of 150 g/min to reach adsorption equilibrium. The filtrates were then analyzed by UHPLC. To desorb the flavonoid C-glycosides from the resins, 5 mL of 95% EtOH was added into each tube, and the mixing was repeated using the same conditions, followed by filtering and the filtrates being analyzed by UHPLC. Three individual experiments were performed. Selection of the optimal MAR for use in subsequent studies was made based on the adsorption and desorption capacities of each MAR.

2.6. Optimization of Sorption Conditions Using Batch Adsorption Tests

In the present study, the four main operating parameters of temperature, pH, equilibrium time point, and initial concentration were optimized. The optimum conditions for the adsorption of flavonoid C-glycosides from OPLAH were performed using batch adsorption tests, where the 15 mL OPLAH solution was mixed with a selected adsorbent (0.3 g) and subjected to continuous agitation using an orbital shaker with a agitation speed of 150 g/min. All experiments were carried out in triplicate. The optimal conditions were selected based on the quantification of orientin, isoorientin, vitexin, and isovitexin using a developed and validated UHPLC-UV/PDA method. Simultaneously, sorption behaviors such as the kinetics and isotherm were assessed.

To select a suitable sorption temperature, the adsorption and desorption were performed at different oscillation temperatures (298 K, 308 K, and 318 K). The OPLAH solution was adjusted to a pH of 5, and the mixture was then agitated for 24 h. EtOH (95%) was used as a desorbing solvent.

To optimize the pH solution of the OPLAH, three different pHs (5, 7, and 9) were adjusted with 1 mol/L HCl or 1 mol/L NaOH. Concurrently, the equilibrium time point was monitored by withdrawing an aliquot of supernatant at 0, 15, 30, 60, 120, 180, 240, 300, 360, 480, and 1440 min. The adsorption kinetics curves for the target flavonoid C-glycosides on the XAD7HP resin were constructed. The kinetic data were subjected to two common kinetic models—*pseudo*-first-order [23] and *pseudo*-second-order models [24]—and one particle diffusion kinetic model [25].

To optimize the suitable initial concentration, different OPLAH concentrations with a known amount of target flavonoid C-glycosides were prepared, whereby the concentrations of isoorientin, orientin, vitexin, and isovitexin, were in ranges of 1.62–45.12 μg/mL, 12.64–89.76 μg/mL, 66.76–863.22 μg/mL and 1.00–12.06 μg/mL, respectively. The OPLAH solution was adjusted to the optimized pH and temperature. Simultaneously, the isotherm data was subjected to two well-known theoretical isotherm models: the Langmuir [26] and Freundlich models [27]. The R_L is a dimensionless constant that was applied to signify the important equilibrium parameter of the Langmuir isotherm [28].

2.7. Dynamic Sorption Experiments on the Chromatography Column

The dynamic sorption procedure was carried out according to our recent publication [9] with modifications. By using a 2.5 cm × 46 cm glass column wet-packed with 4.4 g of dried XAD7HP resin, the dynamic adsorption and desorption experiments were performed. The resin bed volume (BV) was kept at 200 mL. The 750 mg of OPLAH was mixed with 150 mL of deionized water to form a 5 mg/mL solution. The pH of the filtered solution was adjusted to pH 5, applied to the glass column, and allowed to elute at a flow rate of 0.3 mL/min. The eluates were collected every 10 mL for UHPLC analysis. The 5% breakthrough and 95% saturation points were set based on the final to the initial concentration ratio (C/C_o) of each flavonoid C-glycoside. After reaching the saturation point, the desorption process proceeded by first washing the column with 30 mL of deionized water to remove the residue and eluting with EtOH, which acted as desorbing solvent, at a flow rate of 0.3 min/mL. The eluates were collected every 10 mL for UHPLC analysis. All dynamic sorption experiments were carried out in triplicate and at an optimized temperature. The breakthrough and desorption curves were plotted to determine the breakthrough and saturation points.

To select a suitable ethanol concentration for optimal desorption, both the isocratic and gradient elution modes were performed. For the isocratic mode, upon reaching equilibrium, 20% EtOH was loaded to elute the adsorbed flavonoids. The experiment was repeated by using different EtOH concentrations (40%, 60%, 80%, and 95%). For the gradient elution mode, a separate set of experiments was performed by eluting the adsorbed flavonoids using different EtOH concentrations of 20%, 40%, 60%, 80%, and 95%. The collected fractions for both modes were concentrated using a rotary evaporator, freeze-dried, weighed, and subjected to UHPLC analysis.

2.8. Adsorption and Desorption Capacity, Kinetics, and Isotherm Model Equations

Adsorption capacity:

$$q_e = \frac{C_o - C_e}{W} \times V \tag{1}$$

Desorption capacity:

$$q_d = \frac{C_d V_d}{W} \tag{2}$$

Pseudo-first-order:

$$ln(q_e - q_t) = -k_1 t + lnq_e \tag{3}$$

Pseudo-second-order:

$$\frac{t}{q_t} = \frac{1}{k_2 {q_e}^2} + \frac{t}{q_e} \tag{4}$$

Intraparticle diffusion:

$$q_t = k_p \cdot t^{\frac{1}{2}} + C \tag{5}$$

Langmuir:

$$\frac{C_e}{q_e} = \frac{K_L}{q_m} + \frac{C_e}{q_m} \tag{6}$$

Freundlich:

$$q_e = K_f {C_e}^{\frac{1}{n}} \tag{7}$$

R_L:

$$R_L = \frac{1}{1 + K_L C_o} \tag{8}$$

where q_e, q_d, q_m, and q_t are the adsorption capacity, desorption capacity, maximum adsorption capacity, and adsorption capacity at different contact times (t, min), respectively, which are stated as mg/g of dry resin; C_o and C_e, are the initial and equilibrium sample concentrations, respectively, while C_d is the sample concentration in the desorption solution (these concentrations are measured in mg/mL); V, W, and C are the volume of the initial sample solution (mL), weight of the resin (g), and the constant representing the boundary layer diffusion effects (mg/g), respectively; k_1, k_2, and k_p are the *pseudo*-first-order rate constant (1/min), *pseudo*-second-order rate constant (g/mg.min), and particle diffusion rate constant (mg/g.min$^{1/2}$), respectively; K_L is the Langmuir constant (mg/mL), and K_f and 1/n are the Freundlich constant ((mg/g)(mL/mg)$^{1/n}$).

2.9. Determination of the Total Flavonoid Content and Antioxidant Free Radical Scavenging Activities

Evaluation of the total flavonoid content (TFC) was conducted using an aluminum chloride complex colorimetric assay [9]. Briefly, a 125 µL aliquot of 0.1 mg/mL OPL extract was transferred into a 2 mL microcentrifuge tube. Subsequently, 375 µL of 95% EtOH, 25 µL of a 10% aluminum chloride solution, 25 µL of a 1 mol/L sodium acetate solution, and

700 μL of distilled water were added, and the mixture was vortex-mixed (Vortex IKA MS 3 Basic, Selangor, Malaysia). A 200 μL aliquot of the mixture was then transferred into 96 well plates and incubated for 40 min at 25 °C, and the absorbance was recorded at 415 nm on a Tecan Infinite F200 Pro plate reader (Tecan Group Ltd., Männedorf, Switzerland). All tests were performed in triplicate. The TFC values were expressed in milligrams of quercetin equivalents per gram of extract (mg QCE/g extract).

The antioxidant assays, such as 1,1-diphenyl-2-picrylhydrazyl (DPPH), and nitric oxide (NO)-free radical scavenging activities were carried out according to the previous report [9]. The samples were prepared at 1000 μg/mL as a stock solution and serially diluted. For the DPPH assay, aliquots of 50 μL of the sample working solution were pipetted into a microtiter well plate, and each was added with 100 μL of a 59 μg/mL DPPH solution. The reaction mixtures were mixed well and incubated in the dark for 30 min, after which their absorbances were recorded at 515 nm. Similarly, for the NO assay, aliquots of 60 μL of the test concentrations were pipetted into the microtiter well plate, and each was added with 60 μL of a sodium nitroprusside solution. The reaction mixtures were mixed well and incubated for 150 min at 25 °C. Griess reagent (60 μL) was then added to each well, and the absorbance was measured at 550 nm. The scavenging activity (SA) was assessed as $SA\% = [(A_o - A_s)/A_o] \times 100\%$, where A_o and A_s are the absorbances of the blank and test sample, respectively. In this experiment, quercetin was used as a positive control. The experiment was carried out in triplicate, and the results were expressed as IC_{50} values in μg/mL.

2.10. Statistical Analysis

The InStat V2.02 statistical package (GraphPad Software, San Diego, CA, USA) and Minitab statistical software (Version 16, Minitab Inc., State College, PA, USA) were employed for all data analyses. For analysis of significance differences, one-way analysis of variance (ANOVA) done by Tukey's test was employed. The significant level was determined at $p < 0.05$. All data are shown as the mean of three replicates (n = 3).

3. Results and Discussion

3.1. Adsorption and Desorption Capacities of Selected MARs

The sorption capacities of the three MARs (XAD7HP, DAX-8, and XAD4) for the four flavonoid C-glycosides in OPLAH are shown in Figure 2. The adsorption capacity of the XAD7HP resin was 7.62 mg/g, which was considerably higher than those of DAX-8 and XAD4 at 7.41 mg/g and 0.92 mg/g, respectively. With a value of 6.73 mg/g, the desorption capacity of the XAD7HP resin was also higher than that of DAX-8 at 4.89 mg/g. Meanwhile, no desorption of flavonoid C-glycosides was observed for XAD4. Referring to these data, the XAD7HP resin had the best sorption capacities, demonstrating that an acrylic matrix, moderately polar resin, medium surface area, and large average pore diameter are the most suitable characteristics of MARs for the adsorption and desorption of the major OPLAH flavonoid C-glycosides. The findings are consistent with the previous findings, which reported high sorption capacities of the XAD7HP resin and low sorption capacities for XAD4 for grapefruit polyphenols [29] and oleuropein from olive (*Olea europaea*) leaves [30]. Therefore, the XAD7HP resin was selected for further evaluation.

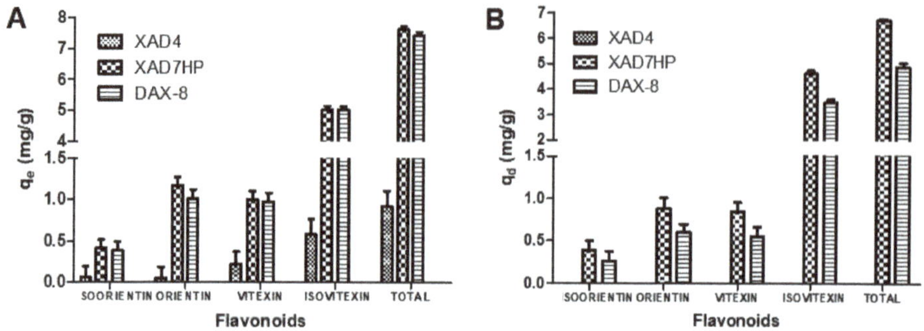

Figure 2. Adsorption (**A**) and desorption (**B**) capacities of three different macroporous resins (MARs) for major flavonoid C-glycosides in acid-hydrolized oil palm leaves (OPLAH). Each value is the average reading from three replicates (n = 3).

3.2. *Effect of Oscillation Temperatures on the Sorption Capacities*

The oscillation temperature is crucial for the optimum sorption properties of the resins, as the intermolecular forces between the adsorbates and adsorbents could be altered by subjection to a suitable temperature. According to the results shown in Figure 3, there was no significant difference in the adsorption capacity of flavonoid C-glycosides at the three different oscillation temperatures. However, the results were different for the desorption capacity, which decreased with an increase in the oscillation temperature. Similar results were reported in a study using MARs to enrich C-glycosyl flavonoids found in trolliflowers and *Abrus mollis* [18,31]. Within the evaluated temperature range, the sorption process is thermopositive [32]. Hence, the optimal oscillation temperature selected was 298 K, due to it demonstrating the highest adsorption and desorption capacities for flavonoid C-glycosides.

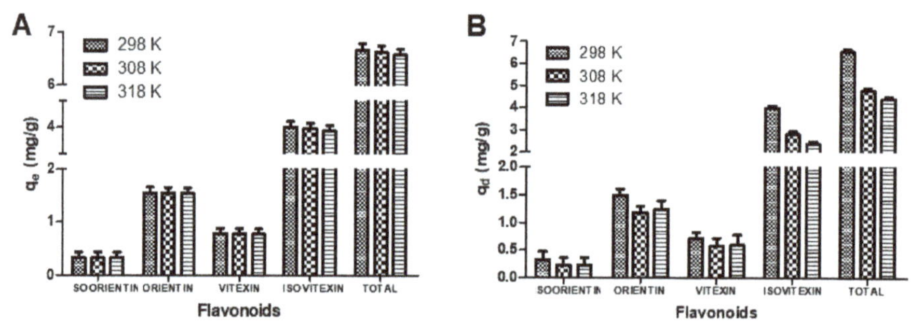

Figure 3. Adsorption (**A**) and desorption (**B**) capacities of XAD7HP resin for major flavonoid C-glycosides in OPLAH at three different temperatures. Each value is the average reading from three replicates (n = 3).

3.3. *Adsorption Kinetics of the XAD7HP Resin*

The pH is a factor affecting the ionization capability of certain compounds in the solvent, which ultimately influences their adsorption affinity. Hence, it is vital to perform the sorption at the right pH [33]. Figure 4 shows that the adsorption capacities (q_e) of the XAD7HP resin for isoorientin, orientin, vitexin, and isovitexin were higher at a pH of 5 than at pHs of 7 and 9. The q_e values for isoorientin, orientin, vitexin, and isovitexin decreased linearly as the pH increased. Based on the observations, hydrogen bonding was deemed to play a significant role in the sorption of the XAD7HP resins. The reduction in adsorption capacity at higher pH values may have been due to the decrease of hydrogen bonding interactions, caused by the deprotonation of hydroxyl groups in the flavonoid C-glycosides and the formation of their corresponding anions [19]. On the other hand, a low pH led to an abundance of hydronium ions at the surface of the resins, which may have enhanced the hydrogen bonding between hydroxyl groups present in the flavonoid C-glycosides with the XAD7HP resin, subsequently enhancing the adsorption capacity. The better adsorption capacity of flavonoid C-glycosides in acidic rather than basic conditions has been reported previously [18,19].

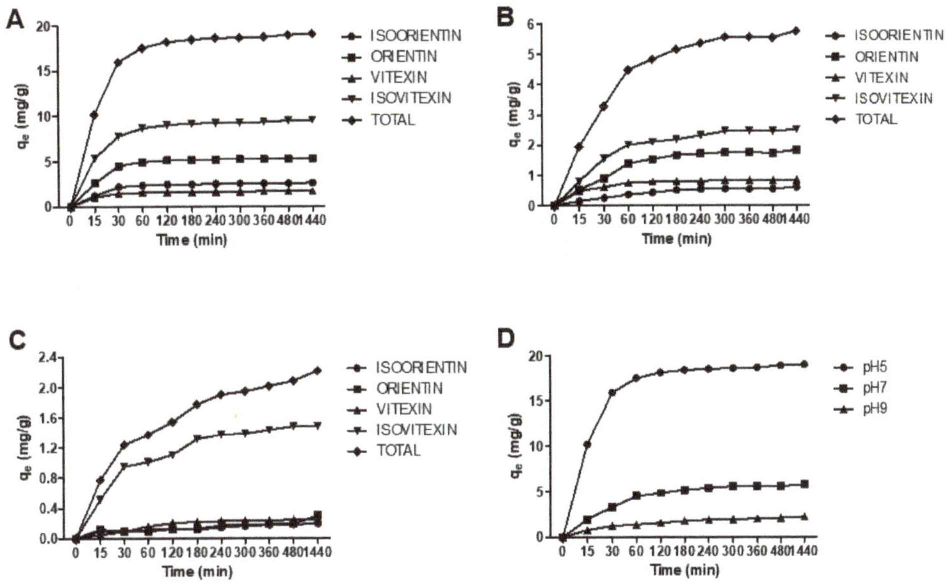

Figure 4. Adsorption kinetic curves of C-glycosyl flavonoids on the XAD7HP resin at pH levels of 5 (**A**), 7 (**B**), and 9 (**C**), and the total of four C-glycosyl flavonoids present in the OPLAH at different pH levels (**D**). Each value is the average reading from three replicates (n = 3).

The kinetics of adsorption, which explains the solute uptake rate governing the contact time of the sorption reaction, is an important characteristic that defines the sorption efficiency [34]. Hence, the adsorption behavior of the XAD7HP resins could be comprehended by accessing the adsorption kinetics of the flavonoid C-glycosides. Figure 4D presents the adsorption capacity q_t versus contact time (t, min) curves for the XAD7HP resin at different pH levels at 298 K. Overall, the q_t values were enhanced with an increase in time before achieving equilibrium [34]. The equilibrium time for the flavonoid C-glycosides was up to 24 h on the XAD7HP resin.

There are three commonly suggested kinetics models for adsorption: *pseudo*-first-order, *pseudo*-second-order, and intraparticle diffusion kinetic models [24,25]. Overall, the correlation coefficient (R^2) values revealed that the adsorption of flavonoid C-glycosides on the XAD7HP resin fit better to a *pseudo*-second order kinetic model compared with a *pseudo*-first-order model. In addition, Table 2 also reveals the multilinear characteristics of adsorption of the flavonoid C-glycosides on the XAD7HP resin, based on the R^2 values of the intraparticle diffusion kinetics model. The intraparticle diffusion curves of the XAD7HP resin show poor linear curves over time. By taking a pH of 5 as an example, the whole process was divided into three major phases: boundary layer diffusion (0–30 min), where the adsorption took place rapidly; a gradual adsorption phase (30–240 min), where the adsorption happened slowly; and finally, the equilibrium phase (240–1440 min), where the adsorption reached equilibrium. Similar results were reported in previous studies that showed intraparticle diffusion took place in the adsorption phase [35]. In the present study, the whole adsorption phase could not be represented by the particle diffusion kinetic models due to weak R^2 values. Nevertheless, it could still explain the adsorption mechanism up to a certain phase [36]. However, it is important to note that different flavonoids, including isoorientin, orientin, vitexin, and isovitexin, will have different ratios of their molecular status to ionic status in various pH environments. This will probably result in a more complex adsorption mechanism and kinetics, which will require more extensive future studies for greater insights into the mechanisms involved.

Table 2. *Pseudo*-first-order and *pseudo*-second-order kinetic equations and the intraparticle diffusion equation for the major C-glycosyl flavonoids in the OPLAH extract on the XAD7HP resin.

Compound	pH	q_e (exp) (mg/g)	*Pseudo*-First Order			*Pseudo*-Second Order			IntraParticle Diffusion		
			R^2	q_e (mg/g)	k_1 (1/min)	R^2	q_e (mg/g)	k_2 (g/mg.min)	R_2	C (mg/g)	k_p (mg/g.min$^{1/2}$)
Isoorientin	5	2.5668 ^{Aa}	0.7935 ^{Aa}	2.0738 ^{Aa}	0.0063 ^{Aa}	0.9995 ^{Aa}	2.5813 ^{Aa}	0.0416 ^{Aa}	0.5133 ^{Aa}	1.6715 ^{Aa}	0.0485 ^{Aa}
	7	0.5916 ^{Ba}	0.8134 ^{Ba}	3.1817 ^{Ba}	0.0050 ^{Ba}	0.9971 ^{Ba}	0.6007 ^{Ba}	0.0403 ^{Ba}	0.8506 ^{Ba}	0.1468 ^{Ba}	0.0218 ^{Ba}
	9	0.1866 ^{Ca}	0.9322 ^{Ca}	8.6789 ^{Ca}	0.0048 ^{Ba}	0.9640 ^{Ca}	0.1851 ^{Ca}	0.1102 ^{Ca}	0.9196 ^{Ca}	0.0701 ^{Ca}	0.0048 ^{Ca}
Orientin	5	5.2571 ^{Ab}	0.8178 ^{Ab}	1.2216 ^{Aa}	0.0085 ^{Ab}	0.9995 ^{Aa}	5.3362 ^{Ab}	0.0238 ^{Ab}	0.5008 ^{Ab}	3.5373 ^{Ab}	0.0978 ^{Ab}
	7	1.8269 ^{Bb}	0.7892 ^{Bb}	1.3399 ^{Bb}	0.0059 ^{Ba}	0.9980 ^{Bb}	1.8643 ^{Bb}	0.0195 ^{Bb}	0.7597 ^{Bb}	0.6503 ^{Bb}	0.0613 ^{Bb}
	9	0.2947 ^{Cb}	0.8829 ^{Cb}	4.9644 ^{Cb}	0.0014 ^{Cb}	0.9641 ^{Ca}	0.2024 ^{Cb}	0.0881 ^{Cb}	0.8223 ^{Cb}	0.0703 ^{Ca}	0.0054 ^{Ca}
Vitexin	5	1.6622 ^{Ac}	0.9158 ^{Ac}	3.7371 ^{Ac}	0.0090 ^{Ab}	0.9999 ^{Aa}	1.6793 ^{Ac}	0.0967 ^{Ac}	0.5653 ^{Ac}	1.2447 ^{Ac}	0.0233 ^{Ac}
	7	0.8356 ^{Bc}	0.8718 ^{Bc}	5.0799 ^{Bc}	0.0075 ^{Bb}	0.9999 ^{Ac}	0.8457 ^{Bc}	0.1137 ^{Bc}	0.8157 ^{Bc}	0.3242 ^{Bc}	0.0235 ^{Ac}
	9	0.2481 ^{Cc}	0.9252 ^{Cc}	6.8244 ^{Cc}	0.0064 ^{Cc}	0.9959 ^{Bb}	0.2683 ^{Cc}	0.0703 ^{Cc}	0.8229 ^{Cb}	0.0518 ^{Cb}	0.0101 ^{Bb}
Isovitexin	5	9.5296 ^{Ad}	0.8763 ^{Ad}	1.9927 ^{Ad}	0.0072 ^{Ac}	0.9998 ^{Aa}	9.5877 ^{Ad}	0.0120 ^{Ad}	0.6099 ^{Ad}	6.4869 ^{Ad}	0.1639 ^{Ad}
	7	2.5159 ^{Bd}	0.8952 ^{Ad}	1.1389 ^{Bd}	0.0078 ^{Ab}	0.9983 ^{Bb}	2.6185 ^{Bd}	0.0143 ^{Bd}	0.7582 ^{Bb}	1.0526 ^{Bd}	0.0772 ^{Bd}
	9	1.4779 ^{Cd}	0.9261 ^{Bc}	1.0830 ^{Cd}	0.0101 ^{Bd}	0.9974 ^{Cc}	1.5564 ^{Cd}	0.0195 ^{Cd}	0.8568 ^{Cc}	0.5758 ^{Cc}	0.0464 ^{Cc}

Values marked with different uppercase letters (A–C) indicate a comparison between pH levels for the same compound. Values marked with different lowercase letters (a–d) indicate a comparison between different compounds for the same pH. Each value is the average reading from three replicates (n = 3).

3.4. Adsorption Isotherms on the XAD7HP Resin

The adsorption isotherms of flavonoid C-glycosides on XAD7HP resins was performed at room temperature (298 K) after taking into consideration several factors, including the practicality and energy conservation. Figure S1 shows the isotherm curves for the individual flavonoid C-glycosides in OPLAH. The adsorption behaviors of flavonoid C-glycosides on the XAD7HP resin were further assessed by using two adsorption isotherm equations, namely Langmuir and Freundlich equations. The equations revealed the interaction between the compounds and the resin [37].

The Langmuir and Freundlich parameters are listed in Table 3. The R^2 values of the two models were relatively higher for isoorientin, orientin, and vitexin. With R^2 values of 0.9977 and 0.9519, the adsorption behavior of isoorientin and orientin, respectively, on the XAD7HP resin followed the Langmuir equation. The results indicated that these two compounds displayed monolayer adsorption on the resin, suggesting that orientin and its isomer were in contact with the surface layer of the XAD7HP resin. Meanwhile, vitexin and isovitexin followed the Freundlich equation with $R^2 = 0.9700$ and 0.8418, respectively, indicating the adsorption of these isomeric compounds followed a multilayer process wherein the XAD7HP resin accommodated more than one layer for the adsorption of vitexin and its pair to take place. This situation could be related to the molecular structures of vitexin/isovitexin and orientin/isoorientin [38]. As shown in Figure 1A, the molecular sizes of the vitexin and isovitexin structures were relatively smaller compared with orientin and isoorientin, due to the lack of one hydroxyl group (-OH). This could have reduced the steric hindrance in the interaction between vitexin/isovitexin and the XAD7HP resin, thus favoring a multilayer adsorption process. In the Freundlich equation, the value of R_L indicates the isotherm shape, which is either unfavorable ($R_L > 1$), linear ($R_L = 1$), favorable ($0 < R_L < 1$), or irreversible ($R_L = 0$) [39]. Hence, the present findings showed that the adsorption of the flavonoid C-glycosides on the XAD7HP resin was favorable. The 1/n value is a measure of the adsorption intensity [38]. A value of 1/n above 2 indicates that adsorption is unlikely to happen [40]. In this study, the 1/n values of the flavonoid

C-glycosides were all above 2, suggesting that the XAD7HP resin was a suitable resin to use for absorbing the flavonoid C-glycosides from OPLAH.

Table 3. Langmuir and Freundlich models for the adsorption of OPLAH flavonoid C-glycosides on XAD7HP resin.

Compound	Langmuir Equation				Freundlich Equation		
	q_m (mg/g)	R_1^2	K_L (mg/mL)	R_L	1/n	K_f ((mg/g) $(mL/mg)^{1/n}$)	R_2^2
Isoorientin	476.1905 [a]	0.9977 [a]	0.2381 [a]	0.2929 [a]	0.3754 [a]	366.090 [a]	0.9260 [a]
Orientin	2000.000 [b]	0.9519 [b]	0.6000 [b]	0.0480 [b]	0.3757 [a]	1075.70 [b]	0.9247 [a]
Vitexin	50000.00 [c]	0.9624 [c]	30.000 [c]	0.0001 [c]	0.9538 [b]	1422.10 [c]	0.9700 [b]
Isovitexin	204.0816 [d]	0.5187 [d]	0.2653 [d]	0.5120 [d]	0.4149 [c]	138.930 [d]	0.8418 [c]

Values marked with different letters (a–d) indicate a comparison between compounds in each tested parameter. Each value is the average reading from three replicates (n = 3).

3.5. Dynamic Sorption Properties of the XAD7HP Resin

On an open column, information of the breakthrough volume is important in estimating the optimum volume of sample-containing compounds of interest that can be loaded onto the column. The breakthrough point was set at 5% of the inlet concentration [9]. As shown in Figure 5A, the dynamic breakthrough curves on the XAD7HP resin were attained for isoorientin, orientin, vitexin, and isovitexin. The breakthrough volume of isoorientin on the XAD7HP resin was 100 mL, while that of orientin, vitexin, and isovitexin was 30 mL. Meanwhile, the saturation point was defined, at which the exit solute concentration reached 95% of the inlet concentration [9]. The saturation volume of isoorientin and isovitexin was 150 mL, while that of orientin and vitexin was 130 mL. The dynamic desorption curves for isoorientin, orientin, vitexin, and isovitexin on the XAD7HP resin are shown in Figure 5B,C. The results indicate that at 150 mL, the flavonoids could be sufficiently desorbed and eluted off of the XAD7HP resin column.

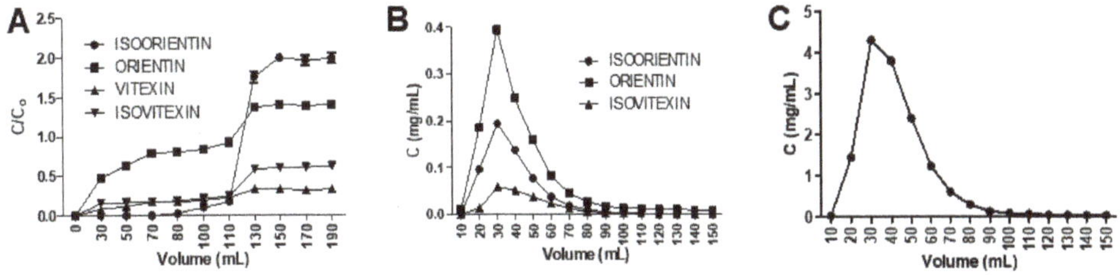

Figure 5. Dynamic breakthrough curves for flavonoid C-glycosides (**A**) and dynamic desorption curves for isoorientin, orientin, and isovitexin (**B**) and vitexin (**C**) in OPLAH on XAD7HP resin. Each value is the average reading from three replicates (n = 3).

3.6. Comparison between Isocratic and Gradient Elution Modes for Optimal Flavonoid C-Glycoside Enrichment

Enrichment of the OPLAH flavonoid C-glycosides was carried out via isocratic and gradient elution modes by using EtOH as a desorbing solvent after considering its low cost, ease of removal, and low toxicity [41]. Previous studies have also used EtOH to desorb flavonoid C-glycosides from other various MARs [20,31,42]. The desorbed fractions from the XAD7HP resin were analyzed qualitatively and quantitatively and compared to the original OPLAH. For the isocratic elution mode, a single desorbing solvent system was applied. As shown in Figure 6A, the amount of desorbed flavonoid C-glycosides increased with an increase in the EtOH concentration (from 20% to 95%). Orientin and vitexin were enriched the most when 80% EtOH was used as a single desorbing solvent system, as a further increment to 95% EtOH gave insignificant changes. Their respective isomers, isoorientin, and isovitexin were found in the highest fold at 95% EtOH. Meanwhile, a multiple desorbing solvent system was employed in the gradient elution mode. Figure 6B shows that orientin, isoorientin, vitexin, and isovitexin started to desorb rapidly from 20% to 40% EtOH concentrations and started to decrease as the EtOH concentration increased from 60% to 95%. Thus, the results revealed that the flavonoid *C*-glycosides found in the OPLAH solution could be desorbed optimally at 95% and 40% for the isocratic and gradient desorption techniques, respectively. The desorption of flavonoid *C*-glycosides from the XAD7HP resin into the solvent was attributed to the competition between the interaction of intermolecular forces and dissolution into the solvent used [9,42].

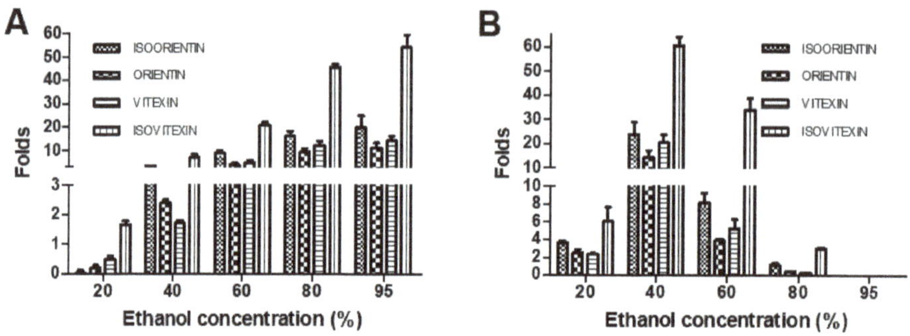

Figure 6. Enrichment of flavonoid *C*-glycosides from OPLAH by the isocratic elution mode (**A**) and gradient elution mode (**B**) on the XAD7HP resin. Each value is the average reading from three replicates (n = 3).

The UHPLC chromatograms of the OPLAH and enriched fractions obtained from the isocratic desorption (95% EtOH) and gradient desorption modes (40% EtOH) are shown in Figure S2A. Comparing the chromatogram of the enriched fractions and OPLAH, it could be observed that some impurities present in the original extract were eliminated, while the relative peak areas of the four major flavonoid *C*-glycosides were increased by different degrees. The compounds assigned to peaks 1–4 were confirmed by commercial standards and characterized by liquid chromatography tandem mass spectrometry (LC-MS/MS) (Figure S2B) [10]. The rest of the unassigned peaks have been comprehensively discussed in our previous publications [7,9,12].

Table 4 summarized the quantitative information of the OPLAH and the enriched fractions obtained through the isocratic (with 95% EtOH) and gradient elution modes (with 40% EtOH). The XAD7HP resin was able to increase the TFC from 88.98 mg QCE/g up to 284.18 mg QCE/g dried extract by 3.2-fold. Among the four flavonoid *C*-glycosides, at the optimum EtOH concentrations, isovitexin was enriched the most, followed by isoorientin, vitexin, and orientin. As illustrated by Figure 6A for the isocratic elution mode, with 95% EtOH, isovitexin was enriched almost 55-fold while isoorientin, vitexin, and orientin were enriched by 11- to 20-fold. A similar trend was also observed in the gradient elution mode, where at 40% EtOH, isovitexin was enriched by 60-fold, isoorientin by 25-fold, vitexin by 20-fold, and orientin by 15-fold (Figure 6B).

Table 4. Quantification and free radical scavenging activities of flavonoid *C*-glycosides in OPLAH, enriched OPLAH fractions and individual compounds.

Extract or Compound	TFC (mg QCE/g)	Flavonoid *C*-Glycosides (μg/mg)				Antioxidant Activities (IC_{50}, μg/mL)	
		Isoorientin	Orientin	Vitexin	Isovitexin	DPPH	NO
OPLAH	88.98 [a]	2.34 [a]	9.35 [a]	84.11 [a]	0.25 [a]	200.00 [a]	44.58 [a]
Enriched OPLAH (Isocratic)	247.28 [b]	46.27 [b]	104.88 [b]	1197.61 [b]	13.03 [b]	69.16 [b]	6.90 [b]
Enriched OPLAH (Gradient)	284.18 [c]	55.98 [c]	136.19 [c]	1726.11 [c]	14.61 [b]	70.63 [b]	7.32 [b]
Isoorientin						14.70 [c]	68.19 [c]
Orientin						57.60 [d]	42.72 [a]
Isovitexin						>1000 [e]	0.73 [d]
Vitexin						>1000 [e]	4.31 [e]

Values marked with different letters (a–e) indicate a comparison between OPLAH, enriched OPLAH and individual compounds for each flavonoid content and antioxidant activities. Each value is the average reading from three replicates (n = 3).

3.7. Antioxidant DPPH and NO Free Radical Scavenging Activities

The antioxidant activities of OPLAH, enriched fractions, and the individual flavonoid C-glycosides were tested using DPPH and NO free radical scavenging assays, and the results are shown in Table 4. The DPPH results revealed that the total flavonoid C-glycoside enriched fractions, obtained using the isocratic and gradient desorption methods, exhibited stronger antioxidant activity, with IC_{50} values of 69.19 and 70.63 μg/mL, respectively. In comparison with the original OPLAH extract, the IC_{50} value was much lower (200 μg/mL). The results for the NO free radical scavenging assay were also similar, as the enriched fractions exhibited significantly improved activity compared with the original extract. This increase in antioxidant activity indicated the substantial contribution of the enriched flavonoid C-glycoside contents to the overall activity. The results highlighted that both the DPPH and NO free radical scavenging assays were in good agreement for evaluating the antioxidant activities in both the original and enriched fractions. Previous studies have also reported the positive correlation between the free radical scavenging activity and the presence of high amounts of phenolic constituents [43,44].

The single flavonoid C-glycoside was also assayed for the free radical scavenging activities. Isoorientin with an IC_{50} value of 14.70 μg/mg exhibited superior DPPH free radical scavenging activity. Its isomer, orientin, was moderately active, with an IC_{50} value of 57.60 μg/mg, while isovitexin and vitexin were weakly active in comparison. The results are in agreement with previous studies that reported vitexin and isovitexin were poor DPPH free radical scavengers [45,46]. The structural differences of these flavonoid C-glycosides, such as the position of glycosidic linkages and the number or position of hydroxyl groups at play, will have a significant effect on the bioactivity. For example, the presence of a single hydroxyl group on the B ring of both vitexin and isovitexin may be the reason for their lower activity, in comparison with isoorientin and orientin, which have two hydroxyl groups on the same ring (Figure 1A). Additionally, the weaker activity of isovitexin could be due to steric hindrance associated with glycosylation on C-6, as compared with on C-8 for its isomer, vitexin [47].

Meanwhile, the NO free radical scavenging activity was quite different from the DPPH free radical scavenging activity. All tested apigenin and luteolin C-glycosides had good antioxidant activity by showing a great ability to inhibit nitric oxide and superoxide anion at low concentrations. In contrast to the DPPH assay results, isovitexin and vitexin exhibited strong NO scavenging activity, with IC_{50} values of 0.73 and 4.31 μg/mg, respectively, whereas orientin and isoorientin exhibited weaker values. These results reflect the superiority of isovitexin and vitexin in scavenging nitric oxide and superoxide radicals, which have also been similarly reported in the study of *Trigonella foenum graecum* L. (fenugreek seeds) [47].

3.8. Adsorption Mechanisms

The efficiency of the adsorption and desorption processes primarily relies on the polarity of the MARs. In this experiment, the tested MARs comprised both nonpolar (XAD4) and moderately polar (XAD7HP and DAX-8) resins. The data obtained showed that moderately polar resins were more appropriate to entrap and release flavonoid C-glycosides from OPL extract. The polarity matching between the extract and resin was related to the multiple interactions between the targeted metabolites and the surface chemistry of the resin [9,19,29,48]. Being a nonpolar resin, XAD4, with its smaller pore size, has low wettability and thus is not well-dispersed in an aqueous solution, which explains the low adsorption and no desorption of polar compounds (flavonoid C-glycosides (29)). Figure 7 displays the possible interactions between the isoorientin, orientin, vitexin, and isovitexin and the moderately polar XAD7HP resin under acidic conditions. The early part of the study indicated that isovitexin showed the highest adsorption and desorption capacities, followed by vitexin, orientin, and isorientin, suggesting multiple interactions, such as electrostatic interaction, intramolecular and intermolecular hydrogen bonding, ion-dipole interactions, cation-π interaction, and Van der Waals forces of the adsorbent. The XAD7HP resin was more favorable in isovitexin compared with the other compounds (Figure 7). More specifically, the hydroxyl groups at C-4′, C-5, or C-7 of the flavones have been reported to be more acidic than the hydroxyl groups attached at other positions [49]. Thus, it is highly likely that the electrostatic interactions of the flavonoids on the surface of the resins could have resulted from the attraction of protons dissociated from the hydroxyl groups at these positions. The hydroxyl groups of the flavonoid C-glycosides can also interact with the resin through the formation of intramolecular and intermolecular hydrogen bonds [50]. Furthermore, polar adsorbates can cause ion-dipole interactions with the polar segment of the moderately polar resin. Other than that, adsorption can also be facilitated by cation-π interaction [29], which can occur between hydronium ions (H_3O^+) surrounding the XAD7HP resin and the benzene ring from flavonoids. Lastly, the main driving force for the sorption process on polymeric XAD7HP resin is the existence of Van der Waals forces in an aqueous solvent system. Therefore, based on the high sorption capacities obtained in the study, the efficient simultaneous sorption of isoorientin, orientin, vitexin, and isovitexin was suggested to be substantially contributed by multiple interactions [9,42].

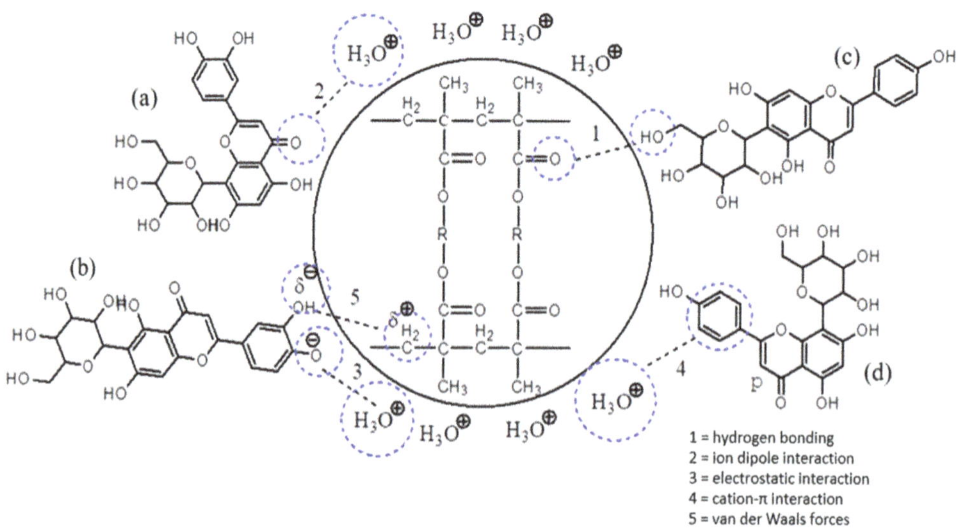

Figure 7. Schematic representation of the postulated interactions between targeted flavonoid C-glycosides for orientin (**a**), isoorientin (**b**), isovitexin (**c**), and vitexin (**d**) and the XAD7HP resin under acidic conditions.

The findings from both the isocratic and gradient elution experiments showed that isovitexin was enriched between 55- and 60-fold, while isoorientin, vitexin, and orientin were enriched by 20–25, 15–20 and 11–15-fold, respectively. This phenomenon could be due to strong hydrogen bonding interactions between the hydroxyl groups attached to the aglycones and the surface of the cross-linked polymeric resins [9,19]. The studied flavonoid C-glycosides shared similar flavone aglycone with differences in the sugar moiety position, as shown in Figure 1A. It was previously reported that a flavonoid molecule with sugar moiety attached to it could reach approximately 2.0 nm in size [50], and with additional hydroxyl groups contributed by glucosides, this collectively enhanced the sorption process to the large pore size and moderately polar resin XAD7HP.

Moreover, in between isomeric compounds, isovitexin/vitexin had higher enrichment than isoorientin/orientin. This could be explained by the steric hindrance [48,51,52]. Referring to Figure 1A, vitexin isomers have one hydroxyl group bonded at C-4′ (para position) of ring B, whereas orientin isomers have two hydroxyl groups attached at C-3′ (meta position) and C-4′ (para position) of the same ring. Since the hydroxyl group is one of the para directing groups, attachment of these hydroxyl groups onto the para position lessened the steric hindrance, resulting in less repulsion between the groups and lastly assisting the interaction of these groups with the resin through hydrogen bonds. However, the addition of one hydroxyl group onto the C-3′ position of the orientin isomers increased the steric hindrance, as the hydroxyl group was larger than the hydrogen atom, causing congestion that may have slowed down the interaction of the atom with the surface of resins.

4. Conclusions

The present study provides experimental data on the enrichment of the total flavonoid C-glycosides content via a process combining acid hydrolysis and adsorption and desorption on MARs. The XAD7HP resin showed the best sorption capacities. The enrichment of the flavonoid C-glycosides content of OPL extract was conducted at optimal conditions, where the leaf extract, prehydrolyzed with acid and adjusted to a pH of 5, was shaken at 298 K for a period of 24 h for static adsorption. The adsorption process of the target flavonoids on the XAD7HP resin could be well-described with the *pseudo*-second-order kinetic model. The equilibrium experimental data of the adsorption of isoorientin and orientin on the XAD7HP resin at 298 K were well fitted to the Langmuir isotherm model, while those of vitexin and isovitexin were well described by the Freundlich isotherm model. The enriched fractions recovered using the isocratic (with 95% EtOH) and gradient elution modes (with 40% EtOH) produced up to 60-fold flavonoid enrichment with excellent antioxidant free radical scavenging activities. The enriched OPLAH contained isoorientin (247.28–284.18 μg/mg), orientin (104.88–136.19 μg/mg), vitexin (1197.61–1726.11 μg/mg), and isovitexin (13.03–14.61 μg/mg), as compared to OPLAH with isoorientin (2.34 μg/mg), orientin (9.35 μg/mg), vitexin (84.11 μg/mg), and isovitexin (0.25 μg/mg). Additionally, the enriched OPLAH also showed excellent antioxidant free radical scavenging activities compared with OPLAH, with IC_{50} values of 6.90–70.63 μg/mL and 44.58–200.00 μg/mL, respectively. Strong hydrogen bonding may explain the efficient enrichment of the target flavonoid C-glycosides. The results indicated the combination of acid treatment and MARs could selectively and effectively enrich flavonoid C-glycosides from OPL. This study presents a simple, rapid, and efficient method for enriching the flavonoid C-glycoside content of oil palm leaf extract, a major type of agriculture waste, which has been underutilized. This method provides several potential applications, such as for further purification of major flavonoid C-glycosides as fine chemicals or pharmaceuticals or the use of the enriched fraction as bioactive ingredients in nutraceutical, cosmeceutical, and other healthcare or personal care products.

Supplementary Materials: The following are available online at https://www.mdpi.com/article/10.3390/pr9040659/s1. Figure S1: Adsorption isotherms of isoorientin (A), orientin (B), vitexin (C) and isovitexin (D) present in OPLAH at 298 K on XAD7HP resin; Figure S2: (A) UHPLC chromatograms (340 nm) of OPLAH and OPLAH-enriched fractions obtained via isocratic and gradient elution desorption modes. (B) Compound confirmation using UHPLC-MS/MS analysis.

Author Contributions: K.S. and M.S.C.Z. conceptualized and designed the experiments; K.S. provided the research materials; M.S.C.Z. performed the experiments and wrote the manuscript; K.S. and M.S.C.Z. analyzed the data; K.S. and C.Y.T. validated the data; K.S. and S.Y.L. edited the manuscript. All authors have read and agreed to the published version of the manuscript.

Funding: This research received no external funding.

Acknowledgments: The first author would like to thank Universiti Putra Malaysia for granting him a graduate research fellowship (GRF) scheme.

Conflicts of Interest: The authors declare no conflict of interest.

References

1. Whitmore, T.C. *The Palms of Malaya*; Longmans: Selangor, Malaysia, 1973.
2. Basiron, Y. Palm oil production through sustainable plantations. *Eur. J. Lipid Sci. Technol.* **2007**, *109*, 289–295. [CrossRef]
3. Malaysian Palm Oil Board (MPOB). Malaysian Oil Palm Industry Performance 2016 and Prospects for 2017. Available online: http://www.mpob.gov.my (accessed on 20 January 2019).
4. Malaysian Palm Oil Board (MPOB). Production of Oil Palm Products 2020. Available online: http://bepi.mpob.gov.my (accessed on 27 January 2021).
5. Be Sustainable Magazine. Malaysia's Biomass Potential. Available online: http://www.besustainablemagazine.com (accessed on 13 February 2021).
6. Owoyele, B.V.; Owolabi, G.O. Traditional oil palm (*Elaeis guineensis* Jacq.) and its medicinal uses: A review. *Tang Humanit Med.* **2014**, *4*, 1–8. [CrossRef]
7. Tahir, N.I.; Shaari, K.; Abas, F.; Parveez, G.K.A.; Ishak, Z.; Ramli, U.S. Characterization of apigenin and luteolin derivatives from oil palm (*Elaeis guineensis* Jacq.) leaf using LC-ESI-MS/MS. *J. Agric. Food Chem.* **2012**, *60*, 11201–11210. [CrossRef] [PubMed]
8. Xiao, J.; Capanoglu, E.; Jassbi, A.R.; Miron, A. Advance on the Flavonoid C-glycosides and Health Benefits. *Crit. Rev. Food Sci. Nutr.* **2016**, *56*, S29–S45. [CrossRef]
9. Zain, M.S.C.; Lee, S.Y.; Teo, C.Y.; Shaari, K. Adsorption and Desorption properties of total flavonoids from oil palm (*Elaeis guineensis* Jacq.) mature leaf on macroporous adsorption resins. *Molecules* **2020**, *25*, 778. [CrossRef]
10. Zain, M.S.C.; Jakariah, N.A.; Yeoh, J.X.; Lee, S.Y.; Shaari, K. Ultrasound-assisted extraction of polyphenolic contents and acid hydrolysis of flavonoid glycosides from oil palm (*Elaeis guineensis* Jacq.) leaf: Optimization and correlation with free radical scavenging activity. *Processes* **2020**, *8*, 1540. [CrossRef]
11. Zain, M.S.C.; Lee, S.Y.; Mad Nasir, N.; Fakurazi, S.; Shaari, K. Metabolite characterization and correlations with antioxidant and wound healing properties of oil palm (*Elaeis guineensis* Jacq.) leaflets via ^{1}H-NMR-based metabolomics approach. *Molecules* **2020**, *25*, 5636. [CrossRef] [PubMed]
12. Zain, M.S.C.; Lee, S.Y.; Sarian, M.N.; Fakurazi, S.; Shaari, K. In vitro wound healing potential of flavonoid c-glycosides from oil palm (*Elaeis guineensis* Jacq.) leaves on 3t3 fibroblast cells. *Antioxidants* **2020**, *9*, 326. [CrossRef]
13. Mordor Intelligence. Flavonoid Market—Growth, Trends and Forecast (2020–2025). Available online: https://www.mordorintelligence.com (accessed on 20 July 2020).
14. Yang, X.; Wei, M.; Tian, H.; Liu, T.; Yang, L. Enrichment and purification of aucubin from *Eucommia ulmoides* ionic liquid extract using macroporous resins. *Materials* **2018**, *11*, 1758. [CrossRef] [PubMed]
15. Kim, J.; Kim, D.; Gwon, Y.J.; Lee, K.W.; Lee, T.S. Removal of sodium dodecylbenzenesulfonate by macroporous adsorbent resins. *Materials* **2018**, *11*, 1324. [CrossRef] [PubMed]
16. Tungmunnithum, D.; Drouet, S.; Kabra, A.; Hano, C. Enrichment in antioxidant flavonoids of stamen extracts from *Nymphaea lotus* L. using ultrasonic-assisted extraction and macroporous resin adsorption. *Antioxidants* **2020**, *9*, 576. [CrossRef]
17. Lin, L.; Zhao, H.; Dong, Y.; Yang, B.; Zhao, M. Macroporous resin purification behavior of phenolics and rosmarinic acid from *Rabdosia serra* (MAXIM.) HARA leaf. *Food Chem.* **2012**, *130*, 417–424. [CrossRef]
18. Du, H.; Wang, H.; Yu, J.; Liang, C.; Ye, W.; Li, P. Enrichment and purification of total flavonoid C-glycosides from *Abrus mollis* extracts with macroporous resins. *Ind. Eng. Chem. Res.* **2012**, *51*, 7349–7354. [CrossRef]
19. Liu, W.; Zhang, S.; Zu, Y.G.; Fu, Y.J.; Ma, W.; Zhang, D.Y.; Kong, Y.; Li, X.-J. Preliminary enrichment and separation of genistein and apigenin from extracts of pigeon pea roots by macroporous resins. *Bioresour. Technol.* **2010**, *101*, 4667–4675. [CrossRef]
20. Wang, X.; Liang, Y.; Zhu, L.; Xie, H.; Li, H.; He, J.; Pan, M.; Zhang, T.; Ito, Y. Preparative isolation and purification of flavone C-glycosides from the leaves of *Ficus microcarpa* L.f by medium-pressure liquid chromatography, high-speed countercurrent chromatography, and preparative liquid chromatography. *J. Liq. Chromatogr. Relat. Technol.* **2010**, *33*, 462–480. [CrossRef]
21. *International Conference on Harmonization of Technical Requirment for Registration of Pharmaceuticals for Human Use, Validation of Analytical Procedures: Text and Methodology Q2 (R1)*; ICH: Geneva, Switzerland, 2005; Volume 7, pp. 1–5.
22. Zain, M.S.C.; Osman, M.F.; Lee, S.Y.; Shaari, K. UHPLC-UV /PDA method validation for simultaneous quantification of luteolin and apigenin derivatives from *Elaeis guineensis* leaf extracts: An application for antioxidant herbal preparation. *Molecules* **2021**, *26*, 1084. [CrossRef]

23. Bhattacharya, A.K.; Venkobachar, C. Removal of Cadmium (II) by low cost adsorbents. *J. Environ. Eng.* **1984**, *110*, 110–122. [CrossRef]
24. Ho, Y.S.; McKay, G. *Pseudo*-second order model for sorption processes. *Process Biochem.* **1999**, *34*, 451–465. [CrossRef]
25. Guibal, E.; Milot, C.; Tobin, J.M. Metal-anion sorption by chitosan beads: Equilibrium and kinetic studies. *Ind. Eng. Chem. Res.* **1998**, *37*, 1454–1463. [CrossRef]
26. Langmuir, I. The adsorption of gases on plane surfaces of glass, mica and platinum. *J. Am. Chem. Soc.* **1918**, *40*, 1361–1403. [CrossRef]
27. Kammerer, J.; Carle, R.; Kammerer, D.R. Adsorption and ion exchange: Basic principles and their application in food processing. *J. Agric. Food Chem.* **2011**, *59*, 22–42. [CrossRef] [PubMed]
28. Hall, K.R.; Eagleton, L.C.; Acrivos, A.; Vermeulen, T. Pore and solid diffusion kinetics in fixed-bed adsorption under constant pattern conditions. *Ind. Eng. Chem. Fundam.* **1966**, *5*, 212–223. [CrossRef]
29. Taktak, F.; Ciğeroğlu, Z.; Öğen, Y.; Kirbaşlar, Ş.İ. Resin-loaded cationic hydrogel: A new sorbent for recovering of grapefruit polyphenols. *Chem. Eng. Commun.* **2018**, *205*, 1442–1456. [CrossRef]
30. Şahin, S.; Bilgin, M. Selective adsorption of oleuropein from olive (*Olea europaea*) leaf extract using macroporous resin. *Chem. Eng. Commun.* **2017**, *204*, 1391–1400. [CrossRef]
31. Sun, Y.; Yuan, H.; Hao, L.; Min, C.; Cai, J.; Liu, J.; Cai, P.; Yang, S. Enrichment and antioxidant properties of flavone C-glycosides from trollflowers using macroporous resin. *Food Chem.* **2013**, *141*, 533–541. [CrossRef]
32. Zhang, L.; Wu, T.; Xiao, W.; Wang, Z.; Ding, G.; Zhao, L. Enrichment and purification of total Ginkgo flavonoid *O*-glycosides from *Ginkgo Biloba* extract with macroporous resin and evaluation of anti-inflammation activities in vitro. *Molecules* **2018**, *23*, 1167. [CrossRef] [PubMed]
33. Huang, J.H.; Huang, K.L.; Wang, A.T.; Yang, Q. Adsorption characteristics of poly(styrene-co-divinylbenzene) resin functionalized with methoxy and phenoxy groups for phenol. *J. Colloid Interface Sci.* **2008**, *327*, 302–307. [CrossRef]
34. Wan, P.; Sheng, Z.; Han, Q.; Zhao, Y.; Cheng, G.; Li, Y. Enrichment and purification of total flavonoids from *Flos Populi* extracts with macroporous resins and evaluation of antioxidant activities in vitro. *J. Chromatogr. B Anal. Technol. Biomed. Life Sci.* **2014**, *945–946*, 68–74. [CrossRef]
35. Lorenc-Grabowska, E.; Gryglewicz, G. Adsorption of lignite-derived humic acids on coal-based mesoporous activated carbons. *J. Colloid Interface Sci.* **2005**, *284*, 416–423. [CrossRef] [PubMed]
36. Ayranci, E.; Hoda, N. Adsorption kinetics and isotherms of pesticides onto activated carbon-cloth. *Chemosphere* **2005**, *60*, 1600–1607. [CrossRef] [PubMed]
37. Chao, L.; Hong, Z.; Li, Z.; Gang, Z. Study on adsorption characteristic of macroporous resin to phenol in wastewater. *Can. J. Chem. Eng.* **2010**, *88*, 417–424. [CrossRef]
38. Liu, Y.; Bai, Q.; Lou, S.; Di, D.; Li, J.; Guo, M. Adsorption characteristics of (−)-epigallocatechin gallate and caffeine in the extract of waste tea on macroporous adsorption resins functionalized with chloromethyl, amino, and phenylamino groups. *J. Agric. Food Chem.* **2012**, *60*, 1555–1566. [CrossRef] [PubMed]
39. Lou, S.; Chen, Z.; Liu, Y.; Ye, H.; Di, D. Synthesis of functional adsorption resin and its adsorption properties in purification of flavonoids from *Hippophae rhamnoides* L. leaves. *Ind. Eng. Chem. Res.* **2012**, *51*, 2682–2696. [CrossRef]
40. Wang, R.; Peng, X.; Wang, L.; Tan, B.; Liu, J.; Feng, Y.; Yang, S. Preparative purification of peoniflorin and albiflorin from peony rhizome using macroporous resin and medium-pressure liquid chromatography. *J. Sep. Sci.* **2012**, *35*, 1985–1992. [CrossRef] [PubMed]
41. Cao, S.Q.; Pan, S.Y.; Yao, X.L.; Fu, H.F. Isolation and Purification of Anthocyanins from Blood Oranges by Column Chromatography. *Agric. Sci. China* **2010**, *9*, 207–215. [CrossRef]
42. Wu, S.; Wang, Y.; Gong, G.; Li, F.; Ren, H.; Liu, Y. Adsorption and desorption properties of macroporous resins for flavonoids from the extract of Chinese wolfberry (*Lycium barbarum* L.). *Food Bioprod. Process.* **2015**, *93*, 148–155. [CrossRef]
43. Koch, W.; Kukuła-Koch, W.; Czop, M.; Helon, P.; Gumbarewicz, E. The role of extracting solvents in the recovery of polyphenols from green tea and its antiradical activity supported by principal component analysis. *Molecules* **2020**, *25*, 2173. [CrossRef]
44. Abdul-Hamid, N.A.; Mustaffer, N.H.; Maulidiani, M.; Mediani, A.; Ismail, I.S.; Tham, C.L.; Shadid, K.; Abas, F. Quality evaluation of the physical properties, phytochemicals, biological activities and proximate analysis of nine Saudi date palm fruit varieties. *J. Saudi Soc. Agric. Sci.* **2010**, *19*, 151–160. [CrossRef]
45. Sinjman, P.W.; Joubert, E.; Ferreira, D.; Li, X.C.; Ding, Y.; Green, I.R.; Gelderblom, W.C. Antioxidant activity of the dihydrochalcones aspalathin and nothofagin and their corresponding flavones in relation to other rooibos (*Aspalathus linearis*) flavonoids, epigallocatechin gallate, and Trolox. *J. Agric. Food Chem.* **2009**, *57*, 6678–6684. [CrossRef]
46. Gerhäuser, C. *Phenolic Beer Compounds to Prevent Cancer. Beer in Health and Disease Prevention*; Elsevier: London, UK, 2009; pp. 669–684.
47. Khole, S.; Panat, N.A.; Suryawanshi, P.; Chatterjee, S.; Devasagayam, T.; Ghaskadbi, S. Comprehensive assessment of antioxidant activities of apigenin isomers: Vitexin and isovitexin. *Free Radic. Antioxid.* **2016**, *6*, 155–166. [CrossRef]
48. Wu, X.; Liu, Y.; Huo, T.; Chen, Z.; Liu, Y.; Di, D.; Guo, M.; Zhao, L. Multiple interactions on macroporous adsorption resins modified with ionic liquid. *Colloids Surfaces A Physicochem. Eng. Asp.* **2015**, *487*, 35–41. [CrossRef]
49. Havsteen, B.H. The biochemistry & medical significance of the flavonoids. *Pharmacol. Ther.* **2002**, *96*, 67–202. [PubMed]

50. Li, J.; Chase, H.A. Development of adsorptive (non-ionic) macroporous resins and their uses in the purification of pharmacologically-active natural products from plant sources. *Nat. Prod. Rep.* **2010**, *27*, 1493. [CrossRef] [PubMed]
51. Xiao, J. Dietary flavonoid aglycones and their glycosides: Which show better biological significance? *Crit. Rev. Food Sci. Nutr.* **2017**, *57*, 1874–1905. [CrossRef]
52. Cao, H.; Chen, X. Structures required of flavonoids for inhibiting digestive enzymes. *Anticancer Agents Med. Chem.* **2012**, *12*, 929–939. [CrossRef]

processes

MDPI

Article

LC-ESI-QTOF-MS/MS Characterization of Phenolic Compounds in Common Commercial Mushrooms and Their Potential Antioxidant Activities

Minghang Chu [1], Rana Dildar Khan [1], Ying Zhou [1], Osman Tuncay Agar [1,2], Colin J. Barrow [3], Frank R. Dunshea [1,4] and Hafiz A. R. Suleria [1,3,*]

[1] School of Agriculture, Food and Ecosystem Sciences, Faculty of Science, The University of Melbourne, Melbourne, VIC 3010, Australia; minghang@student.unimelb.edu.au (M.C.); dildar1943@gmail.com (R.D.K.); yinzhou5@student.unimelb.edu.au (Y.Z.); osman.agar@unimelb.edu.au (O.T.A.); fdunshea@unimelb.edu.au (F.R.D.)

[2] Department of Pharmacognosy, Faculty of Pharmacy, Suleyman Demirel University, Isparta 32000, Türkiye

[3] Centre for Chemistry and Biotechnology, School of Life and Environmental Sciences, Deakin University, Waurn Ponds, Geelong, VIC 3217, Australia; colin.barrow@deakin.edu.au

[4] Faculty of Biological Sciences, The University of Leeds, Leeds LS2 9JT, UK

* Correspondence: hafiz.suleria@unimelb.edu.au; Tel.: +61-470-439-670

Abstract: Mushrooms have a long history of use as food and medicine. They are rich in various nutrients and bioactive compounds, particularly phenolic compounds. In this study, ten mushroom species were selected, and solvent extraction using 80% ethanol was used to extract phenolic compounds. Total phenolic content (TPC), total flavonoid content (TFC) and total condensed tannin content (TCT) were measured to evaluate phenolic content in different mushroom varieties. In the mushroom varieties tested, brown portobello mushroom had the highest TPC (396.78 ± 3.12 μg GAE/g), white cup mushroom exhibited the highest TFC (275.17 ± 9.40 μg CE/g), and shiitake mushroom presented the highest TCT (13.80 ± 0.21 μg QE/g). Antioxidant capacity was evaluated using 2,2-diphenyl-1-picrylhydrazyl (DPPH), ferric reducing antioxidant power (FRAP), 2,2′-azino-bis-3ethylbenzothiazoline-6-sulfonic acid (ABTS) and total antioxidant capacity (TAC) assays. The highest DPPH free radical scavenging ability was found in white cup mushroom (730.14 ± 55.06 μg AAE/g), while the greatest iron-reducing ability (FRAP) was recorded for shiitake mushroom (165.32 ± 10.21 μg AAE/g). Additionally, Swiss brown mushroom showed the highest ABTS antioxidant capacity (321.31 ± 5.7 μg AAE/g), and the maximum TAC value was found in shiitake mushroom (24.52 ± 1.2 μg AAE/g). These results highlight that most of the mushroom varieties studied showed high phenolic contents and demonstrated strong antioxidant activity, with shiitake mushrooms standing out due to their high TCT and FRAP values, and the highest TAC value among the varieties studied. In addition, LC-ESI-QTOF-MS/MS was used to characterize the mushroom samples, and tentatively identified a total of 22 phenolic compounds, including 11 flavonoids, 4 lignans, 3 phenolic acids, 2 stilbenes and 2 other phenolic compounds in all mushroom samples. The research results of this study showed that mushrooms are a good source of phenolic compounds with strong antioxidant potential. The results can provide a scientific basis for the development of mushroom extracts in functional food, health products, and other industries.

Keywords: mushroom; solvent extraction; phenolic compound; antioxidant potential; LC-ESI-QTOF-MS/MS

Citation: Chu, M.; Khan, R.D.; Zhou, Y.; Agar, O.T.; Barrow, C.J.; Dunshea, F.R.; Suleria, H.A.R. LC-ESI-QTOF-MS/MS Characterization of Phenolic Compounds in Common Commercial Mushrooms and Their Potential Antioxidant Activities. *Processes* **2023**, *11*, 1711. https://doi.org/10.3390/pr11061711

Academic Editors: Zongbi Bao and Qianqian Xu

Received: 8 May 2023
Revised: 26 May 2023
Accepted: 1 June 2023
Published: 3 June 2023

1. Introduction

Mushrooms are a common type of fungi found in nature and are one of the most widely consumed foods worldwide. They are popular for their unique flavor, rich nutritional value, and extensive commercial cultivation. According to a report by Expert Market Research (2023), the global mushroom market reached USD 54 billion in 2020. The market is expected to grow at a Compound Annual Growth Rate (CAGR) of 8%, in the forecast period of 2023–2028, to reach a value of USD 86 billion by 2026 [1].

The nutritional value of mushrooms has been extensively studied. In recent years, evidence has shown that mushrooms can reduce the risk of certain chronic diseases, including cancer, cognitive impairment, inflammatory bowel disease, and metabolic-related diseases [2]. The benefit of mushrooms is that they are rich in bioactive compounds, such as vitamins (vitamin B and vitamin D), minerals (selenium, phosphorus, and copper), and polysaccharides (β-glucans) [3]. Moreover, Kalaras et al. showed that mushrooms rich in endogenous ergosterol, which is converted to vitamin D_2 (ergocalciferol) under sunlight or ultraviolet radiation, are a good source of vitamin D_2 [4]. In recent years, researchers have become increasingly interested in studying the antioxidant properties of mushrooms. Some studies have shown that mushrooms are capable of accumulating secondary metabolites, such as phenolic compounds, carotenoids, and steroids, which give them potent antioxidant properties [5,6]. For instance, Barros et al. detected a variety of phenolic acids, such as protocatechuic acid, *p*-hydroxybenzoic acid, *p*-coumaric acid, and vanillic acid, in 16 different wild mushrooms from Portugal [7].

Polyphenols are important secondary metabolites in the plant kingdom. They can be classified into the following four categories based on their structure: flavonoids, phenolic acids, stilbenes, and lignans [8]. Polyphenols are known for their strong antioxidant properties. They have the capacity to inhibit lipoxygenase, chelate metals, and scavenge free radicals [9,10]. In addition, polyphenols play a defensive role in plants and are important substances with antibacterial, antiviral, and antifungal activities [11]. There are many methods for extracting polyphenols from food, with the most common being the traditional solvent extraction method. In addition, new extraction technologies, including pressurized liquid extraction, ultrasound-assisted extraction, microwave extraction, and supercritical fluid extraction, have also been gradually applied to polyphenol extraction [12]. To evaluate the amount of polyphenolic compounds in food, it is useful to measure the contents of total phenols, total flavonoids, and total tannins concurrently. Assays used to identify the potential antioxidant activity of polyphenol compounds in food typically involve free radical scavenging assays, such as ABTS (2,2′-azinobis-(3-ethylbenzothiazoline-6-sulfonic acid)) and DPPH (2,2-diphenyl-1-picrylhydrazyl), as well as TAC (total antioxidant capacity) and FRAP (ferric reducing antioxidant power) assays. Due to the wide variety and complex structures of polyphenol compounds, the liquid chromatography–electrospray ionization quadrupole time-of-flight mass spectrometry (LC-ESI-QTOF-MS/MS) method is commonly used to tentatively characterize extracts [13].

While some studies have explored the antioxidant properties of mushrooms, the majority have focused on the antioxidant properties of multiple wild mushroom species or a single species of cultivated mushrooms. As a result, there is still a significant research gap in the comprehensive analysis and comparative discussion of phenolic compounds across a variety of commercially cultivated mushrooms. This article explores the diverse phytochemicals found in mushrooms, including nutrients, phenolic compounds, and their biological activities, as well as the potential commercial applications of mushroom polyphenols. To this end, the study selected 10 common commercial mushrooms in the Australian market to extract phenolic compounds. Various methods were used to determine the phenolic compound content in mushroom samples and evaluate their antioxidant capacity. The study also tentatively characterized the phenolic compounds in mushrooms using LC-ESI-QTOF-MS/MS, a powerful analytical technique used to identify and characterize phenolic compounds in food samples [14]. It can detect a wide variety of phenolic compounds, including phenolic acids, flavonoids, lignans, stilbenes, and other polyphenols [15]. The goal of this study was to comprehensively compare the phenolic compound content and potential antioxidant capacity of common commercial mushroom varieties in the Australian market. The findings can help optimize their utilization and development for commercial applications.

2. Materials and Methods

2.1. Chemical and Reagents

Almost all of the chemicals used for sample extraction and characterization were of analytical grade and obtained from Sigma Aldrich (Castle Hill, NSW, Australia). Quercetin, catechin, gallic acid, and L-ascorbic acid used for the standard curve were purchased from Sigma Aldrich (St. Louis, MO, USA). In addition, 2,2-diphenyl-1-picrylhydrazyl (DPPH), aluminium chloride, Folin–Ciocalteu's phenol reagent, 2,4,6-tripyridyl-s-triazine (TPTZ), ferric (III) chloride anhydrous, and 2,2′-azino-bis(3-ethylbenz-thiazoline-6-sulphonate) (ABTS) were obtained from Sigma-Aldrich (St. Louis, MO, USA). Sodium carbonate anhydrous was purchased from Gillman, SA, Australia, and sulfuric acid 98% was obtained from RCI Labscan Limited (Bangkok, Thailand). Sodium acetate hydrate was obtained from Ajax Finechem, Scoreby, VIC, Australia. Vanillin was obtained from Glentham Life Science (Wiltshire, United Kingdom), and analytical grade methanol was from Fisher Chemical Company (San Jose, CA, USA). For the mobile phases of LC-ESI-QTOF-MS/MS, LiChrosolv (Darmstadt, Germany) and Sigma-Aldrich (St. Louis, MO, USA) provided acetonitrile and acetic acid, respectively. Standards of phenolic acids and flavonoids used for characterization, including gallic acid, syringic acid, chlorogenic acid, caffeic acid, *p*-hydroxybenzoic acid, coumaric acid, quercetin, catechin, protocatechuic acid, quercetin-3-*O*-glucuronide, kaempferol-3-*O*-glucoside, epicatechin gallate, and kaempferol, were supplied by Sigma-Aldrich (St. Louis, MO, USA).

2.2. Sample Preparation and Extraction of Phenolic Compounds

In this study, a diverse selection of mushrooms was procured from the Australian sales market. These types, including brown portobello, shiitake, white button, organic white, white flat, white cup, portobello flat, Swiss brown, oyster, and needle, were specifically chosen due to their widespread consumption, commercial availability, and to ensure a broad representation of common mushrooms. Fresh mushroom samples were washed and weighed at 500 g. Each mushroom sample was then ground to a slurry using a blender and stored at −20 °C for further study. The polyphenol compounds of the mushroom samples were extracted using a method referenced from Buruleanu et al. with slight modifications [16]. First, the samples were mixed with 80% ethanol and homogenized using an Ultra-Turrax T25 homogenizer (IKA, Staufen, Germany) at 10,000 rpm for 30 s. The mixtures were then incubated in the ZWYR-240 incubator shaker (Labwit, Ashwood, Vic, Australia) at 150 rpm at 4 °C for 16 h. Next, all the samples were centrifuged twice using the Hettich Refrigerated Centrifuge (ROTINA 380R, Tuttlingen, Baden-Württemberg, Germany) at 5000× *g* for 15 min. The supernatant was collected and stored at 4 °C for further antioxidant analysis.

2.3. Polyphenol Estimation

All polyphenol evaluations were performed using the Multiskan® Go microplate photometer (Thermo Fisher Scientific, Waltham, MA, USA) with triplicate measurements. In addition, the standard curves were established with $R^2 > 0.995$.

2.3.1. Determination of Total Phenolic Content (TPC)

The TPC was measured using a modified version of the method outlined by Stojanova, et al. [17]. In a 96-well plate (Costar, Corning, NY, USA), 25 μL of the extract was mixed with 25 μL of 25% (*v*/*v*) Folin–Ciocalteu reagent and 200 μL of Milli-Q water. The mixture was incubated for 5 min at 25 °C, after which 25 μL of 10% (*w*/*w*) sodium carbonate was added. The mixture was then incubated in the darkroom for 1 h at room temperature. The absorbance of the reaction mixture was measured at 765 nm and the results were converted to total polyphenol content using a gallic acid standard calibration curve (ranging from 0 to 200 μg/mL). The results were presented as μg equivalents of gallic acid per gram of the sample (μg GAE/g of raw material) based on fresh weight (FW).

2.3.2. Determination of Total Flavonoids Content (TFC)

The TFC was measured using a modification of the aluminum chloride method developed by Ali, et al. [18]. To perform the assay, 80 µL of the extract, 80 µL of a 2% (*w*/*v*) ethanolic solution of aluminum chloride, and 120 µL of a 50 g/L aqueous solution of sodium acetate were mixed in a 96-well plate. The plate was then incubated for 1 h at 25 °C in a dark room. The reaction mixture was subjected to an absorbance measurement at 440 nm, and the TFC value was determined by converting the absorbance to TFC using the calibration curve prepared with the quercetin standard in a concentration range of 0 to 50 µg/mL. The TPC value was expressed as µg equivalents of quercetin per gram of fresh weight (µg QE/g FW).

2.3.3. Determination of Total Condensed Tannin Content (TCT)

TCT was performed by modifying the method of Ma, et al. [19]. First, 25 µL of the extract and 150 µL of a methanolic vanillin solution (4% *w*/*v*) were mixed in a 96-well plate. Then, 25 µL of 32% sulfuric acid (diluted with methanol) was added. The reaction mixture was incubated in the dark at room temperature for 15 min. The absorbance was measured at 500 nm, and the results were converted to a concentration of tannins (µg CE/g FW) using the calibration curve. The calibration curve was plotted with different concentrations of catechin standards ranging from 0 to 1000 µg/mL.

2.4. *Antioxidant Assays*

2.4.1. 2,2-Diphenyl-1-picrylhydrazyl (DPPH) Antioxidant Assay

The DPPH free radical scavenging capacity was determined by using the method of Wan Mahmood, et al. [20] with some modifications. A concentration of 0.1 mM DPPH methanol solution was used. In addition, 40 µL of the extracts and 260 µL of the DPPH methanol solution were mixed in a 96-well plate. The reaction mixture was incubated for 30 min at room temperature in the dark. The absorbance of the reaction mixture was measured at 517 nm. The results were given as µg of ascorbic acid equivalent per gram of dry weight (mg AAE/g FW) based on the standard curve, which was prepared using ascorbic acid standards with concentrations ranging from 0 to 50 µg/mL.

2.4.2. Ferric Reducing Antioxidant Power (FRAP) Assay

The FRAP of samples was measured by modifying the method of Sogi, et al. [21]. The FRAP reagent was freshly prepared by mixing 300 mM of sodium acetate solution, 10 mM of TPTZ solution (in 40 mM of HCl solution), and 20 mM of $FeCl_3 \cdot 6H_2O$ solution at a ratio of 10:1:1 (*v*:*v*:*v*). To measure the FRAP, 20 µL of the extract was mixed with 280 µL of FRAP reagent and incubated at 37 °C for 10 min. The absorbance of the reaction mixture was then measured at 593 nm. The results were expressed as µg ascorbic acid equivalents per gram of fresh sample weight (µg AAE/g FW) based on the standard curve, which was prepared using the ascorbic acid standard with concentrations ranging from 0 to 50 µg/mL.

2.4.3. 2,2′-Azino-bis-3-ethylbenzothiazoline-6-sulfonic Acid (ABTS) Radical Scavenging Assay

The ABTS radical scavenging capacity was measured according to Queiroz, et al. [22] with some modifications. First, 5 mL of 7 mM ABTS solution and 88 µL of 140 mM potassium persulfate solution were mixed and then the mixture becomes the $ABTS^+$ solution after incubating for 16 h in the dark. The stock $ABTS^+$ solution was further diluted by ethanol until the absorbance of the ABTS cation solution was measured as 0.70 ± 0.02 at 734 nm. In addition, 10 µL of the extract and 290 µL of the ABTS cation solution were mixed in a 96-well plate. After the reaction mixture was incubated at room temperature for 6 min, at the absorbance was measured as 734 nm. The results were expressed as µg equivalents of ascorbic acid per gram of the sample (µg AAE/g FW) based on the standard curve, prepared by the ascorbic acid standard with concentrations ranging from 0 to 200 µg/mL.

2.4.4. Total Antioxidant Capacity (TAC) Assay

The TAC assay has been modified from the method used by Suleria, et al. [23]. First, 260 µL of phosphomolybdate reagent (0.6 M H_2SO_4, 0.028 M sodium phosphate, and 0.004 M ammonium molybdate) was added to each 40 µL sample extract. The mixture was then incubated at 95 °C for 10 min and cooled to 25 °C. The absorbance of the sample solution at 695 nm was measured using a spectrophotometer. Each measurement was repeated three times for each sample. Ascorbic acid (0–200 µg/mL) was used as the standard curve. The TAC of the mushroom samples was expressed as µg equivalents of ascorbic acid per gram of the sample (µg AAE/g FW).

2.5. *LC-ESI-QTOF-MS/MS Characterization of Phenolic Compounds*

The LC-ESI-QTOF-MS/MS assay was carried out by modifying the method of Tang, et al. [24]. The samples were tentatively characterized with an Agilent 1200 HPLC (Agilent Technologies, Santa Clara, CA, USA) equipped with an Agilent 6520 Accurate-Mass Q-TOF LC-MS/MS (Agilent Technologies, CA, USA). Each compound was separated using a Synergi Hydro-RP 80 Å reverse phase column (250 mm × 4.6 mm, 4 µm particle size) with a protected C18ODS (4.0 × 2.0 mm) guard column (Phenomenex, Lane Cove, NSW, Australia). The column was operated at 25 °C and the sample temperature was set at 10 °C. The mobile phase consisted of eluent A (water/acetic acid, 99.5:0.5, *v:v*) and eluent B (acetonitrile/acetic acid/water, 50:49.8:0.2, *v:v:v*). The gradient profile was as follows: 0–10% B (0–5 min), 10–25% B (5–25 min), 25–35% B (25–35 min), 35–40% B (35–45 min), 40–55% B (45–75 min), 55–80% B (75–80 min), 80–90% B (80–82 min), 90–100% B (82–85 min) and isocratic 0% B (85–90 min). Each sample extraction was injected into 6 µL of liquid, and the mobile phase flow rate was 0.8 mL/min. Nitrogen gas atomization was set at 300 °C with a flow rate of 5 L/min at 45 psi, while the sheath gas was set at 11 L/min at 250 °C. The nozzle and capillary voltages were set at 500 V and 3.5 kV, respectively. A complete mass scan ranging from m/z 50 to 1300 was used. Material peaks were identified in positive and negative mode, and processing was performed using LC-ESI-QTOF-MS/MS. MassHunter workstation software (Qualitative Analysis, 152 version B.06.01, Agilent Technologies, Santa Clara, CA, USA) was used for data processing.

2.6. *Statistical Analysis*

All analyses were performed in triplicate. The results of the antioxidant assays and phenolic contents are presented as an average ± standard error (n = 3). One-way analysis of variance (ANOVA) with Tukey's post-hoc test, as provided by Minitab® 19 for Windows (Minitab, NSW, Australia), was used to analyze data and a p-value less than 0.05 indicates a significant difference between samples. Graphs were generated using GraphPad Prism® software (version 9).

3. Results

3.1. *Phenolic Content Estimation (TPC, TFC, TCT)*

Mushrooms are rich in phenolic compounds, such as gallic acid, protocatechuic acid, and cinnamic acid, which have been widely studied as antioxidants [25,26]. Figure 1 shows the results of determining the total polyphenols, total flavonoids, and total tannins in ten different types of mushroom samples.

Ten types of mushrooms were analyzed for their total phenol content, and significant differences were observed. Brown portobello mushroom had the highest content (396.78 ± 3.12 µg GAE/g), followed by shiitake (389.16 ± 5.35 µg GAE/g), Swiss brown mushroom (371.43 ± 3.47 µg GAE/g), and white cup mushroom (370.60 ± 7.14 µg GAE/g). The lowest content was found in organic white mushroom and white flat mushroom (234.57 ± 4.58 and 242.14 ± 5.18 µg GAE/g, respectively). Our results demonstrate that brown portobello and shiitake mushrooms have relatively higher TPC, which aligns with previous studies [27,28]. In the study by Bernaś [27], the TPC for brown portobello mushrooms was reported as 417 mg/100 g DM, which aligns closely with our results. Similarly, the high TPC in shiitake mushrooms is in agreement with the findings by Boonsong et al. [28], who reported a TPC of 24.25 mg GAE/g DW for shiitake mushrooms. However, it should be acknowledged that there are differences in the extraction methods and units of measurement across studies. For instance, the TPC in the cited study was calculated based on dry weight and employed a 50% (v/v) ethanol ratio for extraction. Despite these methodological differences, our results reinforce the idea that brown portobello and shiitake mushrooms are notably rich in phenolic compounds. Indeed, some studies have also shown that the use of different extraction solvents can significantly affect the extraction of polyphenols from mushrooms [29]. On the other hand, the result of the lowest TPC of organic white mushrooms was opposite to the results reported by Popa, et al. [30], who pointed out that organically cultivated foods have higher nutritional value than traditional crops, especially in terms of phenolic compounds, vitamins and minerals. However, there seems to be a lack of research on organic mushrooms in terms of phenolic compounds. Cheung et al. suggested that the total phenol content is closely related to the antioxidant activity [31]. Therefore, the total phenol content plays an important role in the antioxidant properties of mushrooms.

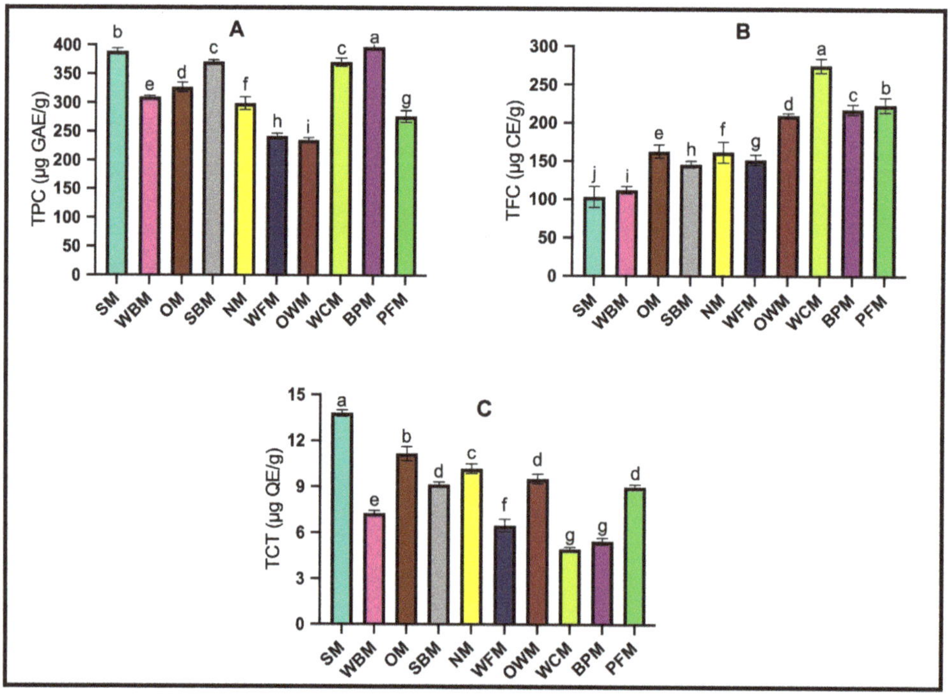

Figure 1. The estimation of phenolic content of ten kinds of mushrooms. (**A**) Total phenolic content (TPC). (**B**) Total flavonoids content (TFC). (**C**) Total condensed tannin content (TCT). The data presented represent the mean ± standard deviation of three replicates from accession. ANOVA and Tukey's test were used to determine the statistically significant difference at $p < 0.05$ as identified by different letters. The codes for the mushroom samples are as follows: SM: shiitake mushroom; WBM: white button mushroom; OM: oyster mushroom; SBM: Swiss brown mushroom; NM: needle mushroom; WFM: white flat mushroom; OWM: organic white mushroom; WCM: white cup mushroom; BPM: brown portobello mushroom; PFM: portobello flat mushroom.

Flavonoids are an important class of secondary metabolites of plant polyphenols. They possess anti-oxidation, anti-inflammatory and anti-cancer properties, and regulate the function of key cell enzymes [32]. As shown in Figure 1, the white cup mushroom had the highest TFC (275.17 ± 9.40 μg CE/g), which was significantly higher than the other nine mushrooms. The portobello flat mushroom (223.88 ± 9.56 μg CE/g), brown portobello mushroom (218.34 ± 6.74 μg CE/g), and organic white mushroom (210.47 ± 3.15 μg CE/g) followed closely behind. In contrast, shiitake mushroom had the lowest total flavonoid content (102.93 ± 13.69 μg CE/g). In a separate study, Buruleanu et al. found that white cap mushroom had the highest TFC with 7.83 ± 4.18 mg QE/g [16]. They also noted that the total flavonoid content varies according to the type of mushroom. Contrastingly, a study by Palacios et al. revealed that *Agaricus bisporus* (white button mushroom) and *Pleurotus ostreatus* (oyster mushroom) had relatively lower flavonoid contents, approximately around 1 mg/g of catechin equivalents [33]. They also observed that the concentration of total flavonoids varies with mushroom species, but the content of total flavonoids does not correlate with the content of phenols. It is speculated that the phenolic compounds contained in mushrooms may vary due to different cultivation substrates and environments.

Tannins are a type of polyphenol compound found widely in plants. They have the ability to bind to and precipitate proteins and other compounds, such as certain amino acids, alkaloids, nucleic acids, and polysaccharides [34]. This study found significant differences in TCT among various mushroom samples. Contrary to the TFC results, shiitake mushrooms had the highest TCT (13.80 ± 0.21 μg QE/g), while white cup mushrooms had the lowest (4.95 ± 0.14 μg QE/g). Following shiitake mushrooms were oyster mushrooms (11.15 ± 0.47 μg QE/g) and needle mushrooms (10.18 ± 0.31 μg QE/g). Organic white mushrooms, Swiss brown mushrooms, and portobello flat mushrooms had no significant differences in their tannin content, which were 9.53 ± 0.31, 9.14 ± 0.17, and 9.01 ± 0.16 μg QE/g, respectively. The TCT of mushrooms has been measured in many studies. For example, the study by Sifat, et al. [35] found that the TCT of oyster mushrooms was 36–40 mg TAE/g. Additionally, a new mushroom species called *Rubroboletus himalayensis* sp. nov. (Boletaceae, Boletales, Basidiomycota), found in the Himalayas of Pakistan, showed high TCT in methanol extraction with 441.0 mg TAE/g [36]. However, no additional research data have been found for the mushroom species studied in this study. Further research is needed to obtain more data on the TCT of mushrooms.

3.2. Antioxidant Activity (DPPH, FRAP, ABTS and TAC)

Based on the estimated results above, it can be concluded that mushrooms are an important source of phenolic active substances. To evaluate the antioxidant capacity of the mushroom samples, this study used DPPH, FRAP, ABTS, and TAC assay methods. The results are shown in Figure 2 and are expressed as μg ascorbic acid/g FW (μg AAE/g FW).

One of the most important functions of antioxidants is to scavenge free radicals. In the current study, the potential radical scavenging activity of the studied mushroom samples was determined using the DPPH method. The research results showed that the white cup mushroom sample had a significantly higher free radical scavenging ability than the other mushroom samples ($p < 0.05$), with a value as high as 730.14 ± 55.06 μg AAE/g. The white button mushroom and portobello flat mushroom had similar results, with values of 309.68 ± 7.48 μg AAE/g and 309.04 ± 42.15 μg AAE/g, respectively. These findings contrast with those of previous studies conducted by Buruleanu et al. [16] and Bach et al. [37]. Both studies indicated that brown *Agaricus bisporus,* especially portobello mushrooms, have a greater ability to scavenge free radicals than white species. Additionally, the DPPH scavenging levels of other mushrooms such as shiitake mushrooms were lower than those of *Agaricus bisporus*. Considering various factors, such as the choice of extraction solvent and the different growth environments of mushroom samples, the reason for these differences may vary. However, in general, white cup mushrooms demonstrate strong DPPH free radical scavenging activity.

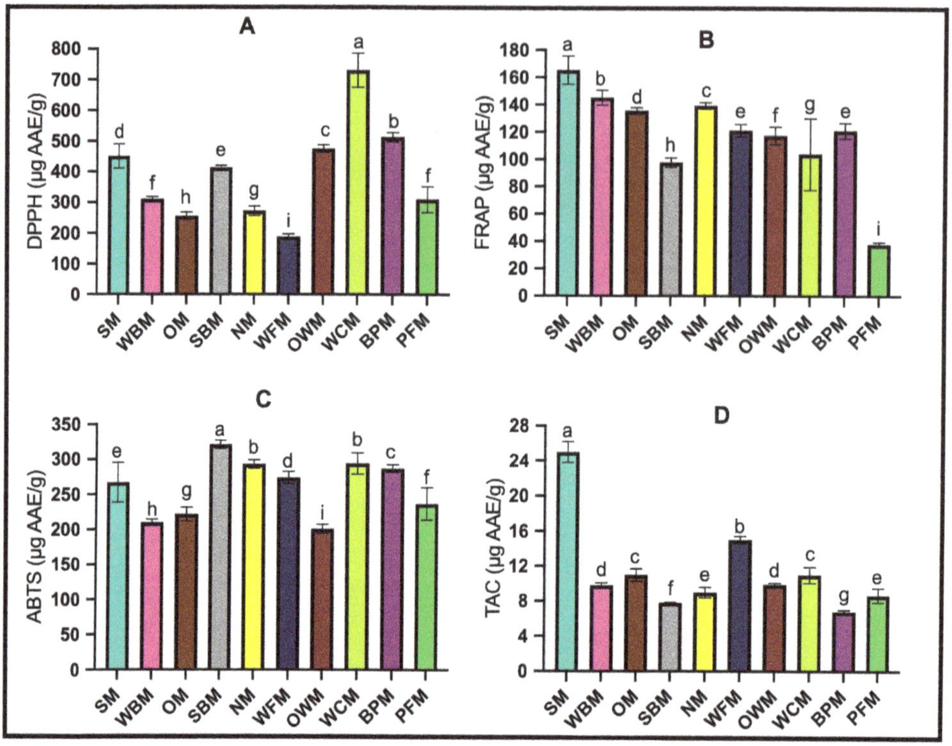

Figure 2. The estimation of antioxidant potential of ten kinds of mushrooms. (**A**) 2,2-diphenyl-1-picrylhydrazyl (DPPH). (**B**) Ferric reducing antioxidant power (FRAP). (**C**) 2,2′-azino-bis-3-ethylbenzothiazoline-6-sulfonic acid (ABTS). (**D**) Total antioxidant capacity (TAC). The data presented represent the mean ± standard deviation of three replicates from accession. ANOVA and Tukey's test were used to determine the statistically significant difference at $p < 0.05$ as identified by different letters. The codes for the mushroom samples are as follows: SM: shiitake mushroom; WBM: white button mushroom; OM: oyster mushroom; SBM: Swiss brown mushroom; NM: needle mushroom; WFM: white flat mushroom; OWM: organic white mushroom; WCM: white cup mushroom; BPM: brown portobello mushroom; PFM: portobello flat mushroom.

The results of FRAP indicate that shiitake mushrooms exhibited the strongest iron-reducing ability (165.32 ± 10.21 µg AAE/g), while the reducing ability of white cup mushrooms was significantly lower than that of other mushroom samples (97.84 ± 3.45 µg AAE/g). These results differed markedly from those of other studies. Sharpe, et al. [38] found that the FRAP value of shiitake mushrooms was lower than that of the other mushroom varieties they examined. Islam, et al. [39] confirmed this point as well. They noted that among the 43 selected Chinese mushrooms, shiitake mushrooms did not demonstrate the strongest antioxidant properties. The FRAP assay measures the ability of an antioxidant in a sample to reduce the Fe^{3+} complex to the Fe^{2+} complex, thus determining the sample's antioxidant capacity. In general, researchers believe that in most cases, the results of antioxidant capacity determination are consistent with the value of TPC [40]. In our study, we found that the portobello flat mushroom has a TPC value of 276.98 ± 10.17 µg AAE/g, but it has the lowest reducibility. On the other hand, the FRAP results of white flat mushrooms, brown portobello mushrooms, and organic white mushrooms are consistent with the TPC value. Further research is needed to understand the causes of these and similar conditions.

ABTS free radical scavenging is a sensitive test method used to evaluate the free radical scavenging ability of phenolic compounds [39]. Antioxidants donate electrons or hydrogen atoms to inactivate $ABTS^+$ ions, causing a color change. According to the ABTS test results, the strongest antioxidant capacity is demonstrated by the Swiss brown mushroom (321.31 ± 5.7 μg AAE/g), followed by white cup mushroom (293.88 ± 14.61 μg AAE/g) and needle mushroom (292.90 ± 6.08 μg AAE/g), and the weakest is the organic white mushroom (201.47 ± 6.57 μg AAE/g). Bach et al. [37] found a strong correlation between the ABTS free radical scavenging activity and the TPC in mushrooms, confirming that the TPC has a direct impact on the antioxidant activity of mushroom samples.

The total antioxidant capacity (TAC) in mushroom samples is a comprehensive parameter that measures the cumulative effect of all antioxidants in the sample, rather than the simple sum of various antioxidants [41]. The results of the TAC assay in this study indicate that shiitake mushrooms have a significantly higher TAC value (24.52 ± 1.2 μg AAE/g) compared to other mushroom varieties. The TAC values of portobello flat mushroom, Swiss brown mushroom, and brown portobello mushroom were found to be similar, at 8.70 ± 0.78, 7.84 ± 0.1, and 6.78 ± 0.17 μg AAE/g, respectively. Some studies suggest that this may be due to the fact that the stems of shiitake mushrooms are rich in true chitosan, which is the source of potential antioxidant properties of shiitake mushrooms [42].

3.3. Tentative Phenolic Characterization by LC-ESI-QTOF-MS/MS

This study used LC-ESI-QTOF-MS/MS to tentatively characterize phenolic compounds in ten mushroom samples. The isolated compounds were tentatively identified using Agilent LC-ESI-QTOF-MS/MS Mass Hunter Qualitative Software and the Personal Compound Database and Library (PCDL), by considering the MS spectra of m/z in the positive and negative ion mode, and retention time (RT). Compounds with screening scores greater than 80 (PCDL scores) and mass errors of ±5 ppm were listed in Table 1 for tentative characterization and m/z validation. Overall, 22 phenolic compounds were identified in all mushroom samples, including 11 flavonoids, 3 phenolic acids, 4 lignans, 2 stilbenes, and 2 other phenolic compounds.

3.3.1. Phenolic Acids

In the studied mushroom samples, three different phenolic acids were tentatively characterized, belonging to two subclasses of phenolic acids, hydroxycinnamic acids and hydroxyphenylpropanoic acids, respectively. Two compounds, cinnamic acid and *p*-coumaroyl malic acid, were tentatively identified in the hydroxycinnamic acids subclass. Cinnamic acid was detected in a variety of mushroom samples, including shiitake mushroom, needle mushroom, white flat mushroom, brown portobello mushroom, portobello flat mushroom, organic white mushroom and oyster mushroom, under the following conditions: retention time RT = 4.53 min, and negative ion mode m/z = 147.0463. In a study to identify phenolic compounds in 26 mushroom species, cinnamic acid was detected in the majority of mushroom samples [43]. However, *p*-coumaroyl malic acid, with its unique retention time of 3.65 min and negative ion mode ($[M-H]^-$ m/z = 279.0511), was mainly identified in white flat mushroom, organic white mushroom, portobello flat mushroom, and shiitake mushroom. This discovery is particularly innovative, as *p*-coumaroyl malic acid has not been previously reported in mushroom studies, but was noted in the North American herb, *Echinacea purpurea* (L.) Moench [44]. Compound 1 has a negative ion mode at m/z = 357.0837 and was tentatively identified as dihydrocaffeic acid 3-*O*-glucuronide, which was previously detected in hops and juniper berries [24].

Table 1. Liquid chromatography–electrospray ionization quadrupole time-of-flight mass spectrometry (LC-ESI-QTOF-MS/MS) and tentative characterization of mushroom polyphenolic compounds.

No.	Proposed Compound	Molecular Formula	RT (min)	Ionization Mode	Molecular Weight	Theoretical (*m/z*)	Observed (*m/z*)	Mass Error	Sample
				Phenolic acids					
				Hydroxyphenylpropanoic acids					
1	Dihydrocaffeic acid 3-*O*-glucuronide	$C_{15}H_{18}O_{10}$	4.16	** $[M-H]^-$	358.0910	357.0837	357.0837	0	* WBM, SM, WFM
				Hydroxycinnamic acids					
2 ***	*p*-Coumaroyl malic acid	$C_{13}H_{12}O_7$	3.65	$[M-H]^-$	280.0586	279.0513	279.0511	−0.7	* WFM, OWM, PFM, SM
3	Cinnamic acid	$C_9H_8O_2$	4.53	$[M-H]^-$	148.0538	147.0465	147.0463	−1.4	* SM, NM, WFM, BPM, PFM, OWM, OM
				Flavonoids					
				Isoflavonoids					
4	Dalbergin	$C_{16}H_{12}O_4$	3.88	$[M-H]^-$	268.0734	267.0661	267.0666	1.9	* SM, PFM, NM
5	2′,7-Dihydroxy-4′,5′-dimethoxyisoflavone	$C_{17}H_{14}O_6$	3.83	** $[M+H]^+$	314.0787	315.0860	315.0860	0	* BPM, NM, WBM
6	6″-*O*-Acetyldaidzin	$C_{23}H_{22}O_{10}$	3.96	$[M-H]^-$	458.1202	457.1129	457.1151	4.8	SM
				Flavonols					
7	Kaempferol 7-*O*-glucoside	$C_{21}H_{19}O_{11}$	3.35	$[M-H]^-$	447.0910	446.0837	446.0858	4.7	WBM
				Flavanones					
8	Eriocitrin	$C_{27}H_{32}O_{15}$	54.24	$[M-H]^-$	596.1713	595.1640	595.1638	−0.3	* WFM, OM
				Flavanols					
9	(+)-Catechin 3-*O*-gallate	$C_{22}H_{18}O_{10}$	3.72	$[M-H]^-$	442.0857	441.0784	441.0783	−0.2	* WBM, OWM, PFM
10	4′-*O*-Methyl-(-)-epigallocatechin 7-*O*-glucuronide	$C_{22}H_{24}O_{13}$	54.71	$[M-H]^-$	496.1233	495.1160	495.1164	0.8	* WCM, WFM
				Dihydroflavonols					
11	Dihydroquercetin 3-*O*-rhamnoside	$C_{21}H_{22}O_{11}$	3.61	$[M-H]^-$	450.1144	449.1071	449.1069	−0.4	* SM, WFM

Table 1. *Cont.*

No.	Proposed Compound	Molecular Formula	RT (min)	Ionization Mode	Molecular Weight	Theoretical (*m*/*z*)	Observed (*m*/*z*)	Mass Error	Sample
12	3-Hydroxyphloretin 2′-*O*-xylosyl-glucoside	$C_{26}H_{32}O_{15}$	59.78	$[M-H]^-$	584.1766	583.1693	583.1692	−0.2	BPM
				Anthocyanins					
13	Cyanidin 3-*O*-(6″-acetyl-glucoside)	$C_{23}H_{23}O_{12}$	3.06	$[M-H]^-$	491.1176	490.1103	490.1102	−0.2	* SBM, NM, SM, PFM, WCM, WBM, WFM, OWM, OM
14 ***	Petunidin 3-*O*-(6″-acetyl-glucoside)	$C_{24}H_{25}O_{13}$	3.80	$[M+H]^+$	521.1297	522.1370	522.1396	5.0	PFM
				Lignans					
15	Todolactol A	$C_{20}H_{24}O_7$	4.43	$[M-H]^-$	376.1519	375.1446	375.1435	−2.9	NM
16	Schisandrin C	$C_{22}H_{24}O_6$	41.09	** $[M+H]^+$	384.1540	385.1613	385.1607	−1.6	* OM
17	Deoxyschisandrin	$C_{24}H_{32}O_6$	3.11	** $[M-H]^-$	416.2196	415.2123	415.2133	2.4	* BPM, SM, WCM
18	Schisandrin	$C_{24}H_{32}O_7$	4.84	** $[M+H]^+$	432.2154	433.2227	433.2222	−1.2	* SM, NM
				Stilbenes					
19	Trans-Resveratrol	$C_{14}H_{12}O_3$	4.58	$[M-H]^-$	228.0780	227.0707	227.0706	−0.4	* WCM, WBM, OWM, NM, WFM
20	Resveratrol 5-*O*-glucoside	$C_{20}H_{22}O_8$	3.81	$[M-H]^-$	390.1311	389.1238	389.1240	0.5	NM
				Other polyphenols					
				Tyrosols					
21	3,4-DHPEA-EDA	$C_{17}H_{20}O_6$	4.87	$[M-H]^-$	320.1276	319.1203	319.1195	−2.5	* PFM, WFM
				Phenolic terpenes					
22	Carnosic acid	$C_{20}H_{28}O_4$	57.83	** $[M-H]^-$	332.1997	331.1924	331.1923	−0.3	* SM, NM, WCM

* The same phenolic compound was detected in more than one mushroom sample and the data shown in this table are from the asterisked samples. ** Compounds were detected in both negative $[M-H]^-$ and positive ion $[M+H]^+$ modes and only single mode data are provided in this table. *** Compounds identified for the first time in this study. As shown in the last column of the table, the codes for the mushroom samples are as follows: BPM: brown portobello mushroom; SM: shiitake mushroom; WBM: white button mushroom; OWM: organic white mushroom; WFM: white flat mushroom; WCM: white cup mushroom; PFM: portobello flat mushroom; SBM: Swiss brown mushroom; OM: oyster mushroom; NM: needle mushroom.

3.3.2. Flavonoids

Flavonoids are the main class of phenolic compounds. In this study, six flavonoid subclasses were tentatively characterized, including anthocyanins, dihydrochalcones, flavanols, flavones, isoflavonoids. Of these, isoflavonoids detected three compounds, flavonols and dihydrochalcones each detected two, and the remaining subclasses each detected one.

Isoflavonoids

Dalbergin was preliminarily characterized in the extracts of shiitake mushroom, portobello flat mushroom and needle mushroom. The compound showed $[M-H]^-$ at an m/z of 267.0666 and the chemical formula was $C_{16}H_{12}O_4$. Dalbergin is a natural flavonoid isolated from *Dalbergia sissoo*, *Machaerium* spp., *Oxytrops falcate*, *Dalbergia odorifera* and some other plants [45]. Compound 5 was tentatively characterized as 2′,7-dihydroxy-4′,5′-dimethoxyisoflavone under the condition of m/z of 315.0860 in the positive ion mode and a retention time of 3.833 min. The compound was detected in brown portobello mushrooms, needle mushrooms and white button mushrooms. It has also been detected in *Lepidium sativum*, as reported by Kadam, et al. [46]. In addition, compound 6 with $[M-H]^-$ at m/z = 457.1151 was identified as 6″-*O*-acetyldaidzin, which was only detected in shiitake mushroom. 6″-*O*-acetyldaidzin is an ortho-glycosylated derivative of isoflavonoids, which was detected in some soybeans and their products by LC-MS [47].

Flavanones and Flavonols

In the $[M-H]^-$ mode, flavonols (kaempferol 7-*O*-glucoside) was tentatively detected in white button mushroom, and flavanones (eriocitrin) were detected om white flat mushroom and oyster mushroom. Kaempferol 7-*O*-glucoside and eriocitrin were characterized at m/z = 446.0858 and 595.1638, with retention times of 3.35 min and 54.24 min, respectively. Research has shown that kaempferol 7-*O*-glucosinolates, which can be isolated from herbaceous plants, have effective anti-HSV-1 activity [48]. Eriocitrin is a disaccharide derivative of eriodictyol. It was previously isolated and identified in lemon peel and pulp vesicles using HPLC, ^{1}H-NMR and ^{13}C-NMR analyses [49].

Flavanols and Dihydroflavonols

Catechin is a flavanol commonly found in fruits and vegetables. Two catechin derivatives were identified in mushroom samples. Compound 9 was initially characterized as (+)-catechin 3-*O*-gallate with a retention time of 3.72 min and $[M-H]^-$ mode at m/z = 441.0783, which was found in white button mushroom, organic white mushroom, and portobello flat mushroom. Compound 10 was tentatively characterized as 4′-*O*-methyl-(-)-epigallocatechin 7-*O*-glucuronide with a retention time of 54.71 min and $[M-H]^-$ mode at m/z = 495.1164. It was detected in white cup mushroom and white flat mushroom. These two compounds were also identified by Peng, et al. [50] in their study of phenolic compounds in Australian mangoes (*Mangifera indica* L.). In addition, two compounds, dihydroquercetin 3-*O*-rhamnoside and 3-hydroxyphloretin 2′-*O*-xylosyl-glucoside, were preliminarily characterized in mushroom samples. Dihydroquercetin 3-*O*-rhamnoside was present in shiitake mushroom and white flat mushroom, while 3-hydroxyphloretin 2′-*O*-xylosyl-glucoside was only detected in brown portobello mushroom. Both compounds showed a negative ion mode at m/z of 449.1069 and 583.1692, respectively. Suprun, et al. [51] reported that dihydroquercetin 3-*O*-rhamnoside is the main dihydroflavonol in white wine, while Ramirez-Ambrosi et al. [5] identified 3-hydroxyphloretin 2′-*O*-xylosyl-glucoside in apple products.

Anthocyanins

Two anthocyanins were tentatively identified in ten mushroom samples, both belonging to anthocyanin cations (compounds 13 and 14). Compound 13, cyanidin 3-*O*-(6″-acetyl-glucoside), exhibits the $[M-H]^-$ mode at m/z of 490.1102 and is detected in almost all mushroom samples, except brown portobello mushroom. Previous research has shown that it is found in multiple species of *Zinnia elegans* [52]. Remarkably, compound 14, identified as petunidin 3-*O*-(6″-acetyl-glucoside), was uniquely characterized in the portobello flat mushroom, displaying an $[M\text{-}H]^+$ mode at an m/z of 522.1396. This is an innovative discovery, as this specific compound has not been previously reported in mushrooms according to the existing literature. Therefore, our study contributes novel insights to the field, augmenting the knowledge of phenolic compounds present in different varieties of mushrooms.

3.3.3. Lignans

In this study, four lignans (compounds 15–18) from ten mushroom samples were tentatively characterized, and compounds 16–18 were detected in both $[M\text{-}H]^-$ and $[M\text{-}H]^+$ modes. Schisandrin C was detected in oyster mushroom as compound 16 with $[M\text{-}H]^-$ at m/z 383.1501 and $[M+H]^+$ at m/z 385.1607. Compound 17 was detected in the $[M+H]^+$ mode at m/z 417.2290 and 417.2282, and $[M\text{-}H]^-$ at m/z 415.2133. It was preliminarily characterized as deoxyschisandrin and detected in brown portobello mushroom, shiitake mushroom and white cup mushroom, while compound 18 was detected in the $[M\text{-}H]^+$ mode at m/z 433.2222 and $[M+H]^-$ at m/z 431.2046 and 431.2100. It was preliminarily characterized as schisandrin and detected in shiitake mushroom and needle mushroom. Schisandrin, schisandrin C and deoxyschisandrin are anti-inflammatory active compounds isolated from the fruit of *Schisandra chinensis* Baill. [53,54]. Compound 15 with $[M\text{-}H]^-$ at m/z 375.1435 was detected in needle mushroom samples and preliminarily characterized as todolactol A. This compound has been previously identified in Norway spruce by Piispanen et al. [55].

3.3.4. Stilbenes and Other Polyphenols

The stilbenes composition identified only two compounds in negative ionization mode, trans-resveratrol (m/z = 227.0706) and resveratrol 5-*O*-glucoside (m/z = 389.1240). Resveratrol is a phenolic compound with various biological activities. Kang, et al. [56] reported that they had cultivated transgenic enoki mushrooms with resveratrol-producing ability. Beekwilder, et al. [57] used transgenic *Saccharomyces cerevisiae* to produce trans-resveratrol in culture. In addition, phenolics were characterized by using UHPLC-ESI/QTOF-MS on *Calligonum azel* Maire, a Tunisian desert plant, and resveratrol 5-*O*-glucoside was detected in this plant [58]. On the other hand, polyphenolic compound 21 exhibited $[M\text{-}H]^-$ at m/z 319.1195 and was preliminarily characterized as 3,4-DHPEA-EDA. Compound 22 also exhibited $[M\text{-}H]^-$ at m/z 331.1923 and was initially identified as carnosic acid. The mushroom samples containing 3,4-DHPEA-EDA were portobello flat mushroom and white flat mushroom, while the mushroom samples containing carnosic acid were shiitake mushroom, needle mushroom, and white cup mushroom. According to Akazawa, et al. [59], 3,4-DHPEA-EDA, the main phenolic compound of olive leaves was detected in cold-water extracts of the leaves, in the form of an aglycone derivative. Carnosic acid is a diterpene compound commonly found in sage and rosemary [60], which has not been previously reported in mushroom species.

4. Conclusions

This study provides a comprehensive analysis of the antioxidant activities, phenolic content, and LC-ESI-QTOF-MS/MS-based characterization of phenolic compounds in ten distinct mushroom varieties. The analysis revealed significant differences in the total polyphenols, flavonoids, and tannins among the tested mushroom varieties, as well as their antioxidant activities. The strength of this study lies in its comprehensive assessment of the phenolic compounds and antioxidant activities in various mushroom samples, contributing to the robustness of the findings and expanding our knowledge of the phenolic content in mushrooms. In addition, 22 phenolic compounds, including flavonoids, phenolic acids, lignans, stilbenes, and other phenolic compounds, some of which have not been previously reported in the literature, have been tentatively identified. Particularly noteworthy in underlining the innovative aspect of this research is the discovery of previously unreported compounds, including *p*-coumaroyl malic acid and petunidin 3-*O*-(6″-acetyl-glucoside), in a diverse selection of mushroom varieties such as organic white mushrooms, thereby filling a significant gap in the existing literature. The findings of this study not only inform consumers, researchers, and industry professionals about the potential health benefits of mushrooms but also pave the way for future research on their potential therapeutic applications or as functional ingredients in food products. However, it is essential to consider the influence of various factors, such as extraction solvents, growth environments, and cultivation substrates, on the phenolic content and antioxidant activities of mushrooms. Further research on mushrooms in terms of phenolic compounds is needed to better understand these effects. Overall, this study highlights the importance of mushrooms as a source of health-promoting phenolic compounds and antioxidants, which can contribute to the development of functional foods and nutraceuticals.

Author Contributions: Conceptualization, methodology, formal analysis, validation and investigation, M.C., Y.Z., R.D.K., O.T.A. and H.A.R.S.; resources, H.A.R.S., C.J.B. and F.R.D.; writing—original draft preparation, M.C., Y.Z., R.D.K., O.T.A. and H.A.R.S.; writing—review and editing, M.C., Y.Z., O.T.A. and H.A.R.S.; supervision, H.A.R.S., O.T.A., C.J.B. and F.R.D.; funding acquisition, H.A.R.S. All authors have read and agreed to the published version of the manuscript.

Funding: Hafiz Suleria is the recipient of an Australian Research Council—Discovery Early Career Award (ARC-DECRA—DE220100055) funded by the Australian Government. This research was funded by the University of Melbourne under the McKenzie Fellowship Scheme (grant no. UoM-18/21), the Future Food Hallmark Research Initiative Funds (grant no. UoM-21/23) and Collaborative Research Development Grant (grant no. UoM-21/23) funded by the Faculty of Veterinary and Agricultural Sciences, the University of Melbourne, Australia.

Data Availability Statement: Data are available from the corresponding authors upon request.

Acknowledgments: We would like to thank researchers of the Hafiz Suleria group from the School of Agriculture, Food and Ecosystem Sciences, Faculty of Science, the University of Melbourne for their incredible supports.

Conflicts of Interest: The authors declare no conflict of interest.

References

1. Expert Market Research. Global Mushroom Market Report and Forecast 2023–2028. 2023. Available online: https://www.expertmarketresearch.com/reports/mushroom-market (accessed on 27 May 2023).
2. Rizzo, G.; Goggi, S.; Giampieri, F.; Baroni, L. A review of mushrooms in human nutrition and health. *Trends Food Sci. Technol.* **2021**, *117*, 60–73. [CrossRef]
3. Ba, D.M.; Gao, X.; Al-Shaar, L.; Muscat, J.; Chinchilli, V.M.; Ssentongo, P.; Zhang, X.; Liu, G.; Beelman, R.B.; Richie, J.P. Prospective study of dietary mushroom intake and risk of mortality: Results from continuous National Health and Nutrition Examination Survey (NHANES) 2003–2014 and a meta-analysis. *Nutr. J.* **2021**, *20*, 80. [CrossRef]
4. Kalaras, M.D.; Beelman, R.B.; Elias, R.J. Effects of Postharvest Pulsed UV Light Treatment of White Button Mushrooms (*Agaricus bisporus*) on Vitamin D2 Content and Quality Attributes. *J. Agric. Food Chem.* **2012**, *60*, 220–225. [CrossRef]
5. Ramirez-Ambrosi, M.; Abad-Garcia, B.; Viloria-Bernal, M.; Garmon-Lobato, S.; Berrueta, L.A.; Gallo, B. A new ultrahigh performance liquid chromatography with diode array detection coupled to electrospray ionization and quadrupole time-of-flight mass spectrometry analytical strategy for fast analysis and improved characterization of phenolic compounds in apple products. *J. Chromatogr. A* **2013**, *1316*, 78–91. [CrossRef] [PubMed]
6. Xiaokang, W.; Brunton, N.P.; Lyng, J.G.; Harrison, S.M.; Carpes, S.T.; Papoutsis, K. Volatile and non-volatile compounds of shiitake mushrooms treated with pulsed light after twenty-four hour storage at different conditions. *Food Biosci.* **2020**, *36*, 100619. [CrossRef]
7. Barros, L.; Duenas, M.; Ferreira, I.C.F.R.; Baptista, P.; Santos-Buelga, C. Phenolic acids determination by HPLC-DAD-ESI/MS in sixteen different Portuguese wild mushrooms species. *Food Chem. Toxicol.* **2009**, *47*, 1076–1079. [CrossRef]
8. Manach, C.; Scalbert, A.; Morand, C.; Rémésy, C.; Jiménez, L. Polyphenols: Food sources and bioavailability. *Am. J. Clin. Nutr.* **2004**, *79*, 727–747. [CrossRef]
9. Ziaullah; Rupasinghe, H.P.V. Chapter 1—Application of NMR Spectroscopy in Plant Polyphenols Associated with Human Health. In *Applications of NMR Spectroscopy*; ur-Rahman, A., Choudhary, M.I., Eds.; Bentham Science Publishers: Sharjah, United Arab Emirates, 2015; pp. 3–92.
10. Decker, E.A. Phenolics: Prooxidants or Antioxidants? *Nutr. Rev.* **1997**, *55*, 396–398. [CrossRef] [PubMed]

11. Wojtunik-Kulesza, K.; Oniszczuk, A.; Oniszczuk, T.; Combrzyński, M.; Nowakowska, D.; Matwijczuk, A. Influence of In Vitro Digestion on Composition, Bioaccessibility and Antioxidant Activity of Food Polyphenols—A Non-Systematic Review. *Nutrients* **2020**, *12*, 1401. [CrossRef] [PubMed]
12. Gogoi, P.; Chutia, P.; Singh, P.; Mahanta, C.L. Effect of optimized ultrasound-assisted aqueous and ethanolic extraction of *Pleurotus citrinopileatus* mushroom on total phenol, flavonoids and antioxidant properties. *J. Food Process Eng.* **2019**, *42*, e13172. [CrossRef]
13. Hamill, L.L.; McRoberts, W.C.; Floyd, S.D.; McKinley, M.C.; Young, I.S.; Woodside, J.V. Analyses of a polyphenol aglycone profile in broccoli and carrots by LC-MS QToF. *Proc. Nutr. Soc.* **2012**, *71*, E93. [CrossRef]
14. Fang, N.; Yu, S.; Prior, R.L. LC/MS/MS Characterization of Phenolic Constituents in Dried Plums. *J. Agric. Food Chem.* **2002**, *50*, 3579–3585. [CrossRef] [PubMed]
15. Rahimi Khoigani, S.; Rajaei, A.; Goli, S.A.H. Evaluation of antioxidant activity, total phenolics, total flavonoids and LC–MS/MS characterisation of phenolic constituents in *Stachys lavandulifolia*. *Nat. Prod. Res.* **2017**, *31*, 355–358. [CrossRef] [PubMed]
16. Buruleanu, L.C.; Radulescu, C.; Georgescu, A.A.; Danet, F.A.; Olteanu, R.L.; Nicolescu, C.M.; Dulama, I.D. Statistical Characterization of the Phytochemical Characteristics of Edible Mushroom Extracts. *Anal. Lett.* **2018**, *51*, 1039–1059. [CrossRef]
17. Stojanova, M.; Pantić, M.; Karadelev, M.; Čuleva, B.; Nikšić, M. Antioxidant potential of extracts of three mushroom species collected from the Republic of North Macedonia. *J. Food Process. Preserv.* **2021**, *45*, e15155. [CrossRef]
18. Ali, A.; Wu, H.; Ponnampalam, E.N.; Cottrell, J.J.; Dunshea, F.R.; Suleria, H.A.R. Comprehensive Profiling of Most Widely Used Spices for Their Phenolic Compounds through LC-ESI-QTOF-MS2 and Their Antioxidant Potential. *Antioxidants* **2021**, *10*, 721. [CrossRef]
19. Ma, C.; Dunshea, F.R.; Suleria, H.A.R. LC-ESI-QTOF/MS Characterization of Phenolic Compounds in Palm Fruits (Jelly and Fishtail Palm) and Their Potential Antioxidant Activities. *Antioxidants* **2019**, *8*, 483. [CrossRef]
20. Wan Mahmood, W.M.A.; Lorwirachsutee, A.; Theodoropoulos, C.; Gonzalez-Miquel, M. Polyol-Based Deep Eutectic Solvents for Extraction of Natural Polyphenolic Antioxidants from *Chlorella vulgaris*. *ACS Sustain. Chem. Eng.* **2019**, *7*, 5018–5026. [CrossRef]
21. Sogi, D.S.; Siddiq, M.; Greiby, I.; Dolan, K.D. Total phenolics, antioxidant activity, and functional properties of 'Tommy Atkins' mango peel and kernel as affected by drying methods. *Food Chem.* **2013**, *141*, 2649–2655. [CrossRef]
22. Queiroz, M.; Oppolzer, D.; Gouvinhas, I.; Silva, A.M.; Barros, A.I.R.N.A.; Domínguez-Perles, R. New grape stems' isolated phenolic compounds modulate reactive oxygen species, glutathione, and lipid peroxidation in vitro: Combined formulations with vitamins C and E. *Fitoterapia* **2017**, *120*, 146–157. [CrossRef]
23. Suleria, H.A.R.; Barrow, C.J.; Dunshea, F.R. Screening and Characterization of Phenolic Compounds and Their Antioxidant Capacity in Different Fruit Peels. *Foods* **2020**, *9*, 1209. [CrossRef]
24. Tang, J.; Dunshea, F.R.; Suleria, H.A.R. LC-ESI-QTOF/MS Characterization of Phenolic Compounds from Medicinal Plants (Hops and Juniper Berries) and Their Antioxidant Activity. *Foods* **2020**, *9*, 7. [CrossRef]
25. Barros, L.; Ferreira, M.-J.; Queirós, B.; Ferreira, I.C.F.R.; Baptista, P. Total phenols, ascorbic acid, β-carotene and lycopene in Portuguese wild edible mushrooms and their antioxidant activities. *Food Chem.* **2007**, *103*, 413–419. [CrossRef]
26. Selli, S.; Guclu, G.; Sevindik, O.; Kelebek, H. Variations in the key aroma and phenolic compounds of champignon (*Agaricus bisporus)* and oyster (*Pleurotus ostreatus*) mushrooms after two cooking treatments as elucidated by GC–MS-O and LC-DAD-ESI-MS/MS. *Food Chem.* **2021**, *354*, 129576. [CrossRef]
27. Bernaś, E. Comparison of the mechanism of enzymatic browning in frozen white and brown *A. bisporus*. *Eur. Food Res. Technol.* **2018**, *244*, 1239–1248. [CrossRef]
28. Boonsong, S.; Klaypradit, W.; Wilaipun, P. Antioxidant activities of extracts from five edible mushrooms using different extractants. *Agric. Nat. Resour.* **2016**, *50*, 89–97. [CrossRef]
29. Xiaokang, W.; Lyng, J.G.; Brunton, N.P.; Cody, L.; Jacquier, J.-C.; Harrison, S.M.; Papoutsis, K. Monitoring the effect of different microwave extraction parameters on the recovery of polyphenols from shiitake mushrooms: Comparison with hot-water and organic-solvent extractions. *Biotechnol. Rep.* **2020**, *27*, e00504. [CrossRef]
30. Popa, M.E.; Mitelut, A.C.; Popa, E.E.; Stan, A.; Popa, V.I. Organic foods contribution to nutritional quality and value. *Trends Food Sci. Technol.* **2019**, *84*, 15–18. [CrossRef]
31. Cheung, L.M.; Cheung, P.C.K.; Ooi, V.E.C. Antioxidant activity and total phenolics of edible mushroom extracts. *Food Chem.* **2003**, *81*, 249–255. [CrossRef]
32. Panche, A.N.; Diwan, A.D.; Chandra, S.R. Flavonoids: An overview. *J. Nutr. Sci.* **2016**, *5*, e47. [CrossRef]
33. Palacios, I.; Lozano, M.; Moro, C.; D'Arrigo, M.; Rostagno, M.A.; Martínez, J.A.; García-Lafuente, A.; Guillamón, E.; Villares, A. Antioxidant properties of phenolic compounds occurring in edible mushrooms. *Food Chem.* **2011**, *128*, 674–678. [CrossRef]
34. Abdelshafy, A.M.; Belwal, T.; Liang, Z.; Wang, L.; Li, D.; Luo, Z.; Li, L. A comprehensive review on phenolic compounds from edible mushrooms: Occurrence, biological activity, application and future prospective. *Crit. Rev. Food Sci. Nutr.* **2021**, *62*, 6204–6224. [CrossRef]
35. Sifat, N.; Lovely, F.; Zihad, S.M.N.K.; Hossain, M.G.; Shilpi, J.A.; Grice, I.D.; Mubarak, M.S.; Uddin, S.J. Investigation of the nutritional value and antioxidant activities of common *Bangladeshi edible* mushrooms. *Clin. Phytosci.* **2020**, *6*, 88. [CrossRef]
36. Sarwar, S.; Siddique, Z.E.B.; Bashir, A.; Khalid, A.N. Rubroboletus himalayensis sarwar & khalid—A new mushroom from Pakistan. *Bangladesh J. Plant Taxon.* **2021**, *28*, 17–26. [CrossRef]
37. Bach, F.; Zielinski, A.A.F.; Helm, C.V.; Maciel, G.M.; Pedro, A.C.; Stafussa, A.P.; Ávila, S.; Haminiuk, C.W.I. Bio compounds of edible mushrooms: In vitro antioxidant and antimicrobial activities. *LWT—Food Sci. Technol.* **2019**, *107*, 214–220. [CrossRef]

38. Sharpe, E.; Farragher-Gnadt, A.P.; Igbanugo, M.; Huber, T.; Michelotti, J.C.; Milenkowic, A.; Ludlam, S.; Walker, M.; Hanes, D.; Bradley, R.; et al. Comparison of antioxidant activity and extraction techniques for commercially and laboratory prepared extracts from six mushroom species. *J. Agric. Food Res.* **2021**, *4*, 100130. [CrossRef]
39. Islam, T.; Yu, X.; Xu, B. Phenolic profiles, antioxidant capacities and metal chelating ability of edible mushrooms commonly consumed in China. *LWT—Food Sci. Technol.* **2016**, *72*, 423–431. [CrossRef]
40. Smolskaitė, L.; Venskutonis, P.R.; Talou, T. Comprehensive evaluation of antioxidant and antimicrobial properties of different mushroom species. *LWT—Food Sci. Technol.* **2015**, *60*, 462–471. [CrossRef]
41. Ghiselli, A.; Serafini, M.; Natella, F.; Scaccini, C. Total antioxidant capacity as a tool to assess redox status: Critical view and experimental data. *Free Radic. Biol. Med.* **2000**, *29*, 1106–1114. [CrossRef]
42. Ming-Tsung, Y.; Yu-Hsiu, T.; Ruei-Chian, L.; Jeng-Leun, M. Antioxidant properties of fungal chitosan from shiitake stipes. *LWT—Food Sci. Technol.* **2007**, *40*, 255–261. [CrossRef]
43. Cayan, F.; Deveci, E.; Tel-Cayan, G.; Duru, M.E. Identification and quantification of phenolic acid compounds of twenty-six mushrooms by HPLC-DAD. *J. Food Meas. Charact.* **2020**, *14*, 1690–1698. [CrossRef]
44. Fu, R.; Zhang, P.; Jin, G.; Wei, S.; Chen, J.; Pei, J.; Zhang, Y. Substrate promiscuity of acyltransferases contributes to the diversity of hydroxycinnamic acid derivatives in purple coneflower. *Plant J.* **2022**, *110*, 802–813. [CrossRef]
45. Wang, X.-Q.; Huang, C.-S.; Liu, H.-X.; He, Y.-Q. Distribution, Synthesis and Biological Activity of Dalbergin. *Nat. Prod. Res. Dev.* **2009**, *21*, 900–904.
46. Kadam, D.; Palamthodi, S.; Lele, S.S. LC–ESI-Q-TOF–MS/MS profiling and antioxidant activity of phenolics from *L. sativum* seedcake. *J. Food Sci. Technol.* **2018**, *55*, 1154–1163. [CrossRef]
47. Wiseman, H.; Casey, K.; Clarke, D.B.; Barnes, K.A.; Bowey, E. Isoflavone Aglycon and Glucoconjugate Content of High- and Low-Soy U.K. Foods Used in Nutritional Studies. *J. Agric. Food Chem.* **2002**, *50*, 1404–1410. [CrossRef] [PubMed]
48. Behbahani, M.; Sayedipour, S.; Pourazar, A.; Shanehsazzadeh, M. In vitro anti-HIV-1 activities of kaempferol and kaempferol-7-O-glucoside isolated from Securigera securidaca. *Res. Pharm. Sci.* **2014**, *9*, 463–469.
49. Miyake, Y.; Yamamoto, K.; Osawa, T. Isolation of Eriocitrin (Eriodictyol 7-rutinoside) from Lemon Fruit (*Citrus limon* BURM. f.) and Its Antioxidative Activity. *Food Sci. Technol. Int. Tokyo* **1997**, *3*, 84–89. [CrossRef]
50. Peng, D.; Zahid, H.F.; Ajlouni, S.; Dunshea, F.R.; Suleria, H.A.R. LC-ESI-QTOF/MS Profiling of Australian Mango Peel By-Product Polyphenols and Their Potential Antioxidant Activities. *Processes* **2019**, *7*, 764. [CrossRef]
51. Suprun, A.R.; Dubrovina, A.S.; Tyunin, A.P.; Kiselev, K.V. Profile of Stilbenes and Other Phenolics in Fanagoria White and Red Russian Wines. *Metabolites* **2021**, *11*, 231. [CrossRef]
52. Qian, J.; Lai, W.; Jiang, L.; Zhan, H.; Zhai, M.; Fu, J.; Zhang, C. Association between differential gene expression and anthocyanin biosynthesis underlying the diverse array of petal colors in *Zinnia elegans*. *Sci. Hortic.* **2021**, *277*, 109809. [CrossRef]
53. Oh, S.-Y.; Kim, Y.H.; Bae, D.S.; Um, B.H.; Pan, C.-H.; Kim, C.Y.; Lee, H.J.; Lee, J.K. Anti-Inflammatory Effects of Gomisin N, Gomisin J, and Schisandrin C Isolated from the Fruit of *Schisandra chinensis*. *Biosci. Biotechnol. Biochem.* **2010**, *74*, 285–291. [CrossRef] [PubMed]
54. Guo, L.Y.; Hung, T.M.; Bae, K.H.; Shin, E.M.; Zhou, H.Y.; Hong, Y.N.; Kang, S.S.; Kim, H.P.; Kim, Y.S. Anti-inflammatory effects of schisandrin isolated from the fruit of *Schisandra chinensis* Baill. *Eur. J. Pharmacol.* **2008**, *591*, 293–299. [CrossRef] [PubMed]
55. Piispanen, R.; Willför, S.; Saranpää, P.; Holmbom, B. Variation of lignans in Norway spruce (*Picea abies* [L.] Karst.) knotwood: Within-stem variation and the effect of fertilisation at two experimental sites in Finland. *Trees* **2008**, *22*, 317–328. [CrossRef]
56. Kang, L.-Z.; Zeng, X.-L.; Ye, Z.-W.; Lin, J.-F.; Guo, L.-Q. Compositional analysis of the fruiting body of transgenic *Flammulina velutipes* producing resveratrol. *Food Chem.* **2014**, *164*, 211–218. [CrossRef]
57. Beekwilder, J.; Wolswinkel, R.; Jonker, H.; Hall, R.; de Vos, C.H.R.; Bovy, A. Production of resveratrol in recombinant microorganisms. *Appl. Environ. Microbiol.* **2006**, *72*, 5670–5672. [CrossRef]
58. Bannour, M.; Fellah, B.; Rocchetti, G.; Ashi-Smiti, S.; Lachenmeier, D.W.; Lucini, L.; Khadhri, A. Phenolic profiling and antioxidant capacity of Calligonum azel Maire, a Tunisian desert plant. *Food Res. Int.* **2017**, *101*, 148–154. [CrossRef]
59. Akazawa, T.; Itami, H.; Furumoto, T.; Nozaki, C.; Koike, H.; Iritani, S.; Amimoto, N.; Ogawa, M. Impact of an Olive Leaf Polyphenol 3,4-DHPEA-EDA on Physical Properties of Food Protein Gels. *J. Agric. Food Chem.* **2021**, *69*, 14250–14258. [CrossRef]
60. Birtić, S.; Dussort, P.; Pierre, F.-X.; Bily, A.C.; Roller, M. Carnosic acid. *Phytochemistry* **2015**, *115*, 9–19. [CrossRef]

processes

Article

A Comprehensive Study on the Acidic Compounds in Gas and Particle Phases of Mainstream Cigarette Smoke

Xinbo Lu [1,2,†], Hongfei Zhang [3,†], Yifeng Cao [2,*], Yongqiang Pang [3], Guojun Zhou [1], Hua Huang [1], Jing Li [2], Jian Jiang [1,*] and Qiwei Yang [2]

1 China Tobacco Zhejiang Industrial Co., Ltd., Hangzhou 310024, China; luxb@zjtobacco.com (X.L.); zhougj@zjtobacco.com (G.Z.); huangh@zjtobacco.com (H.H.)
2 Key Laboratory of Biomass Chemical Engineering of Ministry of Education, College of Chemical and Biological Engineering, Zhejiang University, Hangzhou 310058, China; lijing1006@zju.edu.cn (J.L.); yangqw@zju.edu.cn (Q.Y.)
3 China National Tobacco Quality Supervision & Test Center, Zhengzhou 450001, China; hfzhang1983@126.com (H.Z.); pangyq2726@163.com (Y.P.)
* Correspondence: caoyf@zju.edu.cn (Y.C.); jiangj@zjtobacco.com (J.J.)
† These authors contributed equally to this work.

Abstract: Acidic compounds constitute a group of chemicals present in mainstream cigarette smoke, among which organic acids contribute to flavoring. In order to obtain a comprehensive understanding of the constituents of acidic compounds in both the particulate and gaseous phases of the mainstream smoke of commercial cigarettes, and to delineate the difference between two types of cigarettes, the yields of acidic constituents from nine cigarettes of two commercial brands (L- and M-types) were collected and analyzed in detail by gas chromatography–mass spectrometry (GC-MS). The results identified and quantitatively analyzed 46 compounds, grouped according to the substituent groups. Compositional differences between the two cigarette types were evaluated with statistical approaches. Comparison between individual, grouped, and total acid contents, between the particulate and the gaseous phases, and between the commercial L- and M-type tobaccos were conducted and characterized by the p values obtained from Student's t-test. Multivariate analysis was performed using principal component analysis (PCA) and orthogonal projections to latent structures discriminant analysis (OPLS-DA) models to identify the acids that enable a reliable differentiation of the two types. Seventeen acidic compounds whose $p < 0.05$ and variable importance in projection (VIP) > 1 were identified as key components that could discriminate between the two groups of commercial cigarettes. This study may be beneficial for the development of non-combusted tobacco products, which could serve as alternatives to traditional cigarettes.

Keywords: mainstream cigarette smoke; acidic gas phase; particle phase; acidic compounds; multivariate analysis

Citation: Lu, X.; Zhang, H.; Cao, Y.; Pang, Y.; Zhou, G.; Huang, H.; Li, J.; Jiang, J.; Yang, Q. A Comprehensive Study on the Acidic Compounds in Gas and Particle Phases of Mainstream Cigarette Smoke. *Processes* **2023**, *11*, 1694. https://doi.org/10.3390/pr11061694

Academic Editor: Vladimir S. Arutyunov

Received: 25 April 2023
Revised: 26 May 2023
Accepted: 29 May 2023
Published: 1 June 2023

1. Introduction

In recent decades, a substantial body of evidence has emerged linking cigarette smoking to a range of significant health risks, including lung cancer, other forms of cancer, chronic obstructive pulmonary disease, stroke, liver disease, and coronary heart disease [1–5]. As a preventable cause of death worldwide, cigarette smoking is a matter of considerable public health concern. In addition to the harmful substances found in mainstream cigarette smoke, environmental tobacco smoke, which is composed of exhaled mainstream smoke and sidestream smoke, also contains known carcinogens and toxic compounds [6]. Nonsmokers who are exposed to environmental tobacco smoke over long periods are also at risk of developing lung cancer. In an effort to address these health concerns, alternative smokeless and non-combusted tobacco products, such as e-cigarettes and heat-not-burn devices, have been developed as potential substitutes for traditional cigarettes [7,8]. These products are expected to release fewer harmful chemicals as they do not undergo combustion, which is beneficial to cancer prevention [9–12].

Mainstream cigarette smoke (MCS) is an extremely complex aerosol composed of condensed liquid droplets (particulate phase) and a mixture of gases and volatile compounds (gaseous phase). Thousands of chemicals, including ketones, hydrocarbons, ethers, amides, carboxylic acids, alkaloids, and phenols, most of which are at trace level, have been identified in tobacco smoke [13–15]. In-depth characterization of the chemical composition of MCS is important for evaluating its impact on human health and developing strategies to reduce its harmful effects. However, complicated interactions between these compounds, such as synergistic, additive, or inhibitive effects, make it quite challenging. Knowledge of the characteristic chemical constituents of MCS profiling and the chemical basis of different tobacco types can also be used to determine the processes involved during smoking and the flavor descriptions of tobacco [16]. Flavoring and flavor-enhancing chemicals can be used in the development of non-combusted products [17]. Moreover, it is also critical to characterize both the particulate and gaseous phases separately due to their differing effects on smokers and non-smokers. The particulate phase is more strongly associated with health risks for smokers, whereas non-smokers are more impacted by constituents in the gas phase during passive smoking [18].

Acidic compounds comprise a group of chemicals in mainstream cigarette smoke and tobacco essential oil, of which organic acids contribute to the flavor of tobacco [19]. Despite their potential significance, acidic compounds in tobacco smoke and the differences between different tobacco types have received relatively little attention in the literature. Thus, the aim of this study is to provide a comprehensive analysis of the constituents of acidic compounds in both the particulate and gaseous phases of mainstream cigarette smoke from commercial brands. The present investigation involved the collection of acidic constituent yields from nine cigarettes of two commercial brands, which were subsequently analyzed using gas chromatography–mass spectrometry (GC-MS). Before GC-MS, derivatization of high-polarity and high-boiling-point compounds by trimethylsilylation was performed [20]. The present study offers an in-depth description of the acidic compounds that exist in mainstream cigarette smoke, thus enhancing our comprehension of the chemical constituents of tobacco smokes of two different types.

2. Materials and Methods

2.1. Materials

Nine commercial cigarettes, including five Virginia (L-type) and four blend types (M-type), were used in this study. They were all purchased from local stores in China. Characterization of the cigarette samples, including tobacco type, tar content, and nicotine content, of the samples are listed in Table 1. Standards of the organic acid listed in Table 2, *N*,*O*-bis(trimethylsilyl)trifluoroacetamide (BSTFA), phenethyl phenylacetate, and benzene-d6 were all purchased from Sigma-Aldrich (Merck KGaA, Darmstadt, Germany). The CX-572 (20/45 mesh) cartridge was purchased from Merck Inc. (St. Louis, MO, USA).

Table 1. Characterization of nine commercial cigarette samples used in this study.

Sample	Tobacco Type	TPM [a] (mg/cigarette)	Labelled Tar Content (mg/cigarette)	Measured Tar Content (mg/cigarette)	Water Content (mg/cigarette)	Nicotine Content (mg/cigarette)	CO (mg/cigarette)
L1	Virginia	13.38	10.0	10.81	1.66	0.90	9.93
L2	Virginia	9.50	8.0	8.13	0.69	0.68	10.02
L3	Virginia	13.78	11.0	11.43	1.41	0.95	10.94
L4	Virginia	13.39	11.0	10.80	1.66	0.93	10.13
L5	Virginia	14.80	11.0	12.12	1.68	1.00	11.61
M1	Blend	11.08	10.0	9.13	1.15	0.81	10.07

Table 1. *Cont.*

Sample	Tobacco Type	TPM [a] (mg/cigarette)	Labelled Tar Content (mg/cigarette)	Measured Tar Content (mg/cigarette)	Water Content (mg/cigarette)	Nicotine Content (mg/cigarette)	CO (mg/cigarette)
M2	Blend	10.57	10.0	8.86	1.00	0.71	10.02
M3	Blend	8.67	8.0	7.04	0.97	0.66	8.03
M4	Blend	9.99	8.0	8.34	0.91	0.74	9.27

[a] TPM, total particle matter.

Table 2. Acidic components identified in the mainstream tobacco smoke.

Retention Time/min	CAS Number of the TMS Compound	Compound	Formula	Quantitative Ion	Qualitative Ion	k [a]	Chemical Group [b]
11.80	64-18-6	Formic Acid	CH_2O_2	103	75	3.653	1
14.01	18147-36-9	Acetic Acid	$C_2H_4O_2$	117	75	3.078	1
17.05	13688-55-6	Acrylic Acid	$C_3H_4O_2$	129	75	0.03578	2
17.47	16844-98-7	Propionic Acid	$C_3H_6O_2$	131	75	2.541	1
21.17	16883-61-7	2-methylpropionic Acid	$C_4H_8O_2$	145	117	1.998	3
22.88	55557-14-7	2-methylbutyric Acid	$C_5H_{10}O_2$	159	117	0.03578	3
23.31	86254-80-0	Crotonic Acid/2-butenoic Acid	$C_4H_6O_2$	143	99	0.03578	2
23.39	55557-13-6	Isovaleric Acid	$C_5H_{10}O_2$	159	117	2.092	3
25.39	26429-16-3	Valeric Acid	$C_5H_{10}O_2$	159	177	1.952	2
26.78	55517-33-4	2-methyl-2-ene-butyric Acid	$C_5H_8O_2$	157	83	1.501	8
27.68	88239-45-6	2-pentenoic Acid	$C_5H_8O_2$	157	113	2.256	2
27.77	98983-20-1	3-methylpentanoic Acid	$C_6H_{12}O_2$	173	117	1.744	3
28.09	35707 [c]	4-methylpentanoic Acid	$C_6H_{12}O_2$	173	131	1.729	3
28.68	1529-17-5	Phenol	C_6H_6O	166	151	1.737	4
28.80	17596-96-2	Lactic Acid	$C_3H_6O_3$	147	117	0.03578	5
29.49	14246-15-2	Caproic Acid	$C_6H_{12}O_2$	173	117	1.815	1
29.53	33581-77-0	Glycolic Acid	$C_2H_4O_3$	147	177	4.776	5
30.01	55887-51-9	Pyruvic Acid	$C_3H_4O_3$	147	217	3.671	5
31.63	55133-93-2	2-hydroxybutyric Acid	$C_4H_8O_3$	131	147	2.522	5
31.91	N/A	Levulinic Acid	$C_5H_8O_3$	173	145	0.8642	5
32.00	1009-02-5	*o*-cresol	C_7H_8O	165	180	2.604	4
32.08	55887-53-1	2-furancarboxylic Acid	$C_5H_4O_3$	125	169	2.612	6
32.41	55162-32-8	3-hydroxypropionic Acid	$C_3H_6O_3$	147	219	0.03578	5

Table 2. *Cont.*

Retention Time/min	CAS Number of the TMS Compound	Compound	Formula	Quantitative Ion	Qualitative Ion	k [a]	Chemical Group [b]
32.86	17902-32-8	*p*-cresol	C_7H_8O	165	180	4.457	4
32.95	N/A	3-hydroxy-butyric Acid	$C_4H_8O_3$	147	191	3.552	5
34.94	N/A	Furanacetic Acid	$C_6H_6O_3$	198	154	0.03578	6
35.67	N/A	2-hydroxymethyl-butyric Acid	$C_5H_{10}O_3$	147	247	0.03578	5
36.70	2078-12-8	Benzoic Acid	$C_7H_6O_2$	179	135	2.621	7
38.49	N/A	Phenylacetic Acid	$C_8H_8O_2$	164	193	0.03578	7
39.17	88964 [c]	3-methyl-2-furancarboxylic Acid	$C_6H_6O_3$	139	198	0.03578	6
39.23	5075-52-5	Catechol	$C_6H_6O_2$	254	239	1.763	4
39.56	38191-87-6	Glyceric Acid	$C_3H_6O_4$	189	292	0.3	8
42.31	2117-24-0	Quinol/Resorcinol	$C_6H_6O_2$	239	254	3.44	4
43.79	34084 [c]	2-isopropyl-3-carbonyl-butyric Acid	$C_7H_{12}O_3$	273	183	0.03578	8
45.12	65143-63-7	Malic Acid	$C_4H_6O_5$	147	233	0.9153	5
46.95	17864-23-2	1,2,3-glycinol	$C_6H_6O_3$	239	342	3.895	4
47.41	38191-88-7	Threonic Acid	$C_4H_8O_5$	147	292	0.03578	5
47.83	3782-84-1	*m*-hydroxy-benzoic Acid	$C_7H_6O_3$	267	282	2.109	7
53.70	N/A	Vanillic Acid	$C_8H_8O_4$	297	312	0.9181	7
56.30	18603-17-3	Tetradecanoic Acid/Myristic Acid	$C_{14}H_{28}O_2$	285	285	0.4972	1
61.53	55520-89-3	Palmitic Acid	$C_{16}H_{32}O_2$	313	145	0.163	1
65.59	56259-07-5	Linoleic Acid	$C_{18}H_{32}O_2$	337	262	0.03578	2
65.74	97844-13-8	Linolenic Acid	$C_{18}H_{30}O_2$	335	108	0.03429	2
65.97	96851-47-7	Oleic Acid	$C_{18}H_{34}O_2$	339	117	0.06894	2
66.32	18748-91-9	Octadecanoic Acid	$C_{18}H_{36}O_2$	341	117	0.1029	1
70.72	55530-70-6	Stearic Acid	$C_{20}H_{40}O_2$	369	117	0.03	1

[a]. Slope of the calibration curve $y = k^*x$, whereas y and x corresponding to the peak area and concentration of the compound, respectively. Correlation coefficients R^2 for all the calibration curves are > 0.99. [b]. Chemical class: 1, carboxylate acid with straight alkyl chain; 2, unsaturated acid; 3, carboxylate acid with branched alkyl chain; 4, phenols; 5, acid with a hydroxyl/carbonyl substitution; 6, alkyl chain of the acid bearing a furan ring; 7, alkyl chain of the acid bearing a benzene substitution; 8, acid with more than one substituent group. [c]. CAS No. was not available, therefore NIST library ID was provided.

2.2. Methods

2.2.1. Sample Preparation

The cigarettes were equilibrated under conditions of 22 ± 1 °C, relative humidity 60 ± 3% for 48 h before smoking. Mainstream cigarette smoke was generated under an ISO machine smoking regimen [21] using a linear smoking machine (SM 450, Cerulean, Milton Keynes, UK). The particulate phase of the mainstream cigarette smoke was collected using 44 mm Cambridge filter pads and the gaseous phase was collected using a CX-572 cartridge with 300 mg of carbon molecular sieves. After smoking, the Cambridge filter pad was cut into two pieces and extracted with 6 mL CH_2Cl_2 containing 50 μL of internal standard solution (9.35 mg/mL benzene-d6 and 9.06 mg/mL phenethyl phenylacetate) under sonication for 30 min. After that, the liquid extract was filtered and analyzed. Adsorbent in the CX-572 cartridge was transferred into a 15-mL vial. After adding 4 mL CH_2Cl_2 (at a speed controlled at 1 mL/min) and 50 μL of mixed internal standard solution (9.35 mg/mL benzene-d6 and 9.06 mg/mL phenethyl acetate), the vial was shaken for 2 h and the liquid was collected afterwards.

2.2.2. Derivatization Procedures

Trimethylsilyl (TMS) derivation is routinely used in GC to increase the volatility and stability of the acidic compounds [22]. In this study, trimethylsilylation of the acidic compounds in the collected samples was performed using *N,O*-bis(trimethylsilyl)trifluoroacetamide (BSTFA) as the derivatizing agent prior to GC-MS analysis. The extract solution was filtered through a PTFE membrane with 0.45-μm pore size, 1 mL of which was transferred into a glass gas chromatography vial. Then, 100 μL of BSTFA was added to the vial, which was sealed and heated to 60 ± 1 °C for 40 min using a water bath to complete the derivatization. After that, the sample was cooled to room temperature and analyzed by GC-MS.

2.2.3. GC-MS Conditions

Analysis was carried out on an Agilent 7890B gas chromatography instrument coupled with an Agilent 5977C mass spectrometer (Agilent Technologies, Palo Alto, CA, USA). Separation was achieved with a J&W DB-5ms fused silica capillary column (60 m × 0.25 mm i.d., 1-μm film thickness) from Agilent Technologies, USA. High-purity helium was used as carrier gas and was maintained at 1 mL/min. The injection volume was 1 μL, and the split ratio was 10:1. The inlet temperature and the transfer line temperature were both set at 280 °C. The oven temperature program was as follows: the initial temperature was kept at 40 °C for 3 min, increased to 280 °C at a rate of 4 °C/min, and kept at 280 °C for 20 min. The spectrometer was operated in the electron ionization (EI) mode (70 eV). The scan range was set at 35–450 amu. The ionization source temperature was fixed at 280 °C. Data were collected under both the full scan and selected ion monitoring (SIM) modes. The identification of compounds was based on the National Institute of Standards and Technology (NIST) 14 library (matching score > 70) and by comparison with standards. The slopes of the calibration curves for each of the acids are also provided in Table 2.

2.2.4. Multivariate Analysis

Multivariate analysis of principal component analysis (PCA) and orthogonal projections to latent structures discriminant analysis (OPLS-DA) was performed using the Umetrics SIMC 14.1 software. Unsupervised PCA was firstly performed, followed by supervised OPLS-DA model. Differences between the L- and the M-groups were evaluated.

3. Results

3.1. Constituents of Acids in the Particulate and the Gaseous Phases

Representative total ion chromatograms of the TMS derivatives of acidic compounds in the particulate and gaseous phases of the mainstream cigarette smoke are shown in Figure 1, and the identified peaks are listed in Table 2, where a total number of 46 acidic compounds were identified.

Contents of the individual acids are listed in Tables S1 and S2 and presented in Figure 2. Major acids in the particulate phase are carboxylic acids with straight chains, while those in the gaseous phase are palmitic acid, octadecanoic acid, acetic acid, and formic acid (Figure 2a,b). A number of acids were identified in the particulate phase of iQOS, a commercial heated tobacco product. Acetic acid, palmitic acid, linolenic acid, propanoic acid, oleic acid, linoleic acid, 3-methylpentanoic acid, 3-methyl-2-furancarboxylic acid, 2-methylbutanoic acid, 3-methylbutanoic acid, and myristic acid, which were among the identified acids in this work, were also identified in the total particulate matter of iQOS and in cigarette smoke [23]. Moreover, octadecanoic acid, pentadecanoic acid, arachidic acid, (9Z,12Z)-18-hydroxy-9,12-octadecadienoic acid, palmitoleic acid, butanoic acid, behenic acid, 5-oxo-1-tetradecyl-3-pyrrolidinecarboxylic acid, lignoceric acid, stearidonic acid, and 3-methylpalmitic acid were identified in this work. In another work, acetic acid, formic acid, glycolic acid, lactic acid, and succinic acid were among the major acids in the mainstream and sidestream of four varieties of tobacco [20]. In contrast, formic acid, acetic acid, and levulinic acid were identified in e-cigarette aerosol by high-performance liquid chromatography–high-resolution mass spectrometry (HPLC-HRMS) [24]. Clearly, the acidic constituents of cigarette smoke are highly correlated with the tested samples. It is worth noticing that, given the fact that certain weak organic acids, including but not limited to citric acid, acetic acid, butyric acid, lactic acid, and 2-methyl butyric acid, are among the permitted additives and flavorings to tobacco products [25], their addition could also contribute to the acidic components in the tobacco smoke of the commercial samples.

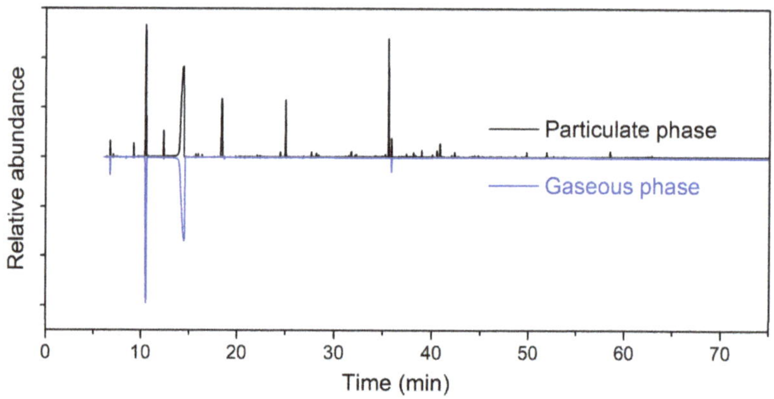

Figure 1. Total ion chromatogram (TIC) of sample M2: the particulate phase (black color) and the gaseous phase (blue color).

3.2. Acid Contents

The total amount of acid compounds in both the particulate and the gaseous phases for all the samples are presented in Figure 3. All the acids were grouped according to the substituent groups on the alkyl chain. No significant difference in the gaseous phase (p = 0.0857) was found between the two groups (Figure 3b). Nevertheless, the high content of acids with straight or unsaturated chains is clear.

According to the chemical structure, the identified acids were categorized into eight groups, including carboxylate acid with straight alkyl chain, unsaturated acid, carboxylate acid with branched alkyl chain, phenols, acid with a hydroxyl/carbonyl substitution, alkyl chain of the acid bearing a furan ring, alkyl chain of the acid bearing a benzene substitution, and acid with more than one substituent group. In the particulate phase (Figure 3a), acids with a straight saturated or unsaturated alkyl chain dominate. For those in the gaseous phase, most have straight alkyl chains with lengths ranging from C1 to C20.

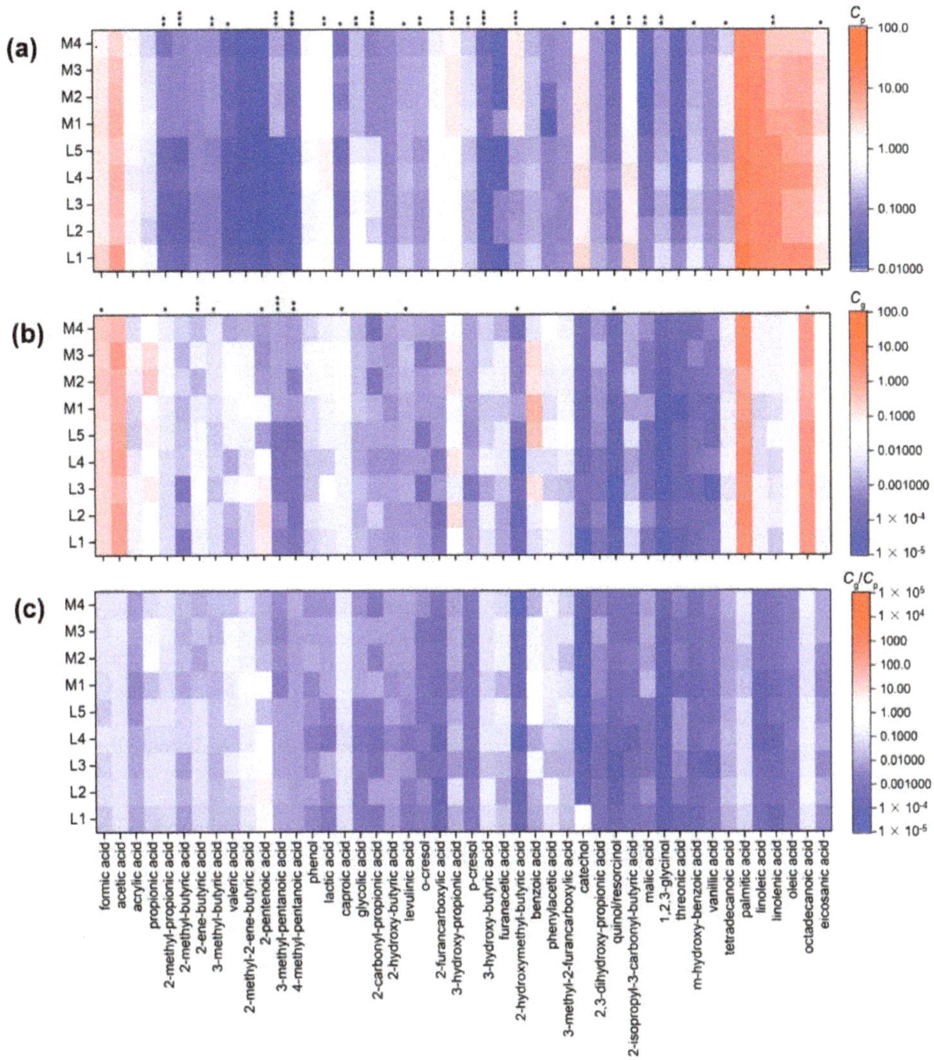

Figure 2. Heat map showing concentrations of the identified acids (μg/mg tar) in the mainstream smoke of different cigarettes: (**a**) particulate phase, (**b**) gaseous phase, as well as (**c**) the ratios between the particulate and the gaseous phases. Acids and sample types are presented on the y- and x-axes, respectively. Contents of the acids (μg/mg tar in a cigarette) are presented in the logarithm form. Student's t-test was performed to test whether the apparent differences between acid components of L- and M- types of tobacco were statistically significant (* $p \leq 0.05$, ** $p \leq 0.01$, *** $p \leq 0.001$).

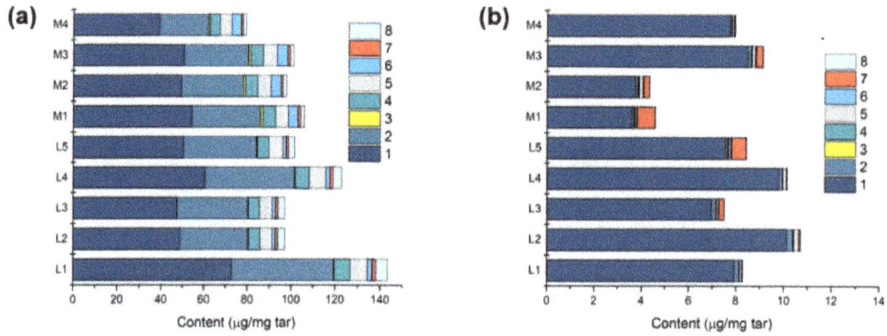

Figure 3. Total acid contents in the (**a**) particulate and (**b**) gaseous phases of different cigarettes, grouped according to the substituent groups on the alkyl chain. The identified acids were categorized into eight groups according to the substituent groups. 1, carboxylate acid with straight alkyl chain; 2, unsaturated acid; 3, carboxylate acid with branched alkyl chain; 4, phenols; 5, acid with a hydroxyl/carbonyl substitution; 6, alkyl chain of the acid bearing a furan ring; 7, alkyl chain of the acid bearing a benzene substitution; and 8, acid with more than one kind of substituent group.

3.3. Acid Content Difference between the Particulate and the Gaseous Phases

Acids in the gaseous phase account for 7.6 ± 1.6% and 6.4 ± 2.6% of all those in the particulate and the gaseous phases for the L- and the M-type, respectively, whereas no significant difference was observed between the two types. Decreased acidity of tobacco emission has been reported to increase the amount of nicotine into blood [26]. In the mainstream smoke, >99% nicotine is in the particulate phase. Strong interaction between the acid compounds and the basic nicotine is expected, which hinders the emission of organic acids to the gaseous phase.

The ratio of individual acid content between the particulate and the gaseous phases of MCS was also calculated and presented in Figure 2c. Overall, organic acids with straight or branched alkyl chain, including those long-chain fatty acids, are relatively more abundant in the gaseous phase, as opposed to those acids bearing hydroxyl, furan, or phenol groups. As the boiling points of long-chain fatty acids are relatively high or equivalent to those preferably accumulating in the particulate phase, their distribution between the gaseous and the particulate phases are not simply dominated by their volatility. One possible explanation is that polycyclic aromatic hydrocarbons (PAHs), tobacco-specific nitrosamines (TSNAs), and phytosterols were found almost exclusively in the particulate phase [27], and hydrogen-bonding interactions between these compounds and acids with the above-mentioned hydroxyl, furan, and phenol groups are stronger than those with straight or branched alkyl groups.

3.4. Acid Content Difference between the L- and the M-Types

The content differences of individual acids and of each acid group between the L- and the M-types are labelled in Figure 2a (individual acid in the particulate phase), Figure 2b (individual acid in the gaseous phase), and Table 3 (acid groups). In the particulate phase, significant difference ($p \leq 0.05$) between the L- and the M-types was observed for 26 acids, including 2-methyl-propionic acid, 2-methyl-butyric acid, 3-methyl-butyric acid, valeric acid, 3-methyl-pentanoic acid, 4-methyl-pentanoic acid, lactic acid, glycolic acid, 2-carbonyl-propionic acid, 3-hydroxy-propionic acid, and 3-hydroxy-butyric acid, etc. Among these acids, the contents of seven acids were found to be highly, significantly different ($p \leq 0.001$). Most of these acids have a substituent group attached to the short alkyl chain. For the gaseous phase, the number was 12 ($p \leq 0.05$) and two ($p \leq 0.001$). Moreover, significant difference in acid content was observed for four acid groups identified in the particulate phase, while the number was two for the gaseous phase (Table 3). For total acid contents, no significant difference was observed between the M- and L-groups, in neither the particulate phase nor the gaseous phase. These results clearly show that the major difference between the L- and the M-types lies in the individual acid/acid groups of the particulate phase.

Table 3. Differences in the grouped acids between cigarettes samples (L- vs. M-types).

	Chemical Group							
	1	2	3	4	5	6	7	8
Particulate Phase		*	***			***		**
Gaseous Phase		*	**					

Note: * $p \leq 0.05$, ** $p \leq 0.01$, *** $p \leq 0.001$.

3.5. Multivariate Analysis

Multivariate analysis with PCA and OPLS-DA methods was performed on the basis of the 46 acids identified in the particulate phase to reduce data dimensionality. PCA results (Figure 4) clearly show clustered L- and the M-groups, confirming that there are differences underlying the acidic components of the two groups. The first two principal components accounted for 88.1% (R^2X) of the total variance. As the first principal component (PC1) and the second one (PC2) account for 50.5% and 37.6% of the variance, respectively, they can represent the majority information of the cigarette samples. The cumulated predictive ability of the model Q^2 (cum) 0.772 (>0.05) indicates the quality of the model.

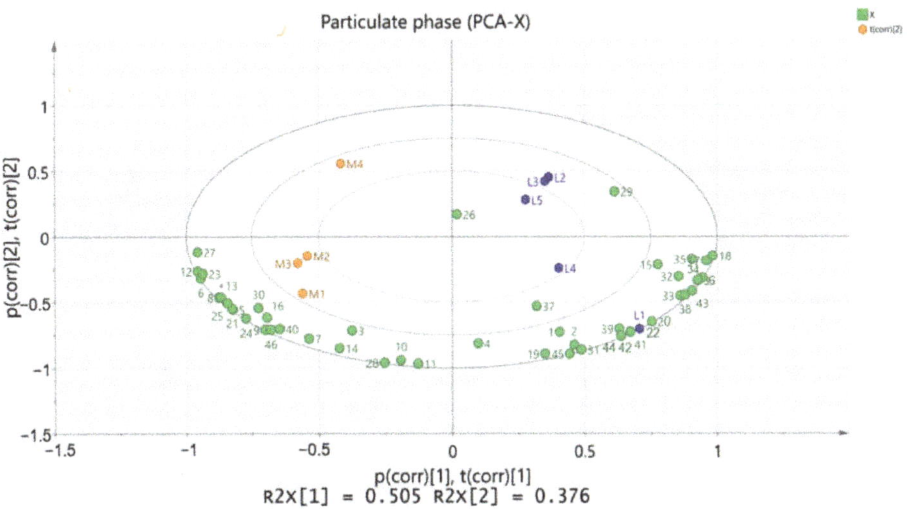

Figure 4. PCA biplot diagram of acid components of the tested cigarette samples (N = 9): blue symbols, M-group; red symbols, L-group; green symbols, individual acid. Cigarette samples were divided into two clusters. The x- and y-axes represent the re-scaled scores and loadings of the first component (PC1) and the second one (PC2), respectively. 1, formic acid; 2, acetic acid; 3, acrylic acid; 4, propionic acid; 5, 2-methyl-propionic acid; 6, 2-methyl-butyric acid; 7, 2-ene-butyric acid; 8, 3-methyl-butyric acid; 9, valeric acid; 10, 2-methyl-2-ene-butyric acid; 11, 2-pentenoic acid; 12, 3-methyl-pentanoic acid; 13, 4-methyl-pentanoic acid; 14, phenol; 15, lactic acid; 16, caproic acid; 17, glycolic acid; 18, 2-carbonyl-propionic acid; 19, 2-hydroxy-butyric acid; 20, levulinic acid; 21, o-cresol; 22, 2-furancarboxylic acid; 23, 3-hydroxy-propionic acid; 24, p-cresol; 25, 3-hydroxy-butyric acid; 26, furanacetic acid; 27, 2-hydroxymethyl-butyric acid; 28, benzoic acid; 29, phenylacetic acid; 30, 3-methyl-2-furancarboxylic acid; 31, catechol; 32, 2,3-dihydroxy-propionic acid; 33, quinol/resorcinol; 34, 2-isopropyl-3-carbonyl-butyric acid; 35, malic acid; 36, 1,2,3-glycinol; 37, threonic acid; 38, m-hydroxy-benzoic acid; 39, vanillic acid; 40, tetradecanoic acid; 41, palmitic acid; 42, linoleic acid; 43, linolenic acid; 44, oleic acid; 45, octadecanoic acid; 46, eicosanic acid.

Compared with the PCA model, the OPLS-DA approach is a powerful statistical modeling tool to discriminate between two groups [28,29]. The results in Figure 5a also demonstrate separation between the two groups. The R^2X of the OPLS-DA was 0.798, indicating that 79.8% of the variation in the cigarette samples could be modeled by the selected components, while R^2Y = 0.976 indicates that the model was well-fitted. Predictability Q^2 = 0.905 is an indication of the good predictivity of the model. In addition, diagnostics such as permutation testing are of high importance to avoid overfitting. The intercept of the blue regression line of the Q^2-points in Figure 5b is (0, −0.838), which strongly indicates that the model is valid. These results overall proved that a combination of the acid contents in the particulate phase with multivariate statistical analysis could effectively differentiate between the two types of cigarettes.

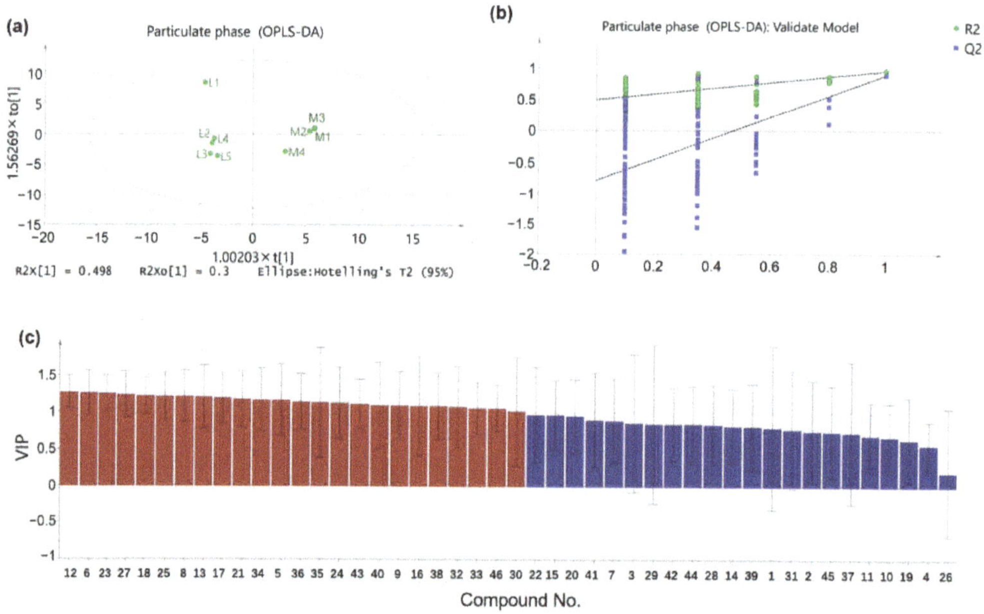

Figure 5. The OPLS-DA model of the two-class separation: (**a**) the scatter plot of the scores t [1] vs. to [1], (**b**) the permutations plot, and (**c**) the variable importance of the projection (VIP) plot. Particulate phase. Intercepts of plots in (**b**): R^2 = (0.0, 0.485), Q^2 = (0.0, −0.838). In (**c**), 1, formic acid; 2, acetic acid; 3, acrylic acid; 4, propionic acid; 5, 2-methyl-propionic acid; 6, 2-methyl-butyric acid; 7, 2-ene-butyric acid; 8, 3-methyl-butyric acid; 9, valeric acid; 10, 2-methyl-2-ene-butyric acid; 11, 2-pentenoic acid; 12, 3-methyl-pentanoic acid; 13, 4-methyl-pentanoic acid; 14, phenol; 15, lactic acid; 16, caproic acid; 17, glycolic acid; 18, 2-carbonyl-propionic acid; 19, 2-hydroxy-butyric acid; 20, levulinic acid; 21, *o*-cresol; 22, 2-furancarboxylic acid; 23, 3-hydroxy-propionic acid; 24, *p*-cresol; 25, 3-hydroxy-butyric acid; 26, furanacetic acid; 27, 2-hydroxymethyl-butyric acid; 28, benzoic acid; 29, phenylacetic acid; 30, 3-methyl-2-furancarboxylic acid; 31, catechol; 32, 2,3-dihydroxy-propionic acid; 33, quinol/resorcinol; 34, 2-isopropyl-3-carbonyl-butyric acid; 35, malic acid; 36, 1,2,3-glycinol; 37, threonic acid; 38, *m*-hydroxy-benzoic acid; 39, vanillic acid; 40, tetradecanoic acid; 41, palmitic acid; 42, linoleic acid; 43, linolenic acid; 44, oleic acid; 45, octadecanoic acid; 46, eicosanic acid.

The variable importance in projection (VIP) scores reflect both the loading weights of each component and the variability of the response explained by this component. Information on the VIP of the acids is presented in Figure 5c and Table S3. In Figure 5c, the VIP plot is sorted from high to low, and the acids whose VIP is higher than one (an indication of 'important' variables) are colored in red, while those with VIP < 1.0 are in blue. A number of 24 acids, including 3-methyl-pentanoic acid, 2-methyl-butyric acid, 3-hydroxy-propionic acid, and 2-hydroxymethyl-butyric acid, were identified as main components in discriminating between the L- and the M-groups.

In general, variables with VIP > 1.0 and $p < 0.05$ are considered significant to contribute to the model and are identified as markers differentiating between the two groups [30]. The p values reflect the compositional difference of individual compound between groups, while the VIP value higher than one suggests that this compound contributes significantly to inter-group difference. These acids include 3-methyl-pentanoic acid, 2-methyl-butyric acid, 3-hydroxy-propionic acid, 2-hydroxymethyl-butyric acid, 3-methyl-butyric acid, glycolic acid, *o*-cresol, 2-isopropyl-3-carbonyl-butyric acid, 2-methyl-propionic acid, malic acid, *p*-cresol, linolenic acid, tetradecanoic acid, caproic acid, *m*-hydroxy-benzoic acid, eicosanic acid, and 3-methyl-2-furancarboxylic acid, in order of decreasing VIP values. These compounds could be useful for establishing fingerprints of the tobacco products.

4. Conclusions

A comprehensive investigation was conducted to determine the concentration of acidic compounds in both the particulate and gaseous phases of mainstream cigarette smoke. The study involved comparative analysis of two commercially available cigarette types, namely the L-type and M-type. A total of 46 acids were analyzed qualitatively and quantitatively using GC-MS. The results indicated that the concentrations of acids in the gaseous phase were much lower than those in the particulate phase. Acids with straight or unsaturated chains were found to be more abundantly present in the gaseous phase, which could be attributed to their high volatility or weaker interaction with compounds in the particulate phase. Significant differences ($p < 0.05$) in acid content between the L- and M-types were observed for 26 acids in the particulate phase, and 12 in the gaseous phase. Multivariate statistical analysis using PCA and OPLS-DA models successfully differentiated between the two tobacco types. These findings may be instrumental in the development of non-combusted tobacco products.

Supplementary Materials: The following supporting information can be downloaded at: https://www.mdpi.com/article/10.3390/pr11061694/s1, Table S1: Contents of the identified acids in the particulate phase (μg/mg tar).; Table S2. Contents of the identified acids in the gaseous phase (μg/mg tar). Table S3: VIP values of individual acid.

Author Contributions: Conceptualization, X.L., Y.C., G.Z., H.H., J.J. and Q.Y.; methodology, X.L., G.Z. and H.H.; data curation, X.L., H.Z., Y.C., Y.P. and J.L.; writing—original draft preparation, X.L. and Y.C.; writing—review and editing, Y.C., J.J. and Q.Y.; project administration, J.J. and Q.Y.; funding acquisition, Y.C., J.L. and Q.Y. All authors have read and agreed to the published version of the manuscript.

Funding: This research was funded by the Zhejiang University- China Tobacco Zhejiang Industrial Co., Ltd. Joint Laboratory, the National Natural Science Foundation of China, Grant Number: 22178306, the Natural Science Foundation of Zhejiang Province, Grant Numbers: LR21B060002 and LGC22B050013, and the Research Funds of Institute of Zhejiang University, Quzhou, Grant Number: IZQ2022RCZX004.

Data Availability Statement: The data presented in this study are available on request from the corresponding authors.

Conflicts of Interest: The authors declare no conflict of interest.

References

1. Kim, S.Y.; Sim, S.; Choi, H.G. Active, passive, and electronic cigarette smoking is associated with asthma in adolescents. *Sci. Rep.* **2017**, *7*, 17789. [CrossRef] [PubMed]
2. Loeb, L.A.; Emster, V.L.; Warner, K.E.; Abbotts, J.; Laszlo, J. Smoking and Lung Cancer: An Overview. *Cancer Res.* **1984**, *44*, 5940–5958. [PubMed]
3. Hjermann, I.; Holme, I.; Byre, K.V.; Leren, P. Effect of diet and smoking intervention on the incidence of coronary heart disease: Report from the Oslo Study Group of a randomised trial in healthy men. *Lancet* **1981**, *318*, 1303–1310.

4. Fagerström, K. The Epidemiology of Smoking Health Consequences and Benefits of Cessation. *Drugs* **2002**, *62*, 1–9.
5. Marti-Aguado, D.; Clemente-Sanchez, A.; Bataller, R. Cigarette smoking and liver diseases. *J. Hepatol.* **2022**, *77*, 191–205. [CrossRef]
6. Sopori, M. Effects of cigarette smoke on the immune system. *Nat. Rev. Immunol.* **2002**, *2*, 372–377. [CrossRef]
7. Simonavicius, E.; McNeill, A.; Shahab, L.; Brose, L.S. Heat-not-burn tobacco products: A systematic literature review. *Tob. Control* **2019**, *28*, 582–594. [CrossRef]
8. O'Connor, R.; Schneller, L.M.; Felicione, N.J.; Talhout, R.; Goniewicz, M.L.; Ashley, D.L. Evolution of tobacco products: Recent history and future directions. *Tob. Control* **2022**, *31*, 175–182. [CrossRef]
9. US Food and Drug Administration. Harmful and Potentially Harmful Constituents in Tobacco Products and Tobacco Smoke: Established List. *Fed. Regist.* **2012**, *77*, 20034–20037. Available online: https://www.fda.gov/tobacco-products/rules-regulations-and-guidance/harmful-and-potentially-harmful-constituents-tobacco-products-and-tobacco-smoke-established-list (accessed on 20 March 2023).
10. Margham, J.; McAdam, K.; Forster, M.; Liu, C.; Wright, C.; Mariner, D.; Proctor, C. Chemical Composition of Aerosol from an E-Cigarette: A Quantitative Comparison with Cigarette Smoke. *Chem. Res. Toxicol.* **2016**, *29*, 1662–1678. [CrossRef]
11. Uchiyama, S.; Noguchi, M.; Takagi, N.; Hayashida, H.; Inaba, Y.; Ogura, H.; Kunugita, N. Simple Determination of Gaseous and Particulate Compounds Generated from Heated Tobacco Products. *Chem. Res. Toxicol.* **2018**, *31*, 585–593. [CrossRef] [PubMed]
12. Hecht, S.S.; Hatsukami, D.K. Smokeless tobacco and cigarette smoking: Chemical mechanisms and cancer prevention. *Nat. Rev. Cancer* **2022**, *22*, 143–155. [CrossRef] [PubMed]
13. Rodgman, A.; Perfetti, T.A. *The Chemical Components of Tobacco and Tobacco Smoke*; CRC Press: Boca Raton, FL, USA, 2013.
14. Ji, H.; Jin, Z. Analysis of six aromatic amines in the mainstream smoke of tobacco products. *Anal. Bioanal. Chem.* **2022**, *414*, 4227–4234. [CrossRef] [PubMed]
15. Guo, Q.; Qin, Y.; Pan, L.; Xie, F.; Liu, S.; Sun, X.; Wang, X.; Cai, J.; Zhao, X.; Liu, H. Accuracy improvement of determination of seven minor tobacco alkaloids in mainstream cigarette smoke using analyte protectants by gas chromatography-mass spectrometry. *J. Chromatogr. A* **2022**, *1684*, 463537. [CrossRef]
16. Kaur, G.; Muthumalage, T.; Rahman, I. Mechanisms of toxicity and biomarkers of flavoring and flavor enhancing chemicals in emerging tobacco and non-tobacco products. *Toxicol. Lett.* **2018**, *288*, 143–155. [CrossRef]
17. Crooks, I.; Neilson, L.; Scott, K.; Reynolds, L.; Oke, T.; Forster, M.; Meredith, C.; McAdam, K.; Proctor, C. Evaluation of flavourings potentially used in a heated tobacco product: Chemical analysis, in vitro mutagenicity, genotoxicity, cytotoxicity and in vitro tumour promoting activity. *Food Chem. Toxicol.* **2018**, *118*, 940–952. [CrossRef]
18. Scherer, G.; Conze, C.; Meyerinck, L.v.; Sorsa, M.; Adlkofer, F. Importance of exposure to gaseous and particulate phase components of tobacco smoke in active and passive smokers. *Int. Arch. Occup. Environ. Health* **1990**, *62*, 459–466. [CrossRef]
19. Sakuma, H.; Kusama, M.; Munakata, S.; Ohsumi, T.; Sugawara, S. The Distribution of Cigarette Smoke Components between Mainstream and Sidestream Smoke I. Acidic Components. *Beiträge Zur Tab. Int.* **1983**, *12*, 63–71. [CrossRef]
20. Karlonas, N.; Padarauskas, A.; Ramanavicius, A.; Ramanaviciene, A. Mixed-mode SPE for a multi-residue analysis of benzodiazepines in whole blood using rapid GC with negative-ion chemical ionization MS. *J. Sep. Sci.* **2013**, *36*, 1437–1445. [CrossRef]
21. *ISO 3308:2012*; Routine Analytical Cigarette-Smoking Machine—Definitions and Standard Conditions. ISO: New York, NY, USA, 2012.
22. Christou, C.; Gika, H.G.; Raikos, N.; Theodoridis, G. GC-MS analysis of organic acids in human urine in clinical settings: A study of derivatization and other analytical parameters. *J. Chromatogr. B* **2014**, *964*, 195–201. [CrossRef]
23. Bentley, M.C.; Almstetter, M.; Arndt, D.; Knorr, A.; Martin, E.; Pospisil, P.; Maeder, S. Comprehensive chemical characterization of the aerosol generated by a heated tobacco product by untargeted screening. *Anal. Bioanal. Chem.* **2020**, *412*, 2675–2685. [CrossRef] [PubMed]
24. Li, Y.; Burns, A.E.; Burke, G.J.P.; Poindexter, M.E.; Madl, A.K.; Pinkerton, K.E.; Nguyen, T.B. Application of High-Resolution Mass Spectrometry and a Theoretical Model to the Quantification of Multifunctional Carbonyls and Organic Acids in e-Cigarette Aerosol. *Environ. Sci. Technol.* **2020**, *54*, 5640–5650. [CrossRef] [PubMed]
25. Winnall, W.R. 12.6 Additives and Flavourings in Tobacco Products. In *Tobacco in Australia: Facts and Issues*; Greenhalgh, E.M., Scollo, M.M., Winstanley, M.H., Eds.; Cancer Council Victoria: Melbourne, VIC, Australia, 2022; Available online: https://www.tobaccoinaustralia.org.au/chapter-12-tobacco-products/12-6-additives-and-flavourings-in-tobacco-products (accessed on 25 May 2023).
26. Pankow, J.F.; Tavakoli, A.D.; Luo, W.; Isabelle, L.M. Percent free base nicotine in the tobacco smoke particulate matter of selected commercial and reference cigarettes. *Chem. Res. Toxicol.* **2003**, *16*, 1014–1018. [CrossRef] [PubMed]
27. Thielen, A.; Klus, H.; Muller, L. Tobacco smoke: Unraveling a controversial subject. *Exp. Toxicol. Pathol.* **2008**, *60*, 141–156. [CrossRef] [PubMed]
28. Pappalardo, L. Pomegranate fruit juice adulteration with apple juice: Detection by UV-visible spectroscopy combined with multivariate statistical analysis. *Sci. Rep.* **2022**, *12*, 5151. [CrossRef]

29. Kang, C.; Zhang, Y.; Zhang, M.; Qi, J.; Zhao, W.; Gu, J.; Guo, W.; Li, Y. Screening of specific quantitative peptides of beef by LC-MS/MS coupled with OPLS-DA. *Food Chem.* **2022**, *387*, 132932. [CrossRef]
30. Peng, Q.; Meng, K.; Zheng, H.; Yu, H.; Zhang, Y.; Yang, X.; Lin, Z.; Xie, G. Metabolites comparison in post-fermentation stage of manual (mechanized) Chinese Huangjiu (yellow rice wine) based on GC-MS metabolomics. *Food Chem.* **2022**, *14*, 100324. [CrossRef]

processes

Article

Carotenoid Extraction from Locally and Organically Produced Cereals Using Saponification Method

Abrar Hussain [1,*], Hans Larsson [2] and Eva Johansson [2,*]

1 Department of Biosciences, Sahiwal Campus, COMSATS University Islamabad, Punjab 57000, Pakistan
2 Department of Plant Breeding, The Swedish University of Agricultural Sciences (SLU), P.O. Box 101, 23053 Alnarp, Sweden; Hans.larsson@slu.se
* Correspondence: Abrar.Hussain@cuisahiwal.edu.pk (A.H.); Eva.Johansson@slu.se (E.J.)

Abstract: Carotenoids are important phytochemicals contributing nutritional health benefits in the human diet, with a significant contribution from cereals as one of the major food component around the world. Different methods have been described and adopted for the extraction and isolation of carotenoid compounds. Saponification can be seen as an option for carotenoid extraction from cereals as it converts retinol esters to retinol and removes other abundant compounds such as triglycerides. Extraction of carotenoids content of locally adapted and organic cereals have been limitedly investigated and was, therefore, evaluated in the present study, with a specific aim to understand genotypic and local cultivation effects and interactions. Therefore, 17 diverse cereal genotypes of local origin were grown organically in four localities and evaluated for carotenoid content and composition by HPLC. The results showed a large variation in content and composition of carotenoids in locally adapted and organically grown cereal genotypes, with lutein as the dominating type in wheat and rye, while zeaxanthin was the dominating type in barley. High-level genotypes showed values (9.9 mg/kg of total carotenoids) similar to the highest values previously reported in specific types of wheat. The barley genotypes showed relatively high stability in carotenoids content within and between cultivation locations, while large interactions were found with the cultivation location for the rest of the genotypes, indicating their local adaptation. The local adaptation of the cereal genotypes evaluated contributes large opportunities for local production of high value, highly nutritious food products, while the direct value of these genotypes for conventional plant breeding for varieties performing similar over broad environmental ranges, are more limited.

Keywords: organic wheat genotypes; saponification; carotenoids; lutein content; environmental effects

Citation: Hussain, A.; Larsson, H.; Johansson, E. Carotenoid Extraction from Locally and Organically Produced Cereals Using Saponification Method. *Processes* **2021**, 9, 783. https://doi.org/10.3390/pr9050783

Academic Editors: Zongbi Bao and Qianqian Xu

Received: 15 April 2021
Accepted: 26 April 2021
Published: 29 April 2021

Publisher's Note: MDPI stays neutral with regard to jurisdictional claims in published maps and institutional affiliations.

1. Introduction

Cereals are the most important type of crops in terms of quantity available for food production to the human population [1–3]. The three major crops contributing to human food are wheat, maize, and rice, all three belonging to the cereal group of crops [1]. The total harvest of these three crops, are 2.5 billion tonnes, which contributes 42% and 37% of the human calorie and protein supply, respectively, thereby making these crops outstanding for human food security [4]. However, within the Nordic hemisphere, only wheat among these three can be grown, but other cereals such as barley, rye, and oats are instead making an important contribution to the cereals share in the human diet [5]. Similarly to other regions of the world, locally adapted cultivars have been developed in the Nordic regions, of crops traditionally in use [6]. Such traditionally grown cultivars (landraces) are often adapted to the local environment in which they have been grown, including an adaptation to the local climatic and soil conditions [7]. Recent studies have also indicated that some traditionally grown cultivars have a good capability for uptake and transportation of, e.g., minerals to the grain, resulting in a high nutritional value of the grains [8,9].

Food from organically produced crops is increasingly desired from consumers, mainly due to an increasing concern for the threat to the global environment and human health issues [10,11]. Concerned consumers are assuming organic production to contribute to less chemicals being used in agriculture, to increased biodiversity in organic production systems and to the production of food that contribute with human health attributes [11,12]. Although the results differ in various studies, some indications are available of organically produced food being richer in minerals and phytochemicals than conventionally produced food [8,9,11].

Cereals contain a number of phytochemicals [13], carotenoids being one of those, present in various amounts in cereals [14,15]. Despite being present in relatively low amounts in cereals, the generally high intake of cereals as being the staple in many countries, make cereals an important contributor of carotenoids in the human diet [16]. Cereals are known to be particularly rich in the carotenoids lutein and xeazanthin [17–20], where previous results have indicated that 40% of the daily requirement of lutein can be obtained from wheat [16]. Both mentioned compounds contribute positively to human health benefits, as they are known to promote the health of skin and eyes and to reduce the risk of cardiovascular diseases [21,22].

Analytical methods used for the extraction, isolation, and quantification of carotenoids are an important factor for accurate estimation of carotenoid compounds [23]. In fact, these compounds are easily degraded by different extraction procedures because of varying affinity between solvents and carotenoid compounds [24]. Due to wide range of polarity among carotenoid compounds different solvents are used, e.g., xanthophylls are extracted by polar solvents, i.e., alcohols and acetones, whereas carotene are extracted by nonpolar solvents [23]. Therefore, content of carotenoid compounds is often underestimated due to extraction procedure and solvent used. Due to these circumstances, saponification may be a suitable option for carotenoid extraction from cereals as it converts retinol esters to retinol and remove the other abundant compounds such as triglycerides. Furthermore, chlorophyll in samples is also destroyed through saponification as it interferes with spectrometric measurements [25].

A number of studies have evaluated the content and composition of carotenoids in cereals [26–30] although the majority of these studies have focused on wheat. Knowledge related to impact of organic production and local adaptation for nutritional value when it comes to carotenoids of cereals is basically lacking. Thus, despite the fact that consumers are willing to cover extra cost for (i) local production, (ii) organic production, and (iii) food from traditional varieties, with the expectation of a healthier food intake [30,31], few scientific studies have been undertaken to evaluate content and composition of nutritional components in such food related crops.

The aim of the present study was to evaluate the content of various carotenoids in locally adapted and organically produced cereals. Effect of genotype and locality on the content, as well as stability over and within locations, were evaluated in order to understand mechanisms for variation in content of carotenoids in traditionally grown cereals in Sweden. Due to recent interest in cereal carotenoids, the importance of wheat and other cereals has been increased during the last few years. Thus, the present study also focused on the role of β-carotene and β-cryptoxanthin in general and lutein in particular with respect to human health and daily intake contribution from different cereals.

2. Materials and Methods

2.1. Chemicals

Standards of α-carotene, β-carotene and lutein were obtained from Sigma (Schnelldorf, Germany). β-cryptoxanthin and zeaxanthin were purchased from CaroteNature (Lupsingen, Switzerland). The stock solution for the standards was prepared at the concentration of 50 μg/mL of n-hexane and stored in the dark at −20 °C. Suitable volume of each stock solution was used to prepare a working solution.

2.2. Samples

A total of 17 genotypes from diverse groups, i.e., landraces, old cultivar and primitive wheat were used in the present study (Table 1). These genotypes were grown in organic trials at Ekhaga (59°49′57″ N, 17°48′58″ E), Krusenberg (59°44′8″ N, 17°38′58″ E), Alnarp (55°39′27″ N, 13°04′51″ E), and Gotland (57°35′52″ N, 18°26′50″ E) in Sweden. At maturity, the cereals were harvested, threshed, and transported to the department in Alnarp where the grains were kept in a seed storage room (dark, dry, and cool—8 °C). Before analysis, the grains were freeze-dried and about six grams of each grain sample was milled to whole meal flour for 20 s by a laboratory mill (Yellow line, A10, IKA-Werke, Staufen, Germany). Afterwards, carotenoids were extracted from the whole meal flour.

Table 1. Name, type of cereal, and origin (year released, NGB number at Nordgen and if selections have been made on the material) of cereals used in the present study.

Name	Type	Cereal Type	Origin Year Released and NGB Number
Algot	Spring wheat	Old cultivar	1953, NGB11611
Atle	Spring wheat	Old cultivar	1936, NGB7455
Atson	Spring wheat	Old cultivar	1954, NGB7460
Aurore	Spring wheat	Old cultivar	1929, NGB9690
Dacke	Spring wheat	Old cultivar	1990, NGB9955
Dalarna	Spring wheat	Landrace	NGB6673, selection
Diamant brun	Spring wheat	Old cultivar	1938, NGB6681, selection
Ella	Spring wheat	Old cultivar	1950, NGB6683
Gengel	Naked 2-row barley	Landrace Czech republic	From Czech growers
Jusso	Spring rye	Finnish landrace	From finish growers
Kärn 1	Spring wheat	Old cultivar	1947, NGB7458
Naket 6-r	Naked 6-row barley	Selection	Diverse origin from Gene bank
Naket korn 2rads	Naked 2-row barley	Selection	Diverse origin from Gene bank
Naket korn 6rads	Naked 6-row barley	Selection	Diverse origin from Gene bank
Ölands	Spring wheat	Landrace	NGB 4798, selection
Prins	Spring wheat	Old cultivar	1962, NGB6688
Våremmer	Spring wheat	Primitive	NGB4499, selection as spring wheat

2.3. Extraction by Saponification

Saponification was carried out for the extraction of carotenoids using the method described by Fratianni et al. [32] with modifications of Hussain et al. [16,33]. Briefly, 1 g of whole meal flour was weighed into a screw cap Teflon tube and saponified with 2.5 mL of ethanol pyrogallol (60 g/L), 1 mL sodium chloride (10 g/L), 1 mL ethanol (95%), and 1 mL potassium hydroxide (600 g/L). The tubes were placed in a 70 °C water bath for 30 min and mixed every 10 min during saponification. Afterwards, the tubes were cooled in an ice-water bath and 7.5 mL of sodium choloride and n-hexane/ethyle acetate (9:1) was added. Then organic layer was separated by centrifuging at 1500 rpm for 5 min. Two additional extractions were carried out by adding 5 mL of n-hexane/ethyle acetate (9:1) in each extraction. The organic layer was evaporated to dryness and residue was dissolved in 2 mL of n-hexane. Each whole meal flour sample was replicated three times.

2.4. HPLC Analysis

Carotenoid compounds were separated by normal phase HPLC according to Panfili et al. [34] with some modifications. Separation was achieved by 250 × 4.6 mm i.d., 5 µm particle size, Phenomenex LUNA Silica column. The mobile phase was n-hexane/isopropyl alcohol (5%) and flow rate was 1.5 mL/min. A fluorescence detector set in the range of 350–500 nm was used for the detection of the carotenoids and peaks were detected at 450 nm. The volume of each injection was 100 µL. Carotenoids compounds were identified by their particular spectra and their retention time compared with respective standards. In addition, the column was reactivated after every 12 injections with a solution of 10%

isopropyl and n-hexane (*v*/*v*). A representative chromatogram of the analyses is shown in Supplementary (Figure S1).

2.5. Data Analysis

Analysis of variance (ANOVA), Spearman rank correlation and Principle Component Analysis (PCA) were carried out using the statistical analysis system (SAS).

Percentage of explanation (obtained from the coefficient of determination—R^2) of different sources (cultivar, location, and their combinations) on content of total and various types of carotenoids were calculated using a simple linear regression analysis, following the procedure described in previous investigations [9,35–37]. By linear regression analysis, an independent variable can be used to predict the value of a dependent variable, and the R^2 value of a linear regression analysis predicts how well a feature (independent variable) can explain a target (dependent variable) [38]. Thus, an R^2 close to 1 means that the proportion is high that the independent variable explains the dependent variable (https://www.colby.edu/biology/BI17x/regression.html) (accessed on 23 March 2021) which makes it possible to select the sources that are of highest importance to determine different traits [38].

To evaluate stability of genotypes within and between localities, coefficient of variation (ratio of standard deviation to the mean) was calculated as a measure of variability. Thus, a lower coefficient of variation indicates higher stability for a genotype either within or across localities.

To secure proper HPLC results, standard solutions were run several times with different batches of samples in order to calibrate and to check the precision of the method. The results of repeated calibrations showed that carotenoid compounds were determined precisely with CV% less than 3 for β-carotene, β-cryptoxanthin, lutein, and zeaxanthin. The calibration curve was plotted between the concentration and peak areas of each carotenoid compound, i.e., β-carotene, β-cryptoxanthin, lutein, and zeaxanthin, which showed linear relationship as indicated by R^2 value > 0.99. The recovery percentage of β-carotene, β-cryptoxanthin, and lutein was 97.8%, 96.6%, and 99.1%, respectively. The limit of detection (LOD) was measured for the HPLC method and it was 0.04 ppm for β-carotene and 0.011 ppm for lutein.

3. Results

3.1. Grain Carotenoid Content in Locally and Organically Produced Cereals and Genotype X Environment Effects

The total carotenoid content in the wheat grain of locally and organically grown cereals varied between 1.96 mg/kg in the spring wheat cultivar 'Ölands' grown in Alnarp to 9.87 mg/kg in the spring wheat cultivar 'Dalarna' grown at Krusenberg (Table 2). The total carotenoid content in the evaluated barley cultivars varied from 4.83 in 'Naket korn 2rads' at Alnarp to 7.32 in 'Gengel' at Krusenberg (Table 2). The spring rye cultivar 'Jusso' showed total carotenoid content from 3.27 when grown in Ekhaga to 8.59 when grown at Krusenberg. Moreover, the different carotenoids evaluated varied widely over cultivars and localities. Contents were found for the different compounds, from 0.06 (Ölands, Alnarp) to 0.48 mg/kg (Jusso, Gotland) for β-carotene, 0.001 (Våremmer, Krusenberg) to 0.019 mg/kg (Jusso, Krusenberg) for β-cryptoxanthin, 1.56 (Ölands, Alnarp) to 8.47 mg/kg (Dalarna, Krusenberg) for lutein, and 0.33 (Ölands, Alnarp) to 4.00 (Gengel, Krusenberg) for zeaxanthin (Table 2).

3.2. Percentage of Explanation for Variation in Carotenoids Content

R square values used to explain the percentage of explanation for the different carotenoids compounds and for the total carotenoids content, clearly showed a higher degree of explanation for the cultivars analyzed as compared to the location used for cultivation. However, the combination of cultivars and localities showed clearly the highest percentage of variation for the compounds, which was especially valid for β-carotene and lutein, as well as for the total content of carotenoids (Table 3).

Table 2. Mean value (mg/kg) ± standard deviation of the different carotenoids per cultivation locality (*italics*) and cultivar.

Locality and Cultivar	β-Carotene (10^{-1})	β-Cryptoxanthin (10^{-3})	Lutein	Zea-Xanthin	Total Carotenoids
		Ekhaga			
Algot	1.13 ± 0.20	4.81 ± 0.72	2.30 ± 0.06	1.12 ± 0.25	3.53 ± 0.30
Dacke	1.94 ± 0.01	7.01 ± 0.40	3.27 ± 0.21	0.90 ± 0.01	4.37 ± 0.20
Dalarna	3.33 ± 0.12	16.4 ± 0.91	7.63 ± 0.16	1.19 ± 0.01	9.17 ± 0.16
Diamant brun	7.93 ± 0.04	3.06 ± 0.11	1.59 ± 0.04	0.67 ± 0.02	2.35 ± 0.06
Ella	1.14 ± 0.03	2.80 ± 0.27	3.21 ± 0.04	1.28 ± 0.03	4.61 ± 0.06
Gengel	2.62 ± 0.12	6.14 ± 0.54	2.49 ± 0.11	3.03 ± 0.10	5.79 ± 0.21
Jusso	1.40 ± 0.22	4.79 ± 0.23	2.36 ± 0.01	0.73 ± 0.06	3.24 ± 0.09
Kärn 1	1.64 ± 0.03	4.32 ± 0.47	4.93 ± 0.05	1.09 ± 0.02	6.19 ± 0.03
Naket 6-r	2.42 ± 0.22	4.27 ± 0.54	2.58 ± 0.14	2.76 ± 0.23	5.58 ± 0.37
Ölands	0.76 ± 0.02	2.91 ± 0.01	1.66 ± 0.01	0.66 ± 0.01	2.40 ± 0.02
Prins	1.48 ± 0.10	4.57 ± 0.01	4.21 ± 0.03	1.09 ± 0.01	5.45 ± 0.03
Våremmer	0.92 ± 0.05	4.02 ± 0.26	3.45 ± 0.13	0.57 ± 0.04	4.12 ± 0.16
		Krusenberg			
Algot	1.58 ± 0.07	3.88 ± 0.72	6.38 ± 0.21	0.98 ± 0.01	7.52 ± 0.22
Aurore	1.42 ± 0.04	6.46 ± 0.14	5.80 ± 0.06	1.13 ± 0.03	7.08 ± 0.08
Dalarna	2.24 ± 0.03	4.94 ± 0.60	8.14 ± 0.44	1.19 ± 0.05	9.56 ± 0.46
Diamant brun	1.84 ± 0.12	4.83 ± 0.22	4.77 ± 0.04	1.34 ± 0.01	6.29 ± 0.06
Ella	1.19 ± 0.02	2.03 ± 0.01	3.89 ± 0.12	1.09 ± 0.03	5.10 ± 0.16
Gengel	2.09 ± 0.07	2.13 ± 0.01	3.06 ± 0.12	3.85 ± 0.14	7.12 ± 0.23
Jusso	3.50 ± 0.12	16.6 ± 2.29	5.46 ± 0.05	2.63 ± 0.06	8.46 ± 0.12
Kärn 1	1.81 ± 0.02	6.75 ± 1.55	6.77 ± 0.11	1.11 ± 0.16	9.07 ± 0.06
Naket 6-r	1.77 ± 0.02	1.98 ± 0.01	2.41 ± 0.24	2.95 ± 0.07	5.54 ± 0.31
Ölands	1.93 ± 0.07	4.74 ± 0.82	4.52 ± 0.24	1.24 ± 0.06	5.97 ± 0.31
Våremmer	1.73 ± 0.07	1.24 ± 0.01	4.14 ± 0.14	0.56 ± 0.03	4.88 ± 0.18
		Alnarp			
Atson	2.12 ± 0.02	4.86 ± 0.36	4.62 ± 0.01	0.67 ± 0.01	5.50 ± 0.01
Diamant brun	1.78 ± 0.04	5.06 ± 0.28	4.18 ± 0.12	0.55 ± 0.01	4.91 ± 0.13
Ella	0.75 ± 0.00	1.29 ± 0.01	1.86 ± 0.01	1.09 ± 0.03	2.24 ± 0.01
Jusso	2.64 ± 0.19	11.3 ± 1.15	3.74 ± 0.16	1.55 ± 0.17	5.57 ± 0.16
Naket korn 2rads	1.84 ± 0.10	3.03 ± 0.19	2.84 ± 0.06	1.88 ± 0.05	4.90 ± 0.12
Naket korn 6rads	2.06 ± 0.07	4.16 ± 0.01	3.18 ± 0.06	1.98 ± 0.05	5.37 ± 0.12
Ölands	0.68 ± 0.06	1.78 ± 0.01	1.69 ± 0.13	0.36 ± 0.03	2.12 ± 0.17
		Gotland			
Atle	2.28 ± 0.09	5.51 ± 00.33	7.21 ± 0.52	1.32 ± 0.07	8.76 ± 0.61
Dalarna	2.74 ± 0.15	4.68 ± 0.58	5.53 ± 0.32	0.85 ± 0.04	6.66 ± 0.37
Diamant brun	1.27 ± 0.12	5.25 ± 0.79	3.64 ± 0.03	0.81 ± 0.01	4.58 ± 0.04
Ella	1.26 ± 0.07	2.25 ± 0.32	3.50 ± 0.03	0.92 ± 0.01	4.55 ± 0.04
Jusso	3.68 ± 0.96	12.4 ± 0.21	4.83 ± 0.12	2.52 ± 0.06	7.73 ± 0.09
Naket korn 2rads	1.84 ± 0.02	4.71 ± 0.51	2.28 ± 0.03	2.71 ± 0.06	5.19 ± 0.09
Naket korn 6rads	2.60 ± 0.03	2.54 ± 0.39	2.76 ± 0.06	2.38 ± 0.02	5.41 ± 0.05
Prins	1.90 ± 0.02	3.09 ± 0.28	4.83 ± 0.05	0.86 ± 0.04	5.88 ± 0.06

Table 3. Percentage of explanation (obtained through the coefficient of determination (R2) from simple linear regression analysis) of cultivar (C), location (L), and their combinations (CL) on amount of different carotenoids compound and for total carotenoids content.

	β-Carotene	β-Cryptoxanthin	Lutein	Zeaxanthin	Total Carotenoids
C	62.8	32.3	48.1	68.6	37.4
L	8.15	0.73	20.2	6.85	28.8
CL	92.6	66.1	92.4	83.5	93.5

3.3. Relationship between Genotype, Environment and Contents of Different Carotenoids

Relationships among the genotypes and environments for carotenoid content and composition was visually revealed by a principal component analysis (PCA; the first principal component explained 61.3% and the second 23.7% of the variation), clearly being able to depict the spring rye cultivar 'Jusso' with high values on the first principal component and positive value on the second principal component, except when Jusso was grown in Ekhaga. Furthermore, the barley cultivar 'Gengel' and naked barley (two and six row types) were found with positive values on the second principal component (Figure 1a). The positive value on the first principal component for spring rye is a result of a general high content of total carotenoids and especially of a high content of β-carotene in spring rye (Figure 1b). However, cultivation conditions were also able to influence the contents of carotenoids in spring rye with lower values when cultivated in Ekhaga. The positive values for the second principal component for barley is the result of high zeaxanthin values in the barley cultivars evaluated (Figure 1b). Among the spring wheats, 'Dalarna' was found with positive values on the first principal component and negative values on the second principal component independent of growing location (Figure 1a), indicating high level of lutein in this cultivar (Figure 1b). Similarly, 'Atle' from Gotland, and 'Algot', 'Aurore', and 'Kärn I' from Krusenberg showed indications of high lutein content while 'Algot' from Ekhaga was found with negative first and positive second principal component values. 'Atle' and 'Aurore' were only evaluated in one location each.

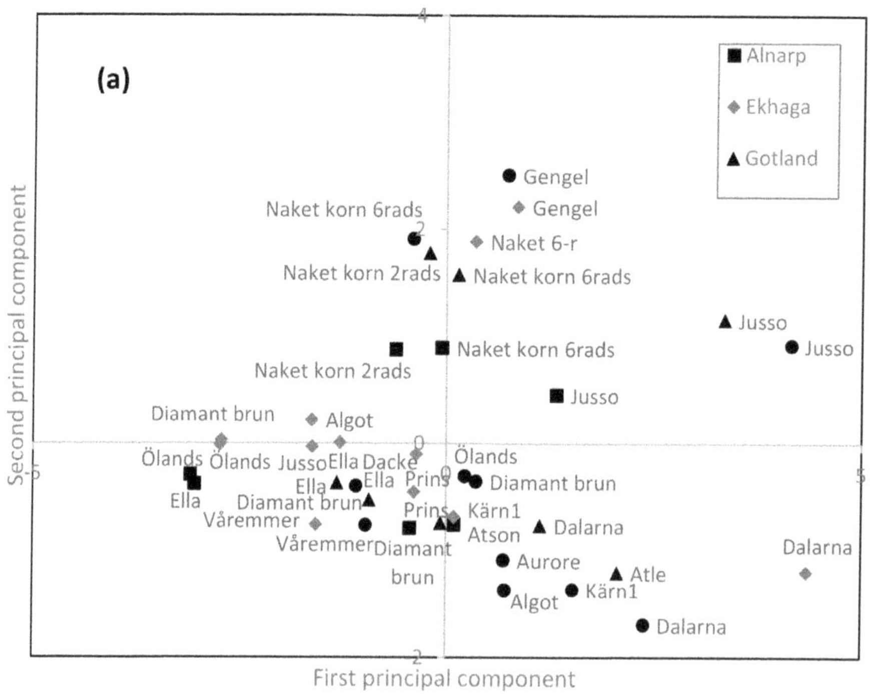

Figure 1. *Cont.*

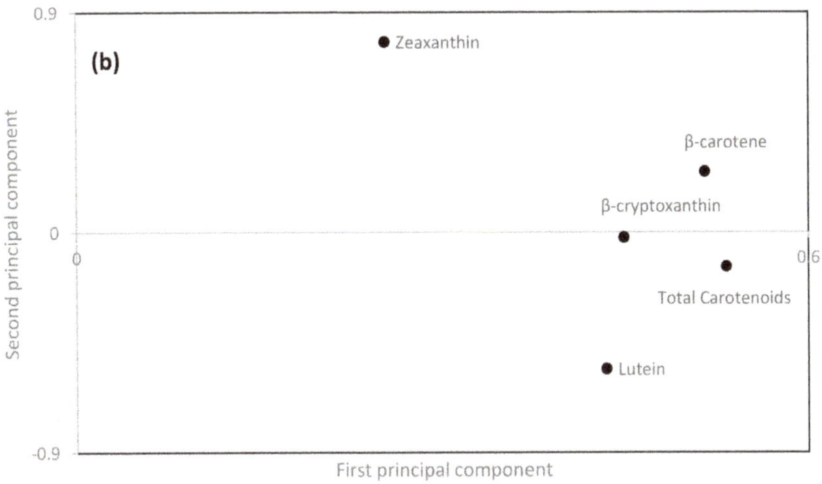

Figure 1. Loading (**a**) and score (**b**) plot from principal component analysis of carotenoid compounds and total carotenoid content in organically grown locally adapted spring cereals. First principal component explained 61.3% of the variation while the second principal component explained 23.7% of the variation.

3.4. Stability Over Environments

The cultivars 'Jusso' and 'Dalarna' showed the highest β-carotene levels across the localities of cultivation for these cultivars although naked barley (2- and 6-rowed) and 'Kärn I' showed the highest stability in β-carotene content over localities (Figure 2a). In Figure 2a,b, percentage of coefficient of variation explain the dispersion of data around mean concentration of carotenoid compounds. It provides a relatively simple and rapid method to understand the level of stability within and across locations of genotypes investigated in the present study. The cultivar 'Jusso' also showed the highest level of β-cryptoxanthin across localities of cultivation although the most stable cultivars over locations was found to be 'Algot', 'Prins', and 'Diamant brun', followed by the naked barley (2 and 6 rows). For lutein, the highest level was found for 'Jusso' and 'Dalarna' while highest stability was found for 'Kärn I' and 'Naket korn 2rads'. Highest and most stable content of zeaxanthin was found for 'Gengel' and naked barley (Naket 6-r) (Figure 2a).

3.5. High Content and Stability per Cultivar within an Environment

The highest content of β-carotene within a locality was found for 'Dalarna' grown in Ekhaga and 'Jusso' grown in Krusenberg while high and stable content was found for 'Atson' grown in Alnarp, 'Dalarna' grown in Krusenberg, and 'Naket korn 6rads' grown in Gotland (Figure 2b). 'Jusso', grown in Gotland was found with high and stable content of β-cryptoxanthin (Figure 2b). High and relatively stable content of lutein was found in 'Dalarna' grown in Ekhaga, and 'Aurore', 'Jusso' and 'Kärn I' grown in Krusenberg (Figure 2b). High and stable zeaxanthin values were found in naked barley (2 and 6 rows) from Alnarp, Krusenberg, and Gotland, 'Jusso' from Gotland and Krusenberg, and 'Gengel' from Krusenberg and Ekhaga (Figure 2b).

3.6. Correlation among Carotenoids and Plant Characters

Significant and positive correlations were found among β-carotene, β-cryptoxanthin, and Lutein, while zeaxanthin only correlated significantly and positively with β-carotene (Table 4). A positive and significant correlation was also found between straw length and content of lutein as well as between straw length and grain protein concentration. No other significant correlations among the content of carotenoids and plant characters were noted in the present study.

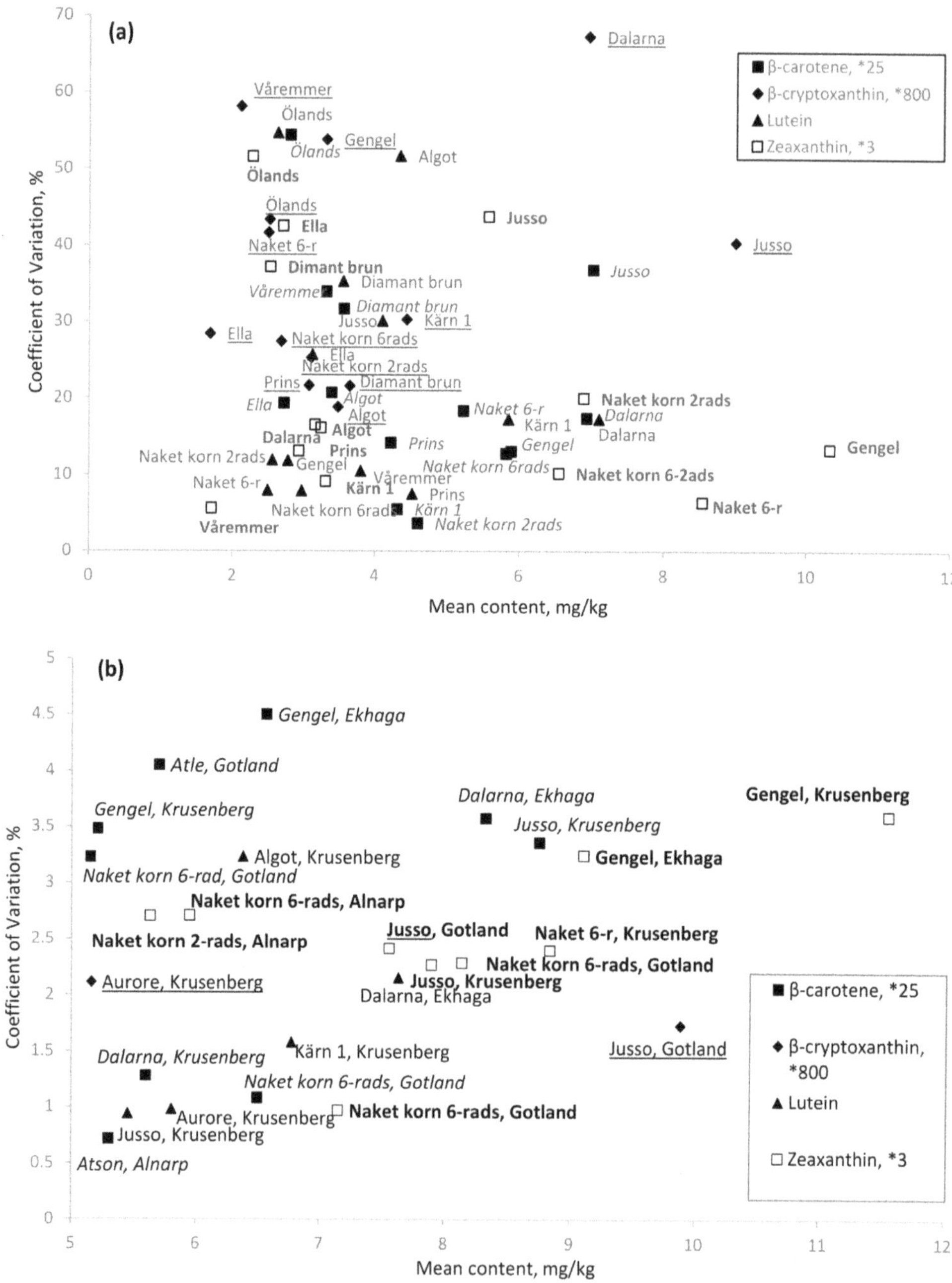

Figure 2. Mean content of carotenoid compounds and stability measured as coefficient of variation of evaluated cultivars over cultivation localities (**a**), and of high mean content cultivars within a specific cultivation locality (**b**). Multiplication numbers given after each carotenoid compound name indicate factor used to obtain similar scale of mean content for the different compounds. Italic, underlined, normal, and bold text indicate mean content of β-carotene, β-cryptoxanthin, lutein, and zeaxanthin, respectively, for the different cultivars (+locations).

Table 4. Pearson correlation coefficients for content of various carotenoids and plant parameters of locally and organically produced cereals.

	β-Cryptoxanthin	β-Carotene	Lutein	Zea-Xanthin	Yield (kg/Ha)	TKV (g)	Protein Content (%)
β-carotene	0.72 ***						
Lutein	0.47 ***	0.48 **					
Zea-xanthin	0.20	0.55 ***	−0.12				
Yield (kg/Ha)	0.14	0.09	0.06	−0.16			
TKV (g)	−0.14	−0.16	−0.14	0.09	−0.27		
Protein Content (%)	−0.17	−0.09	−0.38	0.06	−0.11	0.29	
Straw Length (cm)	0.31	0.11	0.50 **	−0.28	0.24	−0.17	−0.69 ***

, * = significant a $p < 0.01$, $p < 0.005$.

4. Discussion

The locally adapted and organically produced cereals investigated for carotenoids contents extracted through saponification showed high amounts of carotenoid compounds, although with significant differences in content among cereal type, but also among cultivars of the same type of cereal, and cultivation location had a significant impact on the content. Besides variation in mean content, the various cultivars evaluated showed large variation in stability of the content over different cultivation locations and also within the same cultivation location. Thus, the present study revealed the importance of the cultivar and its adaptation to the cultivation location in order to result in a high and stable content of carotenoids, thereby affecting human health positively.

The total carotenoid content in the cereals investigated in this study varied from 1.96 to 9.87 mg/kg, with both highest and lowest content reported for landraces of wheat. Previous studies on wheat have shown large variation in carotenoid contents among wheat cultivars [18,39,40], with values from 0.94 to 13.6 mg/kg [40,41], thereby the levels found in the present study correspond well with previous findings. The large variations in carotenoids content among cereal cultivars highlight opportunities to screen and produce high content batches for specific uses. The highest content of total carotenoids have previously been reported in einkorn, durum wheat, and purple wheat [28], indicating that some of the landraces of wheat evaluated here, showed high content as compared to previously reported values for wheat. Studies on carotenoid content in barley and rye are limited, with reported values of 2.25–4.54 mg/kg for barley [14,28]. Thus, the present study showed high levels of total carotenoid content in the evaluated barley (4.83–7.32) and rye cultivars (3.27–8.59). The high total carotenoid content reported here for locally adapted and organically grown cereals might be an effect of a relative short time from harvest to freezing and analyses of the samples. Lower values were obtained for a wheat material of similar types of landraces when a longer storage time before freezing of the samples was applied [16]. Moreover, a successful saponification may be part of the explanation of the results. Previous studies have indicated that agronomic practices may have an impact on the content of carotenoids in cereals [28,42]. However, large variation in the same material grown at different locations are reported here, indicating also the importance of suitable cultivation locations for specific landraces adapted to different local origins.

The composition of carotenoids varied largely among the investigated cereals. Similarly to what has been reported previously [16,18,34,39], high levels of lutein were found in wheat in the present study. Lutein rich wheat cultivars, with a content of 3.7 mg/kg, have previously been reported as able to contribute 40% of the human daily requirement [16], which would result in the double contribution for the cultivar (i.e., Dalarna) with the highest level in the present study. However, low content of lutein was also found in some of the wheat cultivars in the present study, e.g., in 'Ölands' grown in Alnarp. Previous studies on breeding lines (Öland 5 and Öland 8) from the cultivar 'Ölands' have revealed both high and low content lines [16], which indicate presence of some genetic variation for carotenoids content in this wheat landrace. The reason for the low content of lutein in some of the wheat in this study is, though, a general low content in carotenoids. Thus, for all the evaluated wheat cultivars in the present study, lutein represented 70–95% of the total carotenoids, thereby corresponding to previous results [16]. For spring rye ('Jusso'), the relative content of lutein was 60–70% of the total carotenoids and relatively high contents of β-carotene and β-cryptoxanthin was also noted. Similarly as described previously [28], the barley cultivars evaluated here showed low levels of lutein (40–55% of the total carotenoids), and zeaxanthin was a major component (40–55% of the total carotenoids).

In the present study, R^2 values were used to explain the percentage of explanation of different sources (cultivar, location, and their combinations) on content of total and various types of carotenoids, similarly to what has been reported in previous studies [9,32–34]. Here, the combination of cultivars and localities was found to have the highest percentage of explanation for the amount of all the various carotenoids compounds, i.e., to have an effect of content, the highest effect is obtained by a combination of genotype and the environment. This clearly indicates a local adaptation of the landraces in the present study. Such an adaptation has been developed over long periods of time to a certain region of cultivation, resulting per definition in the specificity of a landrace [43,44]. Such specificity contributes opportunities to produce unique quality grains within a certain region by the use of a specific landrace. However, current plant breeding strategies are often focusing on the production of high yielding genotypes producing similar quality over a large range of environments (cultivation locations and years), thereby contributing food security independent of environment [45]. Thus, unique qualities might be lost on the pace towards high yield and food security. In the present study, we evaluated also the stability in carotenoid content of the landraces between and within localities and the genotypes showed clear variation in this character. Previous studies on stability for minerals content in the same material have also indicated differences in stability [9,46]. Most of the genotypes showed limited stability for all carotenoids compounds between and within cultivation environments (localities) although the different barley genotypes investigated tended to show higher stability than the other cereal genotypes evaluated.

The lipophilic carotenoids (α-carotene, β-carotenes and β-cryptoxanthin) are precursors of vitamin A, which is essential for cell growth, fetal development, and immune functions [47]. Deficiency of vitamin A leads to increased infections risks and pregnancy complications [48] and may even lead to childhood death [48]. The present study, and also previous ones [42], shows that the level of pro-vitamin A compounds in wheat, and also in rye and barley, is generally low and, therefore, consumption of animal products, and some fruits and berries [11,49,50] are recommended for sufficient vitamin A content in the diet. However, cereals are instead rich in other carotenoids than those being precursors of vitamin A. Lutein is the most abundant carotenoid in wheat and barley [18,34,39], which was also confirmed here, while according to this study, zeaxanthin was the pre-dominant carotenoid in rye. Lutein has been ascribed as having an important function for the eye retina and some studies have also indicated an impact in preventing cardiovascular diseases and cancer [51]. Due to the high daily consumption of cereals in large regions in the world, lutein content in the cereals may play an important role for sufficient intake although other sources, e.g., green leafy vegetables [50] may have higher contents. Recommendations of daily intake of lutein is scarce, although a previous study calculating needed intake from serum lutein concentrations in humans, came up with a suggestion of an average intake of lutein + zeaxanthin of 1101 ± 838 μg/day [16]. From these calculations, the consumption of 130 mg wheat per day should be sufficient to cover the daily need of lutein if the highest value genotypes of this study were selected.

5. Conclusions

Locally adapted cereals showed a large variation in carotenoid contents among genotypes and over cultivation locations. Some of the genotypes presented levels of total carotenoids in parity with the highest levels previously reported in einkorn, durum, and purple wheat. Lutein was the major carotenoid in wheat and rye, the latter also with high levels of β-carotene and β-cryptoxanthin, while zeaxanthin was the dominating carotenoid in barley. Most of the locally adapted cereals evaluated showed a clear local adaptation with little stability in carotenoids content especially between cultivation locations. This impair significant opportunities to produce local high value cereals for certain local produce of high value, high nutrition food products. However, the local adaptation also limits conventional breeding opportunities with a variety performing similar over a broad range of environments.

Supplementary Materials: The following are available online at https://www.mdpi.com/article/10.3390/pr9050783/s1, Figure S1: Example of a representative HPLC chromatogram (showing Diamant Brun from Alnarp), with the four peaks of the different carotenoids in the samples indicated by their designations.

Author Contributions: All authors have contributed in the conceptualization, executing, compiling data and write up of the study. All authors have read and agreed to the published version of the manuscript.

Funding: EkoForsk, and the Ekhaga foundation Sweden for project grants and Higher Education Commission, Pakistan for financial support to Abrar Hussain.

Institutional Review Board Statement: Not applicable.

Informed Consent Statement: Not applicable.

Data Availability Statement: Not applicable.

Acknowledgments: We are also thankful to Maria Luisa Prieto-Linde for technical assistance during the chemical analysis.

Conflicts of Interest: The authors declare no conflict of interest.

References

1. Shiferaw, B.; Smale, M.; Braun, H.-J.; Duveiller, E.; Reynolds, M.; Muricho, G. Crops that feed the world 10. Past successes and future challenges to the role played by wheat in global food security. *Food Secur.* **2013**, *5*, 291–317. [CrossRef]
2. FAO Food and Agriculture Organization of the United Nations. *FAOSTAT Database*; FAO: Rome, Italy, 2019.
3. Pena-Bautista, R.J.; Hernandez-Espinosa, N.; Jones, J.M.; Guzman, C.; Braun, H.J. Wheat-based foods: Their global and regional importance in the food supply, nutrition and health. *Cereal Foods World* **2017**, *62*, 231–249. [CrossRef]
4. Save and Grow in Practice: Maize Rice Wheat. A Guide to Sustainable Cereal Production. In *Food and Agriculture Organization of the United Nations (FAO)*; FAO: Rome, Italy, 2016; ISBN ISBN 978-92-5-108519-6.
5. Andersen, V.; Bar, E.; Wirtanen, G. *Nutritional and Health Aspects of Food in Nordic Countries*; Academic Press: Berlin/Heidelberg, Germany, 2018; ISBN 978-0-12-809416-7.
6. Pinheiro, M.A.A.; Bebeli, P.J.; Bettencourt, E.; Costa, G.; Dias, S.; Dos Santos, T.M.M.; Slaski, J.J. Cereal landraces genetic resources in worldwide GeneBanks. A review. *Agron. Sust. Dev.* **2013**, *33*, 177–203.
7. Schmidt, S.B.; George, T.S.; Brown, L.K.; Booth, A.; Wishart, J.; Hedley, P.E.; Martin, P.; Russell, J.; Husted, S. Ancient barley landraces adapted to marginal soils demonstrate exceptional tolerance to manganese limitation. *Ann. Bot.* **2019**, *123*, 831–843. [CrossRef] [PubMed]
8. Hussain, A.; Larsson, H.; Kuktaite, R.; Johansson, E. Mineral composition of organically grown wheat genotypes: Contribution to daily minerals intake. *Int. J. Environ. Res. Public Health* **2010**, *7*, 3442–3456. [CrossRef] [PubMed]
9. Moreira-Ascarrunz, S.G.; Larsson, H.; Prieto-Linde, M.L.; Johansson, E. Mineral nutritional yield and nutrient density of locally adapted wheat genotypes under organic production. *Foods* **2016**, *5*, 89. [CrossRef]
10. Shafie, F.A.; Rennie, D. Consumer perceptions towards organic food. *Procedia Soc. Behav. Sci.* **2012**, *49*, 360–367. [CrossRef]
11. Johansson, E.; Hussain, A.; Kuktaite, R.; Andersson, S.C.; Olsson, M.E. Contribution of organically grown crops to human health. *Int. J. Environ. Res. Public Health* **2014**, *11*, 3870–3893. [CrossRef] [PubMed]
12. Underwood, T.; McCullum-Gomez, C.; Harmon, A.; Roberts, S. Organic agriculture supports biodiversity and sustainable food production. *J. Hunger Environ. Nutr.* **2011**, *6*, 398–423. [CrossRef]
13. Liu, R.H. Whole grain phytochemicals and health. *J. Cereal Sci.* **2007**, *46*, 2017–2219. [CrossRef]
14. Ndolo, V.U.; Beta, T. Distribution of carotenoids in endosperm, germ and aleurone fractions of cereal grain kernels. *Food Chem.* **2013**, *139*, 663–671. [CrossRef] [PubMed]
15. Mellado-Ortega, E.; Hornero-Méndez, D. Carotenoids in cereals: An ancient resource with present and future applications. *Phytochem. Rev.* **2015**, *14*, 873–890. [CrossRef]
16. Hussain, A.; Larsson, H.; Kuktaite, R.; Olsson, M.E.; Johansson, E. Carotenoid content in organically produced wheat: Relevance for human nutritional health on consumption. *Int. J. Environ. Res. Pub. Health* **2015**, *12*, 14068–14083. [CrossRef] [PubMed]
17. Zhou, K.; Yin, J.J.; Yu, L. Phenolic acid, tocopherol and carotenoid compositions, and antioxidant functions of hard red winter wheat bran. *J. Agric. Food Chem.* **2005**, *53*, 3916–3922. [CrossRef]

18. Moore, J.; Hao, Z.; Zhou, K.; Luther, M.; Costa, J.; Yu, L. Carotenoid, tocopherol, phenolic acid and antioxidant properties of Maryland-grown soft wheat. *J. Agric. Food Chem.* **2005**, *53*, 6649–6657. [CrossRef] [PubMed]
19. Leenhardt, F.; Lyan, B.; Rock, E.; Boussard, A.; Potus, J.; Chanliaud, E.; Remesy, C. Genetic variability of carotenoid concentration, and lipoxygenase and perioxidase activities among cultivated wheat species and bread wheat varieties. *Eur. J. Agron.* **2006**, *25*, 170–176. [CrossRef]
20. Konopka, I.; Czaplicki, S.; Rotkiewicz, D. Differences in content and composition of free lipids and carotenoids in flour of spring and winter wheat cultivated in Poland. *Food Chem.* **2006**, *95*, 290–300. [CrossRef]
21. Granado, F.; Olmedilla, B.; Blanco, I. Nutritional and clinical relevance of lutein in human health. *Br. J. Nutr.* **2003**, *90*, 487–502. [CrossRef]
22. Ma, L.; Lin, X.-M. Effects of lutein and zeaxanthin on aspects of eye health. *J. Sci. Food Agric.* **2010**, *90*, 2–12. [CrossRef] [PubMed]
23. Saini, R.K.; Keum, Y.S. Carotenoid extraction methods: A review of recent developments. *Food Chem.* **2018**, *240*, 90–103. [CrossRef]
24. Kopec, R.E.; Cooperstone, J.L.; Cichon, M.J.; Schwartz, S.J. Analysis Methods of Carotenoids. In *Analysis of Antioxidant-Rich Phytochemicals*; Wiley-Blackwell: Oxford, UK, 2012; pp. 105–148. ISBN 9780813823911.
25. Granado, F.; Olmedilla, B.; Gil-Martinez, E.; Blanco, I. A Fast, Reliable and Low-cost Saponification Protocol for Analysis of Carotenoids in Vegetables. *J. Food Comp. Anal.* **2001**, *14*, 479–489. [CrossRef]
26. Abdel-Aal, E.-S.M.; Young, J.C.; Rabalski, I.; Hucl, P.; Frgeau-Reid, J. Identification and quantification of seed carotenois in selected wheat species. *J. Agric. Food Chem.* **2007**, *55*, 787–794. [CrossRef] [PubMed]
27. Hidalgo, A.; Brandolini, A.; Pompei, C.; Piscozzi, R. Carotenoids and tocols of einkorn wheat (*Triticum monococcum* ssp. *monococcum* L.). *J. Cereal Sci.* **2006**, *44*, 182–193. [CrossRef]
28. Paznocht, L.; Kotíková, Z.; Sulc, M.; Lachman, J.; Orsák, M.; Eliášová, M.; Martinek, P. Free and esterified carotenoids in pigmented wheat. Tritordeum and barley grains. *Food Chem.* **2018**, *240*, 670–678. [CrossRef] [PubMed]
29. Tang, Y.; Yerke, A.; Sang, S. *Oats Whole Grains and Their Bioactives: Composition and Health*; Johnson, J., Wallace, T., Eds.; John Wiley & Sons Ltd.: Hoboken, NJ, USA, 2019.
30. Adams, D.C.; Salois, M.J. Local versus organic: A consumer preferences and willingness-to-pay. *Renew. Agric. Food Syst.* **2010**, *25*, 331–341. [CrossRef]
31. Klopcic, M.; Verhees, F.J.H.M.; Kuipers, A.; Kos-Skubic, M. Consumer preceptions of home made, organic, EU certified, and traditional local products in Slovenia. In *Consumer Attitudes to Food Quality Products*; Klopcic, M., Kuipers, A., Hocquette, J.F., Eds.; EAAP—European Federation of Animal Science; Wageningen Academic Press: Wageningen, The Netherlands, 2013; Volume 133.
32. Fratianni, A.; Caboni, M.F.; Irano, M.; Panfili, G.A. Critical comparison between traditional methods and supercritical carbon dioxide extraction for the determination of tocochromanols in cereals. *Eur. Food Res. Technol.* **2002**, *215*, 353–358. [CrossRef]
33. Hussain, A.; Larsson, H.; Olsson, M.E.; Kuktaite, R.; Grausgruber, H.; Johansson, E. Is organically produced wheat a source of tocopherols and tocotrienols for health food? *Food Chem.* **2012**, *132*, 1789–1795. [CrossRef]
34. Panfili, G.; Fratianni, A.; Irano, M. Improved normal-phase high-performance liquid chromatography procedure for the determination of carotenoids in cereals. *J. Agric. Food Chem.* **2004**, *52*, 6373–6377. [CrossRef]
35. Malik, A.H.; Kuktaite, R.; Johansson, E. Combined effect of genetic and environmental factors on the accumulation of proteins in the wheat grain and their relationship to bread-making quality. *J. Cereal Sci.* **2013**, *57*, 170–174. [CrossRef]
36. Vazquez, D.; Berger, A.; Prieto-Linde, M.L.; Johansson, E. Can nitrogen fertilization be used to modulate yield, protein content and bread-making quality in Uruguayan wheat? *J. Cereal Sci.* **2019**, *85*, 153–161. [CrossRef]
37. Mukamuhirwa, A.; Persson Hovmalm, H.; Ortiz, R.; Nyamangyoku, O.; Prieto-Linde, M.L.; Ekholm, A.; Johansson, E. Effect of intermittent drought on grain yield and quality of rice (*Oryza sativa* L.) grown in Rwanda. *J. Agron. Crop Sci.* **2020**, *206*, 252–262. [CrossRef]
38. Ross, S.M. *Introductory Statistics*; Elsevier Inc.: Oxford, UK, 2010.
39. Okarter, N.; Liu, C.S.; Sorrells, M.E.; Liu, R.H. Phytochemical content and antioxidant activity of six diverse varieties of whole wheat. *Food Chem.* **2010**, *119*, 249–257. [CrossRef]
40. Adom, K.K.; Sorrells, M.E.; Liu, R.H. Phytochemicals and antioxidant activity of wheat varieties. *J. Agric. Food Chem.* **2003**, *51*, 7825–7834. [CrossRef] [PubMed]
41. Lachman, J.; Hejtmánková, K.; Kotíková, Z. Tocols and carotenoids of einkorn, emmer and spring wheat varieties: Selection for breeding and production. *J. Cereal Sci.* **2013**, *57*, 207–214. [CrossRef]
42. Fratianni, A.; Giuzio, L.; Di Criscio, T.; Zina, F.; Panfili, G. Response of carotenoids and tocols of durum wheat in relation to water stress and sulfur fertilization. *Food Chem.* **2013**, *61*, 2583–2590. [CrossRef] [PubMed]
43. Ceccarelli, S. Wide adaptation: How wide? *Euphytica* **1989**, *40*, 197–205.
44. Wolfe, M.S.; Baresel, J.P.; Desclaux, D.; Goldringer, I.; Hoad, S.; Kovacs, G.; Löschenberger, F.; Miedaner, T.; Østergård, H.; Van Bueren, E.L. Developments inbreeding cereals for organic agriculture. *Euphytica* **2008**, *163*, 323–346. [CrossRef]
45. Van Bueren, E.L.; Jones, S.S.; Tamm, L.; Murphy, K.M.; Myers, J.R.; Leifert, C.; Messmer, M.M. The need to breed crop varieties suitable for organic farming using wheat, tomato and broccoli as examples: A review. *Njas–Wagen J. Life Sci.* **2010**. [CrossRef]
46. Johansson, E.; Prieto-Linde, M.L.; Larsson, H. Locally adapted and organically grown landrace and ancient spring cereals—A unique source of minerals in the human diet. *Foods* **2021**, *10*, 393. [CrossRef]
47. Rao, A.V.; Rao, L.G. Carotenoids and human health. *Pharmacol. Res.* **2007**, *55*, 207–216. [CrossRef]
48. Zile, M.H. Vitamin A and embryonic development: An overview. *J. Nutr.* **1998**, *128*, 455s–458s. [CrossRef] [PubMed]

49. Andersson, S.C.; Olsson, M.E.; Johansson, E.; Rumpunen, K. Carotenoids in Sea Buckthorn (*Hippophae rhamnoides* L.) Berries during Ripening and Use of Pheophytin a as a Maturity Marker. *J. Agric. Food Chem.* **2009**, *57*, 250–258. [CrossRef] [PubMed]
50. Saini, R.K.; Nile, S.H.; Park, S.W. Carotenoids from fruits and vegetables: Chemistry, analysis, occurrence, bioavailability and biological activities. *Food Res. Int.* **2015**, *76*, 735–750. [CrossRef] [PubMed]
51. Osganian, S.K.; Stampfer, M.J.; Rimm, E.; Spiegelman, D.; Manson, J.E.; Willett, W.C. Dietary carotenoids and risk of coronary artery disease in women. *Am. J. Clin. Nutr.* **2003**, *77*, 1390–1399. [CrossRef] [PubMed]

processes MDPI

Article

Effect of Drying on Nutritional Composition of Atlantic Sea Cucumber (*Cucumaria frondosa*) Viscera Derived from Newfoundland Fisheries

Yi Liu, Deepika Dave *, Sheila Trenholm, Vegneshwaran V. Ramakrishnan and Wade Murphy

Marine Bioprocessing Facility, Centre for Aquaculture and Seafood Development, Fisheries and Marine Institute, Memorial University of Newfoundland, St. John's, NL A1B 4L6, Canada; Yi.Liu@mi.mun.ca (Y.L.); Sheila.Trenholm@mi.mun.ca (S.T.); Vegneshwaran.Ramakrishnan@mi.mun.ca (V.V.R.); Wade.Murphy@mi.mun.ca (W.M.)

* Correspondence: Deepika.Dave@mi.mun.ca

Abstract: *Cucumaria frondosa* is the main sea cucumber species harvested from Newfoundland waters. During processing, the viscera of sea cucumber are usually discarded as waste. As a matter of fact, sea cucumber viscera are abundant in various nutrients and promising for valorization. In the present study, sea cucumber viscera were pretreated by air drying and freeze drying, and the nutritional compositions of the dried products were investigated, including proximate composition, lipid class, fatty acid profile, and amino acid composition. The dried viscera had similar levels of ash, lipids, and proteins compared to fresh viscera. Both air- and freeze-dried viscera had total fatty acid composition similar to fresh viscera, with high levels of omega-3 polyunsaturated fatty acids (PUFAs) (30–31%), especially eicosapentaenoic acid (27–28%), and low levels of omega-6 PUFAs (~1%). The dried samples were abundant in essential amino acids (46–51%). Compared to air-dried viscera, freeze-dried viscera contained a lower content of moisture and free fatty acids, and higher content of glycine and omega-3 PUFAs in phospholipid fraction. The high content of nutritious components in dried viscera of *Cucumaria frondosa* indicates their great potential for valorization into high-value products.

Keywords: marine by-products; *Cucumaria frondosa*; drying; omega-3 fatty acids; eicosapentaenoic acid; amino acids

Citation: Liu, Y.; Dave, D.; Trenholm, S.; Ramakrishnan, V.V.; Murphy, W. Effect of Drying on Nutritional Composition of Atlantic Sea Cucumber (*Cucumaria frondosa*) Viscera Derived from Newfoundland Fisheries. *Processes* **2021**, *9*, 703. https://doi.org/10.3390/pr9040703

Academic Editors: Zongbi Bao and Qianqian Xu

Received: 19 March 2021
Accepted: 14 April 2021
Published: 16 April 2021

1. Introduction

Sea cucumbers are marine invertebrates belonging to the class Holothuroidea. There are approximately 1700 species of sea cucumbers worldwide, which mainly live on the sea floor [1]. Sea cucumbers contain low content of sugar, fat, and cholesterol, but high content of proteins [2]. They are a source of multiple nutritional components, including vitamins, minerals, collagen, amino acids, and polyunsaturated fatty acids. Sea cucumbers also contain a variety of bioactive compounds, such as triterpene glycosides, chondroitin sulfates, glycosaminoglycans, lectins, bioactive peptides, and glycoprotein [3]. During recent decades, therapeutic and medicinal benefits of sea cucumbers have been intensively studied, due to their anticancer, antioxidant, anti-inflammatory, and antimicrobial properties, and wound healing activities [3–5]. Sea cucumbers have been consumed as a type of tonic food in Asia for thousands of years. The market for sea cucumbers was originally developed in Asia and Indo-Pacific regions. With increased awareness about their nutritional value and benefits, consumption of sea cucumbers has increased drastically worldwide, resulting in a global industry of approximately $1 billion [6].

The Canadian market was started in 1980s from the west coast, and spread to Atlantic provinces after two decades [7]. In Newfoundland and Labrador, the most common sea cucumber species is *Cucumaria frondosa* [8]. Although some sea cucumber species worldwide have been included in the International Union for Conservation of Nature

Red List of Threatened Species, *Cucumaria frondosa* is in a sufficient amount, and it is the main sea cucumber species for exportation in Newfoundland fisheries. Sea cucumbers are eviscerated after harvesting. The viscera (all internal organs, including respiratory track, gonad, and intestines) account for approximately 50% of the total body weight of *Cucumaria frondosa*. Sea cucumber viscera (SCV) have been reported to contain various nutrients such as oligosaccharides, saponins, amino acids, polyunsaturated fatty acids, phenols, flavonoids, and trace metals [9]. Mamelona et al. [10] reported that fresh *Cucumaria frondosa* viscera was abundant in essential amino acids (37% of total amino acids) and polyunsaturated fatty acids (45% of total lipids), especially eicosapentaenoic acid (EPA, 17%). However, unlike by-products of many types of finfish and shellfish, which have been intensively valorized for value addition [11,12], there has not been much research performed on valorization of sea cucumber viscera [13–18]. Currently in Newfoundland, the viscera of *Cucumaria frondosa* are entirely discarded as waste. Considering the various bioactive compounds they contain, more research should be performed to valorize them for the industrial production of high-value nutritional products.

Generally, fishery by-products are not immediately processed into value-added products after their collection from processing facilities, especially in remote areas or during peak period of harvesting. Therefore, a proper method is required to preserve the quality and freshness of the raw materials. Drying is a widely used technique in the food industry to maintain product quality and extend their shelf-life, as microbial contamination and subsequent spoilage of products can be effectively slowed down by reducing moisture content in the products [19]. A number of drying methods have been applied in the food industry based on their advantages and the aimed deliverables, such as fluidized bed drying, vacuum drying, hot air drying, and freeze drying [20]. Hot air drying (convective drying) is a traditional drying method, which removes water by a continuous flow of hot air via heat and mass transfer. Hot air drying is the most commonly used method in industry due to its relatively low equipment cost and simple operation. However, it has been claimed to result in product shrinkage and consequent structure damage. The heat-sensitive nutritional components can be destroyed in the hot atmosphere [20]. Freeze drying (lyophilization) is a process that involves sublimation of ice in the frozen product under low temperature and vacuum [21]. Compared to air drying, freeze drying requires higher investment in equipment installation and operation. However, freeze drying has been reported to be more effective in preserving quality and retaining nutritional properties of products [22–25]. In the present study, *Cucumaria frondosa* viscera were dried using two methods, including air drying and freeze drying, and the dried products were analyzed for nutritional composition. The effect of two drying methods was investigated in terms of dehydrating sea cucumber waste and retaining their nutritional profiles, and the potential of valorization of dried sea cucumber waste was evaluated (Figure 1).

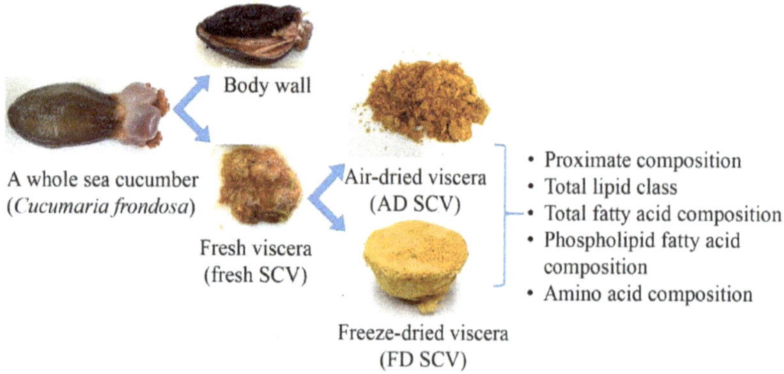

Figure 1. Experimental design of drying sea cucumber (*Cucumaria frondosa*) viscera, analyzing their nutritional profiles.

2. Materials and Methods

2.1. Chemicals

Hexane (≥98.5%), acetanilide (99%), glycine (98.5–101.0%), Kjeltabs Cu−3.5, concentrated sulfuric acid (95.0–98.0%), and concentrated hydrochloric acid (36.5–38.0%) were purchased from Fisher Scientific (Waltham, MA, USA). Nicotinic acid (99.5%) was obtained from Acros Organics (Fair Lawn, NJ, USA). Boric acid (4%) with methyl red/methylene blue indicator was from RICCA Chemical (Arlington, TX, USA). Sodium hydroxide (40%) was from VWR (Radnor, PA, USA). Amino acid standard solution (containing 17 amino acids), acetonitrile (≥99.9%) and N-tert-butyldimethylsilyl-N-methyltrifluoroacetamide (MTBSTFA, ≥99.0%) were purchased from Sigma-Aldrich (St. Louis, MO, USA). All chemicals were of analytical grade.

2.2. Preparation of Samples

Approximately 10 kg of fresh SCV (sea cucumber viscera) was supplied in August 2018 from a local seafood processing plant in Newfoundland, Canada. The samples were immediately stored at 4–8 °C until further processing.

2.2.1. Air Drying

A portion of the fresh SCV (approximately 3 kg) was spread evenly on parchment paper and placed on trays, which were loaded into a drying oven at 34 °C for 24 h (Figure 2a). Afterwards, the samples were removed from the drying oven and allowed to cool (Figure 2b). The partially dried samples were scraped off the parchment paper and flipped over to expose the undersurface, which was not fully dried (Figure 2c). Then the samples were returned to the drying oven for complete drying (Figure 2d). To prevent surface hardening, the oven temperature was lowered to 27 °C. After 48 h, the samples were removed and reweighed. The air-dried SCV were ground for 2 min in a Ninja blender on high speed to obtain powder with particle size ≤5 mm.

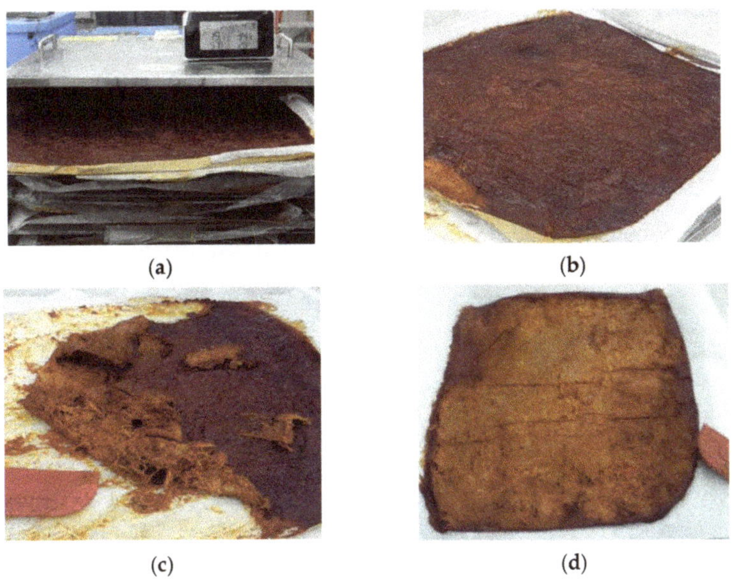

Figure 2. (**a**) Air drying of SCV at 34 °C; (**b**) partially dried SCV after 24 h at 34 °C; (**c**) flipping over the partially dried material; (**d**) SCV after flipped over, ready for 2nd round of drying.

2.2.2. Freeze Drying

A portion of the fresh SCV (approximately 3 kg) was freeze dried using a Labconco 2.5 L Lab Scale freeze dryer. The fresh SCV were frozen in silicone trays at −1 °C overnight. Afterwards, the samples were transferred to freeze drying flasks and weighed. The flasks were loaded onto the freeze dryer and dried between −48 and −52 °C at ≤0.133 mBar for 48 h (Figure 3). After freeze drying, the flasks were removed from the freeze dryer and reweighed. The samples were ground for 2 min in a Ninja blender on high speed to obtain powder with particle size ≤5 mm.

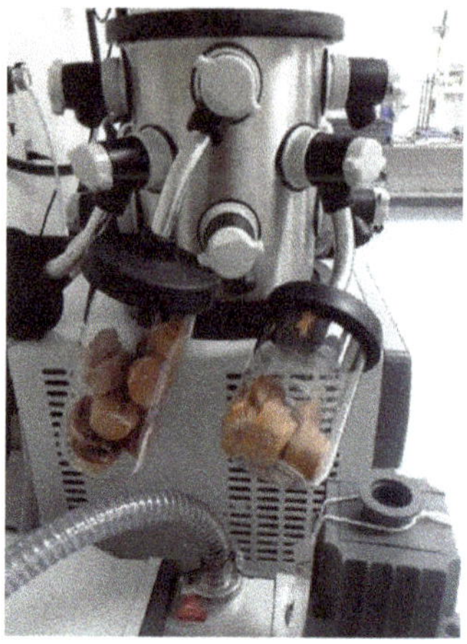

Figure 3. Freeze drying of SCV.

2.3. Proximate Analysis

Proximate analysis was performed on fresh, air-dried, and freeze-dried SCV following the standard method of the Association of Official Analytical Chemists International (AOAC) [26]. Moisture and ash content were determined following AOAC 930.15 and AOAC 938.08. The Kjeldahl method (AOAC 954.01) was applied for protein content analysis and Soxhlet extraction (AOAC 948.15) was applied to determine total lipid content.

2.4. Total Lipid Class and Fatty Acid Composition

Total lipid class, and fatty acid composition of total lipids and phospholipids of fresh and dried SCV were analyzed at Oceans Sciences Centre of Memorial University of Newfoundland (St. John's, NL, Canada).

2.4.1. Lipid Extraction

Lipids in sea cucumber viscera were extracted according to the method by Parrish [27]. An aliquot of approximately 0.25 g sample was weighed in a test tube containing 2 mL of chloroform. Prior to addition of the sample, the test tubes and Teflon® lined caps were rinsed three times with methanol and chloroform, respectively, to remove any lipids. 1 mL of ice-cold methanol, 1 mL of chloroform: methanol (2:1 v/v), and 0.5 mL of chloroform-extracted water were added to the test tube. (Chloroform-extracted water was prepared by adding 1 l of distilled water and 30 mL of chloroform to a separatory funnel. The funnel was manually shaken for 2 min, and the chloroform was allowed to settle and removed from the bottom of the funnel. This procedure was repeated twice to remove any lipids present in the distilled water.) The test tube was then recapped and sonicated for 10 min, followed by centrifugation for 3 min at 3000 rpm. The entire lower organic lipid layer was removed by a double pipetting technique and transferred to a 15 mL lipid-free vial [28]. The double pipetting technique was performed in three steps. Firstly, a 14 cm pipette was passed through the top aqueous layer in the test tube to the bottom of the test tube, by bubbling air with the pipette bulb to prevent the aqueous layer from entering the pipette. Secondly, the pipette bulb was removed, and a 27 cm pipette was placed inside the 14 cm pipette until it reached the bottom of the test tube. Finally, the lipid layer was extracted using the long pipette and transferred to another lipid-free vial. Each of the short and long pipettes was washed with 3 mL ice-cold chloroform, and the wash was collected. The samples were again sonicated, centrifuged, double pipetted, and the pipettes were rinsed three times as previously described, and all the organic layers were collected together. The collected lipid fraction was evaporated under a gentle stream of nitrogen, sealed with Teflon® tape and stored at −20 °C until use.

2.4.2. Lipid Class Determination

Lipid class was determined using an Iatroscan Mark VI thin-layer chromatograph-flame ionization detector (Iatron Laboratories, Inc., Tokyo, Japan) and silica-coated Chromarods® through a three-step development method [29]. The Chromarods® were calibrated (0.2–20 µg) using lipid standards including: nonadecane (aliphatic hydrocarbon), cholesteryl palmitate (wax esters/steryl ester), 3-hexadecanone (ketone), glyceryl tripalmitate (triacylglycerol), 1,2-dihexadecanoylglycerol (diglyceride), 1-hexadecanol (free aliphatic alcohol), cholesterol (free sterol), 1-monopalmitoyl-rac-glycerol (acetone mobile phase lipids), and 1,2-di-o-hexadecyl-sn-glycerol-3-phosphocholine (phospholipids) purchased from Sigma-Aldrich. The lipid extracts and standards were applied to the Chromarods® and focused on a narrow band using 100% acetone.

Four different solvent systems were used to obtain three chromatograms per rod. The first development system was hexane: diethyl ether: formic acid (99.95:1:0.05). The rods were developed for 25 min, removed from the system for 5 min, and placed again in the system for 20 min for double development. The first chromatograms were obtained by scanning each rod to the lowest point behind the ketone peak. The second development was developed for 40 min in hexane: diethyl ether: formic acid (79:20:1). The second chromatogram was obtained by scanning each rod to the lowest point behind the diglyceride peak. The final development was carried out in two steps. The lipid extract was first developed using 100% acetone for two 15 min periods, and then two 10 min periods in chloroform: methanol: chloroform-extracted water (5:4:1). The third chromatogram was obtained as a complete scan after two double developments. Before each solvent system, the rods were dried in a constant humidity chamber. After each development system, the rods were scanned in the Iatroscan and the data were collected using Peak Simple software (ver 3.67, SRI Instruments, Torrance, CA, USA).

2.4.3. Preparation of Fatty Acid Methyl Esters (FAMEs)

An aliquot of 40 µL lipid extract was transferred to a lipid-free vial, in which 1.5 mL of methylene chloride and Hilditch reagent was added. The Hilditch reagent was prepared by adding 1.5 mL of concentrated sulfuric acid to 100 mL of dry methanol. The sample was capped and vortexed for approximately 5 s, followed by sonication for 4 min. The vial was flushed with nitrogen, capped, sealed with Teflon® tape and heated at 100 °C for 1 h in a VWR drying oven (VWR international, Mississauga, ON, Canada). The vial was cooled to room temperature. Approximately 0.5 mL of a saturated solution of sodium bicarbonate in chloroform-extracted water was slowly added to the vial, followed by the addition of 1.5 mL of hexane and vortexing for 10 s. The top organic layer was removed to a new vial, and the hexane was evaporated with a gentle stream of nitrogen. The fatty acids were re-suspended by adding approximately 0.5 of hexane, sealing the vial with Teflon® tape under nitrogen, and sonicating for an additional 4 min.

2.4.4. Fatty Acid Composition Analysis

Fatty acid composition of the FAME samples was analyzed using a gas chromatograph-flame ionization detector (Hewlett Packard 6890 Series II, Agilent Technologies, Mississauga, ON, Canada) following the method developed by [27]. A ZB wax+ polar capillary column (Phenomenex, Torrance, CA, USA, 30 m in length, 0.32 mm of internal diameter and 0.25 μm of film thicknesses) was used. An aliquot of 10 μL FAME sample was injected directly into the column. The initial oven temperature was set at 65 °C and held for 0.5 min, followed by the ramped temperature of 195 °C at a rate of 40 °C/min and held for 15 min, and further ramped to a final temperature of 220 °C at a rate of 2 °C/min and held for 0.75 min. The flame ionization detector was operated at 260 °C with hydrogen as the carrier gas at a flow rate of 2 mL/min. The injector temperature was started at 150 °C and ramped to a final temperature of 250 °C at a rate of 120 °C/min. The analyte peaks were identified using retention times of standards purchased from Supelco, including 37 component FAME mix, bacterial acid methyl ester mix, PUFA No. 1 and PUFA No. 3. Chromatograms were integrated using Agilent OpenLAB Data Analysis-Build 2.203.0.573.

2.4.5. Separation of Phospholipids

The phospholipids were separated using Strata SI-1 silica tubes (Phenomenex 8B-S012-JDG, Phenomenex, Torrance, CA, USA) with the 12-position vacuum manifold set (Phenomenex AH0-6023, Phenomenex, Torrance, CA, USA). The aspirator was started by pulling a very light vacuum. The column was washed by eluting 2 × 3 mL methanol and 2 × 3 chloroform. Afterwards, 3 mL of a 98:1:0.5 mixture of chloroform: methanol: formic acid was eluted. Once the solvent mixture reached the top of the silica, the stop cocks at the base of the tubes were closed and the sample extract was applied directly to the silica using a long pipette. The sample vial was rinsed with a small amount of chloroform as quickly as possible and the rinsing was applied to the silica. A lipid-cleaned 15 mL vial was located at the end of the column for collection. The stop cocks were opened and once the last of the rinse reached the top of the silica, 8 mL of the 98:1:0.5 mixture of chloroform: methanol: formic acid was eluted through the pipette and all the resulting neutral lipid-containing eluent was collected. Afterwards, the silica gel was rinsed with 2 × 3 mL acetone to recover the acetone mobile polar lipids in a second 15 mL vial. Then the vial containing the acetone mobile polar lipids fraction was replaced with a 40 mL vial. 3 mL of chloroform was passed through the column to return the column to a more neutral polarity. Phospholipids were eluted with 6 mL methanol followed by 9 mL of a mixture of chloroform: methanol: water (5:4:1). The phospholipid fraction was transferred to a 50 mL round bottom flask and dried completely in a flash-evaporator. The lipids were then washed into a 15 mL vial using methanol and chloroform.

2.4.6. Preparation of FAMEs of Phospholipids and Fatty Acid Composition Analysis

The same procedure was followed as described in Section 2.4.3 and Section 2.4.4.

2.5. *Amino Acid Analysis*

2.5.1. Hydrolysis of Dried SCV

Acidic hydrolysis of dried SCV was carried out following the method by Fountoulakis and Lahm [30]. Ten milligrams of dried viscera sample was reacted with 1 mL 6 N hydrochloric acid at 110 °C for 24 h. The resulting solution was centrifuged at 13,000 rpm for 15 min. The supernatant was collected, and 0.5 mL of the internal standard solution (0.5 mg/mL norleucine) was added. The solution was thoroughly mixed and ready for further analysis.

2.5.2. Derivatization and Measurement of Amino Acids

The derivatization and analysis of amino acids were performed following the method by Stenerson [31]. Fifty microliters of the solution from 2.5.1 was transferred into a 10 mL test tube and completely dried at 70 °C under nitrogen for 5 min. One hundred microliters of MTBSTFA (N-tert-butyldimethylsilyl-N-methyltrifluoroacetamide) was added to the test tube, followed by 100 μL of acetonitrile. The test tube was tightly capped and heated at 100 °C for 2 h. Afterwards, the sample was allowed to cool at room temperature, and

200 µL of acetonitrile was added. The solution was thoroughly mixed and transferred into a GC (gas chromatograph) vial for analysis.

The analysis was carried out using a gas chromatograph-mass spectrometer (ThermoFisher Trace 1300 GC/ISQ-LT MS) equipped with a SLB-5ms, 20 m × 0.18 mm column (internal diameter of 0.18 µm). The inlet temperature was 280 °C. The splitless flow and time was maintained at 100 mL/min and 0.3 min, respectively. The helium flow was 0.5 mL/min. The initial temperature of GC oven was 60 °C, ramped up to 100 °C at a rate of 20 °C/min and held for 1 min, followed by the ramped temperature of 290 °C at a rate of 10 °C/min and held for 3 min, and finally ramped to 340 °C at a rate of 10 °C/min and held for 2 min. The temperature of MS transfer line (the instrumental component where the eluate from the column of the gas chromatograph is transferred to the ion source of the mass spectrometer) and ion source was maintained at 320 °C and 280 °C, respectively. The standards were initially scanned using MS in the range of m/z 40–639. Once the target compounds were identified, the data were exported and analyzed in SIM acquisition mode (Selected Ion Monitoring mode, in which the mass spectrometer is set to only collect data of compounds that possess the selected mass fragments, instead of a wide range of masses).

The amino acid standard solution was diluted in 0.1 N hydrochloric acid at a concentration of 2.5 µmoles/mL of each amino acid, except cystine at a concentration of 1.25 µmoles/mL. Samples for the calibration curve were prepared by taking 10, 20, 30, 40, and 50 µL from the standard solution and diluted using 400 µL 0.1 N hydrochloric acid. The samples were processed and analyzed following the same procedure as described above.

2.6. Statistical Analysis

The proximate composition, lipid class, and fatty acid composition analyses were performed in triplicates. The amino acid analysis was performed in duplicates. The data were analyzed with analysis of variance (ANOVA) at 95% confidence level using Minitab 17.3.1.

3. Results and Discussion

3.1. Proximate Analysis

The proximate composition of fresh (fresh SCV), air-dried (AD SCV), and freeze-dried sea cucumber viscera (FD SCV) are shown in Table 1 and Figure 4. So far, there have been only two studies performed about proximate composition of *Cucumaira frondosa* and its by-products. Mamelona et al. [10] reported 92.3% moisture, 0.7% ash, 2.0% lipids, and 4.5% proteins in fresh viscera of *Cucumaira frondosa* from Québec, QC, Canada. The differences in proximate composition between reported results from their and the present studies are likely due to variation of living conditions of the sea cucumber, such as the habitat and climate, and the harvesting location and time [32,33]. Zhong et al. [34] reported that body wall of *Cucumaira frondosa* from Newfoundland waters contained 87.4% moisture, 2.97% ash, 0.50% lipids, and 8.34% proteins. In comparison to body wall, viscera have similar content of moisture, ash, and proteins, but much higher amount of lipids, possibly due to the better ability of internal organs to store fat.

In comparison to fresh SCV (82.07% moisture), both drying methods removed most water from the fresh sample. However, freeze drying (1.79% moisture in FD SCV) was comparatively more effective than air drying (7.84% moisture in AD SCV). For food materials with high content of moisture (e.g., above 80%), freeze drying might be more effective in removing the high amount of water. Although Tukey Test indicated that the content of ash, lipids, and proteins of the dried samples were statistically different from the fresh SCV (dry weight basis), the actual differences between the dried samples and the fresh SCV were within a range of approx. 0.1% for ash, 0.04–3% for lipids and 1–3% for proteins (Table 1). Considering the large amounts of ash (approx. 12%), lipids (approx. 23–26%), and proteins (approx. 46–48%) these samples contained, the differences are very small. Therefore, the drying methods did not significantly change the proximate composition of fresh SCV except removing water.

Table 1. Proximate composition of fresh, air-dried (AD SCV), and freeze-dried (FD SCV) sea cucumber viscera (wwb: wet weight basis, dwb: dry weight basis).

Parameter	Fresh SCV	AD SCV	FD SCV
	Percent of wet tissue (wwb, Mean ± SD)		
Moisture (%)	82.07 ± 0.03	7.84 ± 0.14	1.79 ± 0.07
Ash (%)	2.14 ± 0.05	11.51 ± 0.14	12.31 ± 0.05
Lipid (%)	4.68 ± 0.01	23.68 ± 0.33	22.77 ± 0.12
Protein (%)	8.65 ± 0.17	42.20 ± 0.29	46.12 ± 0.20
	Percent of dry matter (dwb, Mean ± SD)		
Ash (%)	11.95 ± 0.29 [a]	12.49 ± 0.15 [b]	12.53 ± 0.05 [b]
Lipid (%)	26.12 ± 0.03 [a]	25.69 ± 0.36 [a]	23.19 ± 0.12 [b]
Protein (%)	48.26 ± 0.17 [a]	45.79 ± 0.31 [b]	46.96 ± 0.20 [a,b]

[a,b] sample means of proximate composition compared by Tukey Test at 95% confidence level was performed for dry matter; means in the same line that do not share the same letter are significantly different ($p < 0.05$).

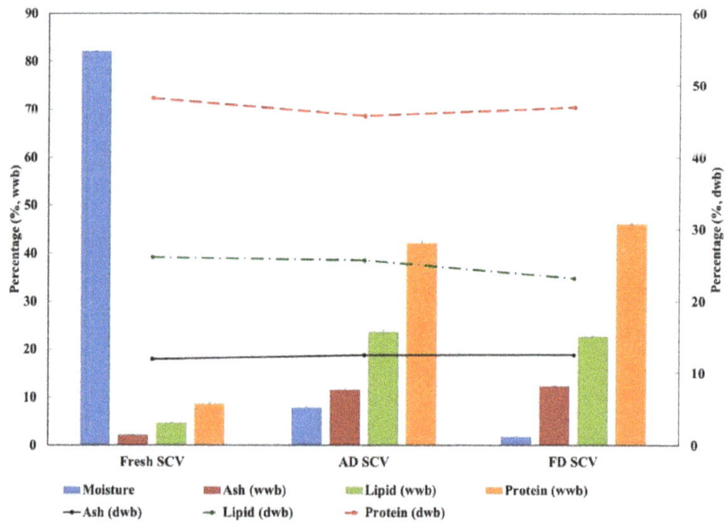

Figure 4. Proximate composition of fresh, air-dried (AD SCV) and freeze-dried (FD SCV) sea cucumber viscera (wwb: wet weight basis, dwb: dry weight basis).

3.2. *Total Lipid Class*

As indicated in Table 2 and Figure 5, the largest component of total lipids in fresh SCV was triacylglycerols (TAGs) (32.93%). TAGs are composed of one glycerol chain linked to three long-chain fatty acids via esterification. Each TAG molecule usually contains one EPA or DHA. So far, TAGs in plant oil and fish oil are the primary source of omega-3 PUFAs for human consumption. In the present study, after air drying, the TAG content was significantly decreased by 19.73% (13.20% in AD SCV), along with a notable increase of free fatty acids (FFAs) by 21% from fresh SCV (7.75%) to AD SCV (28.75%). After freeze drying, the TAG content was slightly decreased by 6.63%, but the FFA content in FD SCV (6.84%) was similar to fresh SCV (7.75%). FFAs are formed due to hydrolysis of lipids catalyzed by hydrolytic enzymes in raw materials [35]. In the present study, the high amount of FFAs in AD SCV was likely due to hydrolysis of TAGs in fresh SCV resulting from the higher temperature (27 and 34 °C) of air drying for longer drying time (72 h) compared to freeze drying (in the range of −48 ~ −52 °C, 48 h).

Table 2. Total lipid class of fresh, air-dried (AD SCV), and freeze-dried (FD SCV) sea cucumber viscera.

Lipid Class	Fresh SCV (%)	AD SCV (%)	FD SCV (%)
Hydrocarbons	0.05 ± 0.09	0.24 ± 0.28	0.30 ± 0.43
Steryl esters/wax esters	0.10 ± 0.17	0.00 ± 0.00	0.00 ± 0.00
Ethyl esters	0.00 ± 0.00	0.00 ± 0.00	0.00 ± 0.00
Methyl esters	0.00 ± 0.00	0.00 ± 0.00	0.00 ± 0.00
Ethyl ketones	0.16 ± 0.27	0.00 ± 0.00	0.00 ± 0.00
Methyl ketones	0.00 ± 0.00	0.00 ± 0.00	0.00 ± 0.00
Glyceryl ethers	26.87 ± 1.93	22.15 ± 0.64	23.14 ± 2.48
Triacylglycerols	32.93 ± 1.96	13.20 ± 0.90	26.30 ± 3.29
Free fatty acids	7.75 ± 1.21	28.75 ± 0.20	6.84 ± 0.36
Alcohols	0.00 ± 0.00	4.01 ± 0.35	0.51 ± 0.72
Sterols	1.92 ± 0.15	1.04 ± 0.18	0.86 ± 0.21
Diacylglycerols	0.00 ± 0.00	0.00 ± 0.00	0.00 ± 0.00
Acetone mobile polar lipids	3.60 ± 1.70	2.85 ± 2.61	4.91 ± 3.30
Phospholipids	26.63 ± 1.31	27.77 ± 3.20	37.13 ± 8.78

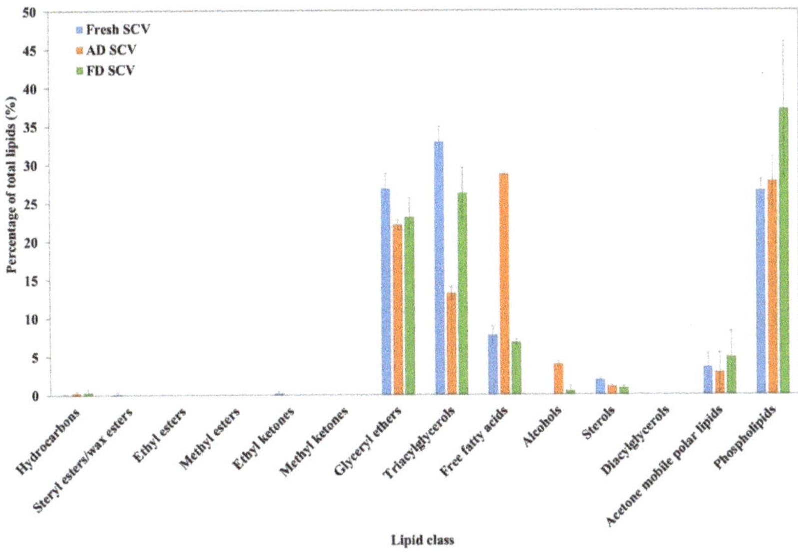

Figure 5. Total lipid class of fresh, air-dried (AD SCV), and freeze-dried (FD SCV) sea cucumber viscera.

In comparison to fresh SCV and AD SCV, FD SCV contained less moisture (Table 1) so they remained in more compact form. Therefore, during lipid extraction, it was more difficult for the organic solvents (chloroform and methanol, see Supplementary Information; for detailed procedure) to penetrate the solid material to extract non-polar TAGs. By comparison, phospholipids (PLs) are amphiphilic due to the hydrophobic fatty acid chains and hydrophilic phosphate moieties in their structures. Therefore, part of PLs in the dried sample might be dissolved in the water added during lipid extraction, which facilitated their extraction into the organic solvents. Consequently, the percentage of extracted TAGs in FD SCV was decreased (26.30%), while extracted PLs was increased (37.13%) compared to fresh SCV (32.93% TAGs and 26.63% PLs). However, both TAG and PL content in FD SCV was higher compared to AD SCV. Since TAGs and PLs are the lipid components of high interest when utilizing marine oil, freeze drying would be more effective for valorization of *Cucumaria frondosa* viscera for nutraceutical products compared to air drying. For drying of other food materials that are abundant in lipids, freeze drying might be selected to prevent the undesired degradation of TAGs to FFAs during the drying process, and retain the high amounts of TAGs and PLs in the materials.

To the authors' best knowledge, no published reports are available on lipid class of *Cucumaria frondosa* viscera. Gianasi et al. [32] reported that fresh *Cucumaria frondosa* gonads (part of viscera) contained 19.0–22.5% TAGs and 17.0–21.2% PLs, which are consistent with the results in the present study. Vaidya and Cheema [36] reported extremely high content of PLs (75.59%) and low content of TAGs (2.27%) for freeze-dried *Cucumaria frondosa* derived from Newfoundland. However, they did not indicate whether the raw material was obtained from a specific body part or whole sea cucumber. Their reported results are quite different from the present study, possibly because they might have used whole sea cucumber (mixture of different parts) with abundant PLs in some parts. Moreover, lipid class of sea cucumber can be significantly influenced by harvesting season and location.

3.3. Fatty Acid Composition

3.3.1. Total Fatty Acid Composition

As indicated in Table 3 and Figure 6, fatty acid compositions of total lipids in both dried samples were similar to fresh SCV, indicating that both drying methods had no effect on total fatty acid composition of the material. For all three samples, polyunsaturated fatty acids (PUFAs) accounted for the highest fraction (33.01–40.14%), followed by monounsaturated fatty acids (MUFAs, 27.01–28.73%) and saturated fatty acids (SFAs, 25.93–26.97%). The content of omega-3 PUFAs was 30.58–31.65% and omega-6 PUFAs was 1.02–1.26%. The amounts of SFAs, MUFAs, and omega-3 PUFAs in the present study are consistent with the values reported by Mamelona et al. [10] for viscera of *Cucumaria frondosa* in Québec, Canada (26.4% SFAs, 28.2% MUFAs, and 29.5% omega-3 PUFAs), but their omega-6 PUFA content was higher (15.4%), possibly due to different living and harvesting conditions of the sea cucumber.

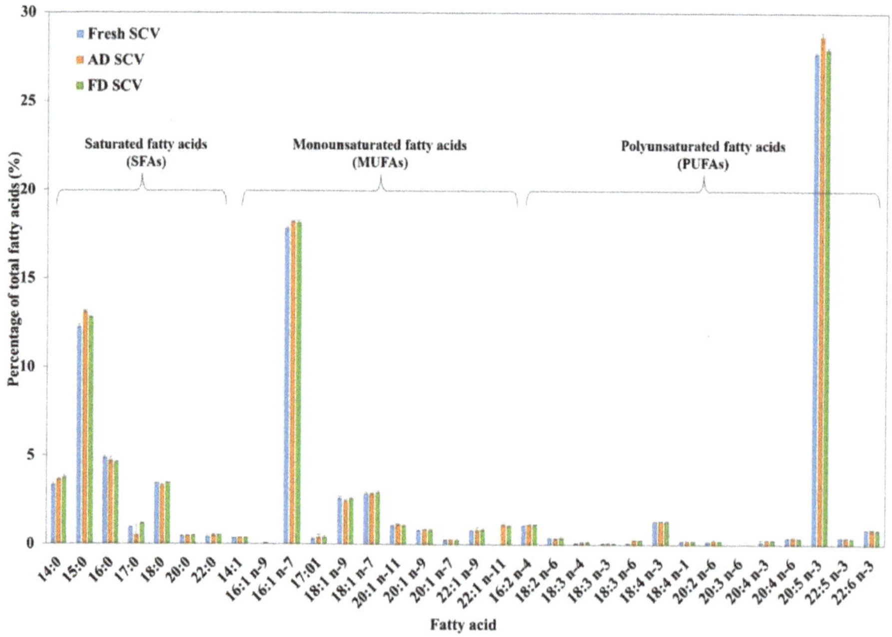

Figure 6. Total fatty acid composition of fresh, air-dried (AD SCV), and freeze-dried (FD SCV) sea cucumber viscera.

Table 3. Total fatty acid composition of fresh, air-dried (AD SCV), and freeze-dried (FD SCV) sea cucumber viscera.

Type	Isomer	Systematic Name	Fresh SCV	AD SCV	FD SCV
Saturated fatty acids (SFAs)	14:0	Tetradecanoic acid	3.39 ± 0.04	3.66 ± 0.04	3.79 ± 0.09
	15:0	Pentadecanoic acid	12.27 ± 0.10	13.12 ± 0.06	12.81 ± 0.04
	16:0	Hexadecanoic acid	4.90 ± 0.09	4.73 ± 0.17	4.66 ± 0.02
	17:0	Heptadecanoic acid	0.99 ± 0.01	0.55 ± 0.54	1.19 ± 0.02
	18:0	Octadecanoic acid	3.45 ± 0.03	3.37 ± 0.05	3.46 ± 0.02
	20:0	Eicosanoic acid	0.48± 0.01	0.50 ± 0.04	0.52 ± 0.01
	22:0	Docosanoic acid	0.45 ± 0.08	0.53 ± 0.06	0.54 ± 0.01
Subtotal (SFAs)			25.93	26.46	26.97
Monounsaturated fatty acids (MUFAs)	14:1	Tetradecenoic acid	0.37 ± 0.00	0.40 ± 0.00	0.40 ± 0.00
	16:1 n-9	cis-7-Hexadecenoic acid	0.00 ± 0.00	0.06 ± 0.01	0.00 ± 0.00
	16:1 n-7	Hexadecenoic acid	17.86 ± 0.05	18.20 ± 0.05	18.16 ± 0.08
	17:1	Heptadecenoic acid	0.33 ± 0.05	0.43 ± 0.14	0.42 ± 0.08
	18:1 n-9	Octadecenoic acid	2.61 ± 0.08	2.47 ± 0.04	2.59 ± 0.06
	18:1 n-7	cis-Vaccenic acid	2.87 ± 0.04	2.89 ± 0.02	2.94 ± 0.06
	20:1 n-11	Gadoleic acid	1.07 ± 0.04	1.14 ± 0.03	1.07 ± 0.03
	20:1 n-9	Eicosenoic acid	0.82 ± 0.03	0.84 ± 0.01	0.85 ± 0.05
	20:1 n-7	Paullinic acid	0.26 ± 0.00	0.30 ± 0.00	0.26 ± 0.05
	22:1 n-9	Erucic acid	0.81 ± 0.01	0.81 ± 0.14	0.77 ± 0.06
	22:1 n-11	Docosenoic acid	0.00 ± 0.00	1.14 ± 0.03	1.07 ± 0.03
Subtotal (MUFAs)			27.01	28.73	28.60
Polyunsaturated fatty acids (PUFAs)	16:2 n-4	Hexadecadienoic acid	1.10 ± 0.00	1.13 ± 0.01	1.13 ± 0.02
	18:2 n-6	Octadecadienoic acid	0.38 ± 0.00	0.34 ± 0.02	0.39 ± 0.05
	18:3 n-4	Octadecatrienoic acid	0.08 ± 0.02	0.13 ± 0.04	0.15 ± 0.00
	18:3 n-3	15-Octadecatrienoic acid	0.06 ± 0.01	0.09 ± 0.02	0.10 ± 0.01
	18:3 n-6	12-Octadecatrienoic acid	0.09 ± 0.02	0.25 ± 0.01	0.25 ± 0.02
	18:4 n-3	6,9,12,15-Octadecatetraenoic acid	1.30 ± 0.01	1.31 ± 0.01	1.31 ± 0.03
	18:4 n-1	Octadeca-9,11,13,15-tetraenoic acid	0.23 ± 0.01	0.23 ± 0.00	0.23 ± 0.01
	20:2 n-6	11,14-Eicosadienoic acid	0.17 ± 0.03	0.24± 0.03	0.22 ± 0.01
	20:3 n-6	8,11,14-Eicosatrienoic acid	0.00 ± 0.00	0.00 ± 0.01	0.00 ± 0.01
	20:4 n-3	8,11,14,17-Eicosatetraenoic acid	0.18 ± 0.07	0.25 ± 0.05	0.25 ± 0.02
	20:4 n-6	5,8,11,14-Eicosatetraenoic acid	0.39 ± 0.01	0.42 ± 0.07	0.38 ± 0.01
	20:5 n-3	Eicosapentaenoic acid (EPA)	27.76 ± 0.07	28.71 ± 0.20	27.97 ± 0.11
	22:5 n-3	Docosapentaenoic acid (DPA)	0.41 ± 0.03	0.41 ± 0.04	0.38 ± 0.01
	22:6 n-3	Docosahexaenoic acid (DHA)	0.88 ± 0.01	0.87 ± 0.01	0.85 ± 0.00
Other PUFAs			6.53 ± 0.27	5.74 ± 0.18	6.31 ± 0.12
Subtotal (PUFAs)			33.01	40.14	39.91
Total Omega-3 PUFAs			30.58	30.58	31.65
Total Omega-6 PUFAs			1.02	1.02	1.26
Other fatty acids			14.05	4.68	4.52

In the present study, eicosapentaenoic acid (EPA) was the predominant fatty acid (27.76–28.71%), which is in agreement with all reported studies about total fatty acid composition of different parts of *Cucumaria frondosa* [10,32,34,36]. EPA has been proved with a variety of health benefits, such as prevention of inflammation, promotion of fetal development, reduction of cardiovascular risk, and enhancement of cognitive function [37]. Therefore, *Cucumaria frondosa* viscera can be potentially valorized for production of omega-3 PUFA (especially EPA) products.

The content of omega-3 PUFAs, omega-6 PUFAs, and EPA in total lipids of fresh and dried SCV were compared to crude oil extracted from some marine by-products that have been widely commercialized for nutritional supplements [38] (Table 4). In comparison to crude oil from salmon, cod liver, and seal, both fresh and dried SCV have higher amounts of omega-3 PUFAs, and at the same time lower amounts of omega-6 PUFAs. As reported by Simopoulos [39], the omega-6/3 ratio in western diets is as high as 15~20:1. An excessive intake of omega-6 PUFAs might promote pathogenesis of many diseases, such as inflammation, obesity, cancer, and cardiovascular diseases [39–41]. An increased intake of omega-3 PUFAs can reduce the risks, and a recommended dietary ratio of omega-6/3 is 1~5:1 [42,43]. In fresh and dried SCV, the omega-6/3 ratio is as low as 0.03 and 0.04, respectively, along with high omega-3 PUFA content (above 30%). Furthermore, the EPA content in SCV is significantly higher (27–28%) in comparison to crude oil from other marine by-products (4–9%, Table 4). Therefore, it seems promising to process *Cucumaria frondosa* viscera into omega-3 PUFAs and EPA enriched nutritional products without the concern of excessive intake of omega-6 PUFAs.

Table 4. Comparison of omega-3 PUFAs, omega-6 PUFAs, and EPA in total lipids from the present study and literature.

Species	Total Omega-3 PUFAs (%)	Total Omega-6 PUFAs (%)	Omega-6 /Omega-3	EPA (%)	Reference
Fresh SCV	30.58	1.02	0.03	27.76	The present study
AD SCV	31.65	1.26	0.04	28.71	
FD SCV	30.85	1.24	0.04	27.97	
Crude farmed Atlantic salmon oil	9.93	15.11	1.52	4.63	Dave et al. [38]
Crude seal oil	16.27	2.03	0.12	7.12	
Crude cod liver oil	20.77	2.29	0.11	8.52	
Crude wild Pacific salmon oil	21.18	2.26	0.11	9.54	

3.3.2. Phospholipid Fatty Acid Composition

Phospholipids (PLs) are the main component of cellular membranes of various living organisms, and essential in gene expression, lipoprotein formation, and signaling systems [44]. PLs have been reported with numerous health benefits, including relief of inflammatory reactions, inhibition of certain types of cancer, regulation of blood lipid profiles, improvement of neurological development, enhancement of immunological functions, and reduction of liver diseases [45]. Compared to TAGs, PLs are more biocompatible and bioavailable [46,47]. In the present study, PLs were the most abundant lipid component in both air- and freeze-dried SCV (27.77% in AD SCV and 37.13% in FD SCV, Table 2). Therefore, fatty acid composition of PLs of dried SCV were analyzed and compared to fresh SCV (Table 5 and Figure 7).

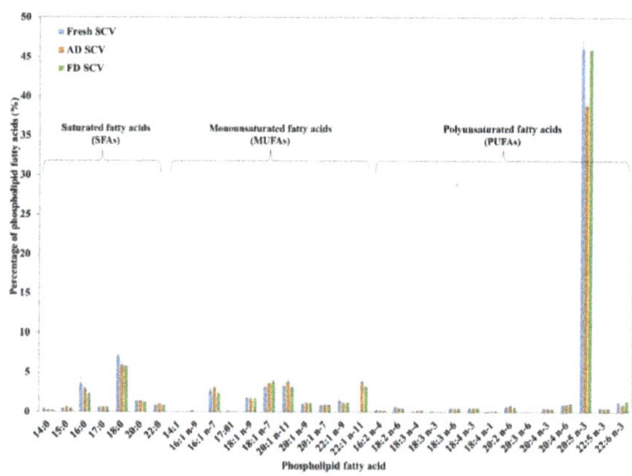

Figure 7. Phospholipid fatty acid composition of fresh, air-dried (AD SCV), and freeze-dried (FD SCV) sea cucumber viscera.

Table 5. Phospholipid fatty acid composition of fresh, air-dried (AD SCV), and freeze-dried (FD SCV) sea cucumber viscera.

Type	Isomer	Systematic Name	Fresh SCV	AD SCV	FD SCV
Saturated fatty acids (SFAs)	14:0	Tetradecanoic acid	0.35 ± 0.11	0.25 ± 0.01	0.18 ± 0.03
	15:0	Pentadecanoic acid	0.44 ± 0.03	0.65 ± 0.03	0.43 ± 0.07
	16:0	Hexadecanoic acid	3.62 ± 0.68	2.96 ± 0.60	2.34 ± 0.13
	17:0	Heptadecanoic acid	0.62 ± 0.03	0.66 ± 0.02	0.64 ± 0.01
	18:0	Octadecanoic acid	7.08 ± 0.32	5.93 ± 0.38	5.81 ± 0.02
	20:0	Eicosanoic acid	1.37 ± 0.12	1.34 ±0.02	1.29 ± 0.03
	22:0	Docosanoic acid	0.93 ± 0.18	1.07 ± 0.03	0.87 ± 0.04
Subtotal (SFAs)			14.41	12.85	11.56
Monounsaturated fatty acids (MUFAs)	14:1	Tetradecenoic acid	0.00 ± 0.00	0.04 ± 0.01	0.03 ± 0.00
	16:1 n-9	cis-7-Hexadecenoic acid	0.19 ± 0.06	0.00 ± 0.00	0.01 ± 0.02
	16:1 n-7	Hexadecenoic acid	2.68 ± 0.29	3.09 ± 0.13	2.40 ± 0.14
	17:1	Heptadecenoic acid	0.12 ± 0.18	0.07 ± 0.00	0.03 ± 0.00
	18:1 n-9	Octadecenoic acid	1.81 ± 0.10	1.69 ± 0.07	1.75 ± 0.03
	18:1 n-7	cis-Vaccenic acid	3.15 ± 0.07	3.63 ± 0.02	3.84 ± 0.08
	20:1 n-11	Gadoleic acid	3.40 ± 0.32	3.86 ± 0.30	3.15 ± 0.07
	20:1 n-9	Eicosenoic acid	1.03 ± 0.02	1.15 ± 0.02	1.16 ± 0.04
	20:1 n-7	Paullinic acid	0.84 ± 0.11	0.91 ± 0.03	0.94 ± 0.03
	22:1 n-9	Erucic acid	1.49 ± 0.14	1.82 ± 0.05	1.76 ± 0.06
	22:1 n-11	Docosenoic acid	0.00 ± 0.00	3.86 ± 0.30	3.15 ± 0.07
Subtotal (MUFAs)			14.71	19.46	17.60
Polyunsaturated fatty acids (PUFAs)	16:2 n-4	Hexadecadienoic acid	0.28 ± 0.05	0.22 ± 0.01	0.25 ± 0.00
	18:2 n-6	Octadecadienoic acid	0.65 ± 0.14	0.48 ± 0.06	0.44 ± 0.03
	18:3 n-4	Octadecatrienoic acid	0.09 ± 0.03	0.20 ± 0.00	0.24 ± 0.01
	18:3 n-3	15-Octadecatrienoic acid	0.10 ± 0.03	0.09 ± 0.01	0.08 ± 0.00
	18:3 n-6	12-Octadecatrienoic acid	0.50 ± 0.11	0.42 ± 0.00	0.42 ± 0.01
	18:4 n-3	6,9,12,15-Octadecatetraenoic acid	0.49 ± 0.03	0.48 ± 0.01	0.52 ± 0.01
	18:4 n-1	Octadeca-9,11,13,15-tetraenoic acid	0.08 ± 0.03	0.14 ±0.01	0.13 ± 0.03
	20:2 n-6	11,14-Eicosadienoic acid	0.67 ± 0.13	0.84 ± 0.02	0.58 ± 0.02
	20:3 n-6	8,11,14-Eicosatrienoic acid	0.00 ± 0.00	0.00 ± 0.00	0.00 ± 0.00
	20:4 n-3	8,11,14,17-Eicosatetraenoic acid	0.39 ± 0.15	0.40 ± 0.02	0.38 ± 0.02
	20:4 n-6	5,8,11,14-Eicosatetraenoic acid	0.93 ± 0.07	0.97 ± 0.03	1.10 ± 0.02
	20:5 n-3	Eicosapentaenoic acid (EPA)	46.13 ± 1.39	38.88 ± 0.36	45.97 ± 0.50
	22:5 n-3	Docosapentaenoic acid (DPA)	0.46 ± 0.12	0.39 ± 0.01	0.42 ± 0.02
	22:6 n-3	Docosahexaenoic acid (DHA)	1.26 ± 0.27	0.97 ± 0.13	1.32 ± 0.01
Other PUFAs			14.14 ± 4.74	10.46 ± 0.69	8.89 ± 0.27
Subtotal (PUFAs)			52.02	54.93	60.72
Total Omega-3 PUFAs			48.82	48.82	41.21
Total Omega-6 PUFAs			2.75	2.75	2.71
Other fatty acids			18.87	12.77	10.12

For all three samples, the content of PUFAs (52.02–60.72%) in PLs were significantly higher than SFAs (11.56–14.41%) and MUFAs (14.71–19.46%). The amounts of omega-3 PUFAs in PLs of FD SCV (48.69%) were similar to fresh SCV (48.82%), which were slightly higher than AD SCV (41.21%). EPA was the most abundant fatty acid, accounting for 46.13% and 45.97 % in fresh and FD SCV, respectively, and 38.88% in AD SCV, which was relatively lower. Vaidya and Cheema [36] reported similar content of omega-3 PUFAs (49.59%) and EPA (44.4%) in PLs of freeze-dried *Cucumaria frondosa* derived from Newfoundland.

Araujo et al. [48] reported 49.4% omega-3 PUFAs, 2.5% omega-6 PUFAs, and 28.5% EPA for PLs in krill oil. Krill oil is a type of marine oil abundant in PLs, and has been widely commercialized as nutritional supplements. In comparison to fish oil, which is mainly composed of TAGs, omega-3 PUFAs in krill oil can be absorbed by humans more safely and efficiently with low side-effects, due to the high efficiency of PLs to deliver the fatty acid residues into cell membranes [45,49]. The amounts of total omega-3 and 6 PUFAs in PLs of fresh and dried SCV in the present study are quite similar to krill oil, and the EPA content in SCV is even higher. Therefore, there is a great potential for *Cucumaria frondosa* viscera to be valorized for value-added PL-rich nutritional products.

3.4. Total Amino Acid Composition

In the present study, sixteen amino acids were identified, including eight essential amino acids (EAAs) and eight non-essential amino acids (NEAAs). As indicated in Table 6, the two drying methods afforded similar yields of total EAAs and NEAAs from fresh SCV, with the EAAs/NEAAs ratio of approximately 1. As claimed by Mamelona et al. [50], marine by-products commonly have a ratio of EAAs/NEAAs higher than 0.5, indicating the high quality of marine proteins and their potential as a source of balanced dietary proteins.

Table 6. Amino acid composition of air-dried (AD SCV) and freeze-dried (FD SCV) sea cucumber viscera.

Type	AD SCV (mg/100 g)	AD SCV (%)	FD SCV (mg/100 g)	FD SCV (%)
		Essential amino acids (EAAs)		
Histidine	396.18 ± 163.86	1.04	225.99 ± 135.30	0.62
Threonine	3815.16 ± 766.14	10.00	2878.91 ± 377.70	7.95
Valine	64.00 ± 35.21	0.17	62.26 ± 6.95	0.17
Methionine	3100.43 ± 44.01	8.12	2854.84 ± 290.10	7.88
Isoleucine	5938.32 ± 263.54	15.56	5385.95 ± 181.72	14.87
Leucine	295.86 ± 92.82	0.78	185.63 ± 44.80	0.51
Phenylalanine	5841.59 ± 295.02	15.31	5056.60 ± 229.27	13.96
Lysine	224.32 ± 82.05	0.59	223.01 ± 42.83	0.62
Total EAAs	19,675.86	51.57	16,873.19	46.58
		Non-essential amino acids (NEAAs)		
Aspartic acid	191.85 ± 22.15	0.50	140.87 ± 5.67	0.39
Glutamic acid	9897.19 ± 4.90	25.93	7926.57 ± 136.11	21.89
Serine	647.22 ± 479.46	1.70	263.47 ± 91.24	0.73
Glycine	1210.25 ± 861.02	3.17	6275.05 ± 98.92	17.33
Alanine	1877.57 ± 270.81	4.92	673.43 ± 342.47	1.86
Proline	4121.35 ± 11.38	10.80	3755.68 ± 184.38	10.37
Tyrosine	407.29 ± 136.74	1.07	196.90 ± 95.39	0.54
Cystine	134.40 ± 4.06	0.35	105.48 ± 9.33	0.29
Total NEAAs	18,487.12	48.44	19,337.45	53.40
EAAs/NEAAs	1.06		0.87	

Glutamic acid was the predominant amino acid in both dried samples (9897.19 mg/100 g in AD SCV and 7926.57 mg/100 g in FD SCV, Table 6 and Figure 8), which agrees with amino acid profiles reported by Mamelona et al. [10] for *Cucumaria frondosa* viscera and Zhong et al. [34] for body wall. The most abundant EAA in both dried samples was isoleucine, followed by phenylalanine. In comparison to AD SCV, FD SCV contained a significantly higher amount of glycine (6275.05 mg/100 g). Heat treatment has been claimed possibly changing compositions of nitrogenous compounds [51]. Similar observations that different drying methods resulted in different amino acid compositions of dried marine materials have been reported by Deng et al. [52] and Kim et al. [53] in their studies about squid and yellow croaker, respectively. As observed in the present study, for drying of food materials that are abundant in glycine, freeze drying might be more effective in retaining the high level of glycine compared to air drying.

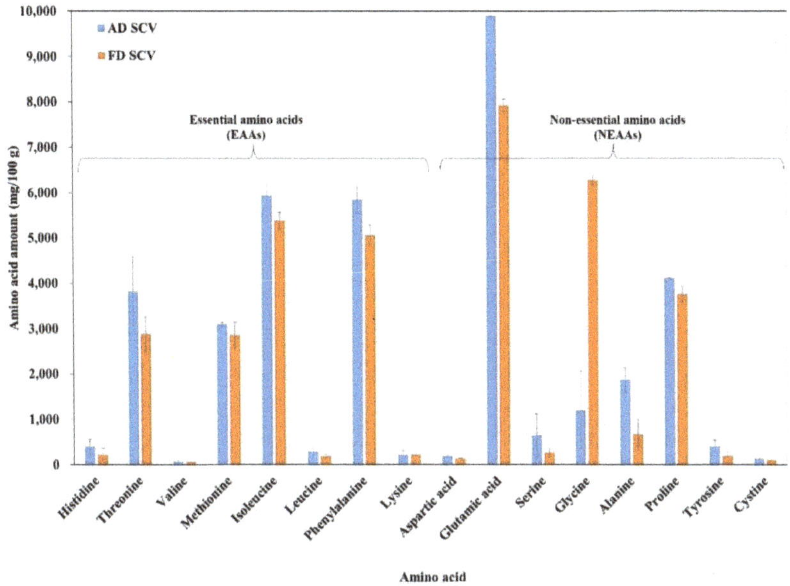

Figure 8. Amino acid composition of air-dried (AD SCV) and freeze-dried (FD SCV) sea cucumber viscera.

Glutamic acid and its derivative glutamine have been reported with various health benefits, including anticancer activity, cell proliferation, wound healing, improvement of protein metabolism, enhancement of immune system, and prevention of bacterial translocation [54,55]. Isoleucine plays a crucial role in formation of hemoglobin, stabilization and regulation of blood sugar, and energy levels [56]. It also promotes muscle recovery after physical exercise and tissue repair after surgery [57]. Phenylalanine is an important precursor for production of tyrosine, which can be converted to proteins and signaling molecules that are closely related to learning ability and stress control [58,59]. Glycine is important for living organisms as it plays the role of a neurotransmitter, and is an essential component in many biological molecules and metabolic reactions [60]. It promotes production and release of interleukin-2 and B lymphocytes in the immune system, thus facilitating phagocytosis [61]. Glycine is crucial in synthesis of glutathione, which is a primary antioxidant enzyme in the human body and promotes generation and activation of natural killer cells [62]. Moreover, glycine has been proved with multiple cytoprotective effects for brain, heart, liver, and kidney [60]. In spite of slightly higher amounts of glutamic acid, isoleucine, and phenylalanine in AD SCV, FD SCV had a significantly higher content of glycine, therefore freeze drying might be more effective in processing *Cucumaria frondosa* viscera for production of functional foods and nutritional supplements.

4. Conclusions

This study creates a paradigm for future research on using drying technology to pretreat sea cucumber viscera for production of value-added products. Compared to air drying, freeze drying was more effective in removing moisture and retaining the nutritional components (TAGs, PLs and glycine) of *Cucumaria frondosa* viscera. The present study indicates the high amounts of nutrients (omega-3 PUFAs and glycine) in dried *Cucumaria frondosa* viscera, and addresses their potential for further processing into functional food, nutraceuticals and pharmaceuticals.

Supplementary Materials: The following are available online at https://www.mdpi.com/article/10.3390/pr9040703/s1, Experimental methodologies of determination of total lipid class and fatty acid composition of total lipids and phospholipids.

Author Contributions: Conceptualization, Y.L., D.D., and W.M.; methodology, Y.L. and D.D.; validation, S.T. and V.V.R.; formal analysis, Y.L., S.T., and V.V.R.; investigation, Y.L., S.T., and V.V.R.; resources, W.M.; data curation, Y.L.; writing—original draft preparation, Y.L.; writing—review and editing, Y.L. and D.D.; visualization, Y.L.; supervision, D.D.; project administration, D.D.; funding acquisition, D.D. All authors have read and agreed to the published version of the manuscript.

Funding: This work was funded by the National Research Council Canada; and Natural Sciences and Engineering Research Council of Canada under Discovery Grants program (RGPIN-2015-06121).

Institutional Review Board Statement: This study doesn't involve humans or animals.

Informed Consent Statement: This study doesn't involve humans.

Data Availability Statement: This study doesn't report any data.

Conflicts of Interest: The authors declare no conflict of interest. The funders had no role in the design of the study; in the collection, analyses, or interpretation of data; in the writing of the manuscript, or in the decision to publish the results.

References

1. Conand, C. Tropical sea cucumber fisheries: Changes during the last decade. *Mar. Pollut. Bull.* **2018**, *133*, 590–594. [CrossRef] [PubMed]
2. Wen, J.; Hu, C.; Fan, S. Chemical composition and nutritional quality of sea cucumbers. *J. Sci. Food Agric.* **2010**, *90*, 2469–2474. [CrossRef] [PubMed]
3. Hossain, A.; Dave, D.; Shahidi, F. Northern sea cucumber (*Cucumaria frondosa*): A potential candidate for functional food, nutraceutical, and pharmaceutical sector. *Mar. Drugs* **2020**, *18*, 274. [CrossRef] [PubMed]
4. Shi, S.; Feng, W.; Hu, S.; Liang, S.; An, N.; Mao, Y. Bioactive compounds of sea cucumbers and their therapeutic effects. *Chin. J. Oceanol. Limnol.* **2016**, *34*, 549–558. [CrossRef]
5. Pangestuti, R.; Arifin, Z. Medicinal and health benefit effects of functional sea cucumbers. *J. Tradit. Complement. Med.* **2018**, *8*, 341–351. [CrossRef]
6. Tridge Sea Cucumber. Available online: https://www.tridge.com/intelligences/sea-cucumber/production (accessed on 18 August 2020).
7. Hamel, J.-F.; Mercier, A. Population status, fisheries and trade of sea cucumbers in temperate areas of the Northern Hemisphere. *Sea Cucumbers A Glob. Rev. Fish. Trade* **2008**, *516*, 257–291.
8. Fisheries and Oceans Canada. *Fisheries and Oceans Canada Sea Cucumber Stock Status Update in NAFO Subdivision 3PS*; Fisheries and Oceans Canada: St. John's, NL, Canada, 2018.
9. Zhang, Y.; Chang, H. Research progress on the structural components and function of sea cucumber viscera and their application. *Sci. Technol. Food Ind.* **2014**, *13*, 382–386.
10. Mamelona, J.; Saint-Louis, R.; Pelletier, É. Proximate composition and nutritional profile of by-products from green urchin and Atlantic sea cucumber processing plants. *Int. J. Food Sci. Technol.* **2010**, *45*, 2119–2126. [CrossRef]
11. Hou, Y.; Shavandi, A.; Carne, A.; Bekhit, A.A.; Ng, T.B.; Cheung, R.C.; Bekhit, A.E. Marine shells: Potential opportunities for extraction of functional and health-promoting materials. *Crit. Rev. Environ. Sci. Technol.* **2016**, *46*, 1047–1116. [CrossRef]
12. Shavandi, A.; Hou, Y.; Carne, A.; McConnell, M.; Bekhit, A.E.D.A. Marine waste utilization as a source of functional and health compounds. In *Advances in Food and Nutrition Research*; Toldra, F., Ed.; Academic Press: Cambridge, MA, USA, 2019; Volume 87, pp. 187–254. ISBN 9780128160497.
13. Mamelona, J.; Pelletier, É. Producing high antioxidant activity extracts from Echinoderm byproducts by using pressured liquid extraction. *Biotechnology* **2010**, *9*, 523–528. [CrossRef]
14. Zhou, X.; Zhou, D.-Y.; Yin, F.-W.; Song, L.; Liu, Y.-X.; Xie, H.-K.; Gang, K.-Q.; Zhu, B.-W.; Shahidi, F. Glycerophospholipids in sea cucumber (*Stichopus japonicus*) and its processing by-products serve as bioactives and functional food ingredients. *J. Food Bioact.* **2018**, *1*, 134–142. [CrossRef]
15. Mamelona, J.; Pelletier, É.; Girard-Lalancette, K.; Legault, J.; Karboune, S.; Kermasha, S. Quantification of phenolic contents and antioxidant capacity of Atlantic sea cucumber, *Cucumaria frondosa*. *Food Chem.* **2007**, *104*, 1040–1047. [CrossRef]
16. Zhong, C.; Sun, L.C.; Yan, L.J.; Lin, Y.C.; Liu, G.M.; Cao, M.J. Production, optimisation and characterisation of angiotensin converting enzyme inhibitory peptides from sea cucumber (*Stichopus japonicus*) gonad. *Food Funct.* **2018**, *9*, 594–603. [CrossRef] [PubMed]
17. Zheng, J.; Wu, H.T.; Zhu, B.W.; Dong, X.P.; Zhang, M.M.; Li, Y.L. Identification of antioxidative oligopeptides derived from autolysis hydrolysates of sea cucumber (*Stichopus japonicus*) guts. *Eur. Food Res. Technol.* **2012**, *234*, 895–904. [CrossRef]

18. Yan, M.; Tao, H.; Qin, S. Effect of enzyme type on the antioxidant activities and functional properties of enzymatic hydrolysates from sea cucumber (*Cucumaria frondosa*) viscera. *J. Aquat. Food Prod. Technol.* **2016**, *25*, 940–952. [CrossRef]
19. Moses, J.A.; Norton, T.; Alagusundaram, K.; Tiwari, B.K. Novel drying techniques for the food industry. *Food Eng. Rev.* **2014**, *6*, 43–55. [CrossRef]
20. Guiné, R.P.F. The drying of foods and its effect on the physical-chemical, sensorial and nutritional Properties. *ETP Int. J. Food Eng.* **2018**, *4*, 93–100. [CrossRef]
21. Oetjen, G.-W.; Haseley, P. *Freeze-Drying*, 2nd ed.; Wiley-VCH: Weinheim, Germany, 2004.
22. Mejia-Meza, E.I.; Yanez, J.A.; Davies, N.M.; Rasco, B.; Younce, F.; Remsberg, C.M.; Clary, C. Improving nutritional value of dried blueberries (*Vaccinium corymbosum L.*) combining microwave-vacuum, hot-air drying and freeze drying technologies. *Int. J. Food Eng.* **2008**, *4*. [CrossRef]
23. Asami, D.K.; Hong, Y.J.; Barrett, D.M.; Mitchell, A.E. Comparison of the total phenolic and ascorbic acid content of freeze-dried and air-dried marionberry, strawberry, and corn grown using conventional, organic, and sustainable agricultural practices. *J. Agric. Food Chem.* **2003**, *51*, 1237–1241. [CrossRef]
24. Barbosa, J.; Borges, S.; Amorim, M.; Pereira, M.J.; Oliveira, A.; Pintado, M.E.; Teixeira, P. Comparison of spray drying, freeze drying and convective hot air drying for the production of a probiotic orange powder. *J. Funct. Foods* **2015**, *17*, 340–351. [CrossRef]
25. Ratti, C. Hot air and freeze-drying of high-value foods: A review. *J. Food Eng.* **2001**, *49*, 311–319. [CrossRef]
26. Association of Official Analytical Chemists. *Official Methods of Analysis of AOAC International*, 17th ed.; Association of Official Analytical Chemists: Gaithersburg, MD, USA, 2000.
27. Parrish, C.C. Determination of total lipid, lipid classes, and fatty acids in aquatic samples. In *Lipids in Freshwater Ecosystems*; Arts, M.T., Wainman, B.C., Eds.; Springer: New York, NY, USA, 1999; pp. 4–20.
28. Hooper, T.; Parrish, C.C. Profiling neutral lipids in individual fish larvae by using short-column gas chromatography with flame ionization detection. *Limnol. Oceanogr. Methods* **2009**, *7*, 411–428. [CrossRef]
29. Parrish, C.C. Separation of aquatic lipid classes by chromarod thin-layer chromatography with measurement by latroscan flame ionization detection. *Can. J. Fish. Aquat. Sci.* **1987**, *44*, 722–731. [CrossRef]
30. Fountoulakis, M.; Lahm, H.-W. Hydrolysis and amino acid composition analysis of proteins. *J. Chromatogr. A* **1998**, *826*, 109–134. [CrossRef]
31. Stenerson, K.K. The derivatization and analysis of amino acids by GC-MS. *Rep. US* **2011**, *25*, 1–3.
32. Gianasi, B.L.; Parrish, C.C.; Hamel, J.F.; Mercier, A. Influence of diet on growth, reproduction and lipid and fatty acid composition in the sea cucumber *Cucumaria frondosa*. *Aquac. Res.* **2017**, *48*, 3413–3432. [CrossRef]
33. Dave, D.; Liu, Y.; Clark, L.; Dave, N.; Trenholm, S.; Westcott, J. Availability of marine collagen from Newfoundland fisheries and aquaculture waste resources. *Bioresour. Technol. Rep.* **2019**, 100271. [CrossRef]
34. Zhong, Y.; Khan, M.A.; Shahidi, F. Compositional characteristics and antioxidant properties of fresh and processed sea cucumber (*Cucumaria frondosa*). *J. Agric. Food Chem.* **2007**, *55*, 1188–1192. [CrossRef]
35. Karlsdottir, M.; Arason, S.; Thorarinsdottir, K.; Nguyen, M.V.; Kristinsson, H. Lipid degradation of cod liver during frozen storage as influenced by temperature, packaging method, and seasonal variation. *J. Aquat. Food Prod. Technol.* **2016**, *25*, 802–810. [CrossRef]
36. Vaidya, H.; Cheema, S.K. Sea cucumber and blue mussel: New sources of phospholipid enriched omega-3 fatty acids with a potential role in 3T3-L1 adipocyte metabolism. *Food Funct.* **2014**, *5*, 3287–3295. [CrossRef] [PubMed]
37. Swanson, D.; Block, R.; Mousa, S.A. Omega-3 fatty acids EPA and DHA: Health benefits throughout life. *J. Adv. Nutr.* **2012**, *3*, 1–7. [CrossRef] [PubMed]
38. Dave, D.; Ramakrishnan, V.V.; Trenholm, S.; Manuel, H.; Pohling, J.; Murphy, W. Marine oils as potential feedstock for biodiesel production: Physicochemical characterization. *J. Bioprocess. Biotech.* **2014**, *4*, 10001678. [CrossRef]
39. Simopoulos, A.P. An increase in the Omega-6/Omega-3 fatty acid ratio increases the risk for obesity. *Nutrients* **2016**, *8*, 128. [CrossRef] [PubMed]
40. Simopoulos, A.P. The importance of the ratio of omega-6/omega-3 essential fatty acids. *Biomed. Pharmacother.* **2002**, *56*, 365–379. [CrossRef]
41. Simopoulos, A.P. Evolutionary aspects of diet: The omega-6/omega-3 ratio and the brain. *Mol. Neurobiol.* **2011**, *44*, 203–215. [CrossRef] [PubMed]
42. Dinicolantonio, J.J.; O'Keefe, J.H. Importance of maintaining a low omega-6/omega-3 ratio for reducing inflammation. *Open Heart* **2018**, *5*, 1–4. [CrossRef]
43. Simopoulos, A.P. The importance of the omega-6/omega-3 fatty acid ratio in cardiovascular disease and other chronic diseases. *Exp. Biol. Med.* **2008**, *233*, 674–688. [CrossRef]
44. Burri, L.; Hoem, N.; Banni, S.; Berge, K. Marine omega-3 phospholipids: Metabolism and biological activities. *Int. J. Mol. Sci.* **2012**, *13*, 15401–15419. [CrossRef]
45. Küllenberg, D.; Taylor, L.A.; Schneider, M.; Massing, U. Health effects of dietary phospholipids. *Lipids Health Dis.* **2012**, *11*, 3. [CrossRef] [PubMed]
46. Li, J.; Wang, X.; Zhang, T.; Wang, C.; Huang, Z.; Luo, X.; Deng, Y. A review on phospholipids and their main applications in drug delivery systems. *Asian J. Pharm. Sci.* **2015**, *10*, 81–98. [CrossRef]

47. Ahmmed, M.K.; Ahmmed, F.; Tian, H.; Carne, A.; Bekhit, A.E.D. Marine omega-3 (n-3) phospholipids: A comprehensive review of their properties, sources, bioavailability, and relation to brain health. *Compr. Rev. Food Sci. Food Saf.* **2020**, *19*, 64–123. [CrossRef]
48. Araujo, P.; Zhu, H.; Breivik, J.F.; Hjelle, J.I.; Zeng, Y. Determination and structural elucidation of triacylglycerols in krill oil by chromatographic techniques. *Lipids* **2014**, *49*, 163–172. [CrossRef] [PubMed]
49. Ramprasath, V.R.; Eyal, I.; Zchut, S.; Jones, P.J. Enhanced increase of omega-3 index in healthy individuals with response to 4-week n-3 fatty acid supplementation from krill oil versus fish oil. *Lipids Health Dis.* **2013**, *12*, 1–11. [CrossRef]
50. Mamelona, J.; Saint-Louis, R.; Pelletier, É. Nutritional composition and antioxidant properties of protein hydrolysates prepared from echinoderm byproducts. *Int. J. Food Sci. Technol.* **2010**, *45*, 147–154. [CrossRef]
51. Sun, L.; Xia, W. Effect of steam cooking on muscle and protein heatdenature of tuna. *Food Mach.* **2010**, *26*, 22–25.
52. Deng, Y.; Luo, Y.; Wang, Y.; Zhao, Y. Effect of different drying methods on the myosin structure, amino acid composition, protein digestibility and volatile profile of squid fillets. *Food Chem.* **2015**, *171*, 168–176. [CrossRef]
53. Kim, B.-S.; Oh, B.-J.; Lee, J.-H.; Yoon, Y.S.; Lee, H.-I. Effects of various drying methods on physicochemical characteristics and textural features of yellow croaker (*Larimichthys polyactis*). *Foods* **2020**, *9*, 196. [CrossRef]
54. Zaenuri, M.; Anggoro, S.; Kusumaningrum, H.P.S. Nutritional value of sea cucumber [*Paracaudina Australis* (Semper, 1868)]. *Aquat. Procedia* **2016**, *7*, 271–276. [CrossRef]
55. Dutta, S.; Ray, S.; Nagarajan, K. Glutamic acid as anticancer agent: An overview. *Saudi Pharm. J.* **2013**, *21*, 337–343. [CrossRef] [PubMed]
56. Bajwa, M.A.; Zahoor, T.; Butt, T.M.; Atiq, M.; Sahi, S.T. Microbial production of L-isoleucine from different substrates using locally isolated bacteria. *Int. J. Agric. Biol.* **2010**, *12*, 668–672.
57. Moorthi, P.P.; Gunasekaran, S.; Ramkumaar, G.R. Vibrational spectroscopic studies of isoleucine by quantum chemical calculations. *Spectrochim. Acta—Part A Mol. Biomol. Spectrosc.* **2014**, *124*, 365–374. [CrossRef]
58. Fernstrom, J.D.; Fernstrom, M.H. Tyrosine, phenylalanine, and catecholamine synthesis and function in the brain. *J. Nutr.* **2007**, *137*, 1539S–1547S. [CrossRef]
59. Meyers, S. Use of neurotransmitter precursors for treatment of depression. *Altern. Med. Rev.* **2000**, *5*, 64–71. [PubMed]
60. Gundersen, R.Y.; Vaagenes, P.; Breivik, T.; Fonnum, F.; Opstad, P.K. Glycine—An important neurotransmitter and cytoprotective agent. *Acta Anaesthesiol. Scand.* **2005**, *49*, 1108–1116. [CrossRef]
61. Bordbar, S.; Anwar, F.; Saari, N. High-value components and bioactives from sea cucumbers for functional foods—A review. *Mar. Drugs* **2011**, *9*, 1761–1805. [CrossRef]
62. Liguori, I.; Russo, G.; Curcio, F.; Bulli, G.; Aran, L.; Della-Morte, D.; Gargiulo, G.; Testa, G.; Cacciatore, F.; Bonaduce, D.; et al. Oxidative stress, aging, and diseases. *Clin. Interv. Aging* **2018**, *13*, 757–772. [CrossRef]

processes

Article

Volatile Profiling Aided in the Isolation of Anti-Proliferative Lupeol from the Roots of *Clinacanthus nutans* (Burm. f.) Lindau

Angelina Ying Fang Cheng [1], Peik Lin Teoh [1], Lalith Jayasinghe [2] and Bo Eng Cheong [1,*]

1 Biotechnology Research Institute, Universiti Malaysia Sabah, Jalan UMS, Kota Kinabalu 88400, Malaysia; angelinacheng16@gmail.com (A.Y.F.C.); peiklin@ums.edu.my (P.L.T.)

2 National Institute of Fundamental Studies, Hantana Road, Kandy 20000, Sri Lanka; lalith.ja@nifs.ac.lk

* Correspondence: becheong@ums.edu.my; Tel.: +60-88-320000 (ext. 8530)

Abstract: Isolation of anti-proliferative compounds from plants is always hindered by the complexities of the plant's nature and tedious processes. *Clinacanthus nutans* (Burm. f.) Lindau is a medicinal plant with reported anti-proliferative activities. Our study aimed to isolate potential anti-proliferative compounds present in *C. nutans* plant. To start with, for our study, we came up with a strategy by first profiling the volatile compounds present in the leaf, stem and root of *C. nutans* using GC-MS. Comparing the plant's volatile profiles greatly narrowed down our target of study. We decided to start with the isolation and characterization of a pentacyclic terpenoid, i.e., lupeol from the roots of *C. nutans*, as this compound was found to present abundantly in the roots compared to the leaf or stem. We developed a simple maceration and re-crystallization method, without the necessity to go through the fractionation or column chromatography for the isolation of lupeol. Characterizations of the isolated compound identified the compound as lupeol. The anti-proliferative activity of the isolated lupeol was further investigated against the MCF-7 cell line, which showed comparable anti-proliferative activity with the authentic lupeol and camptothecin. Our strategy to profile every part of the plant first, followed by selection of the most suitable plant part and targeted compound proved useful for further isolation and characterization bioactive compound from *C. nutans*.

Keywords: *Clinacanthus nutans*; profiling; root; GC-MS; lupeol; anti-proliferative; MCF-7 cells

Citation: Cheng, A.Y.F.; Teoh, P.L.; Jayasinghe, L.; Cheong, B.E. Volatile Profiling Aided in the Isolation of Anti-Proliferative Lupeol from the Roots of *Clinacanthus nutans* (Burm. f.) Lindau. *Processes* **2021**, *9*, 1383. https://doi.org/10.3390/pr9081383

Academic Editor: Zongbi Bao

Received: 29 June 2021
Accepted: 30 July 2021
Published: 9 August 2021

1. Introduction

Clinacanthus nutans (Burm. f.) Lindau from the Acanthaceae family is a small shrub which is naturally found in tropical and subtropical Asian countries, including Malaysia, Thailand, Indonesia, Vietnam, and China [1]. In Malaysia, the plant is commonly known as Belalai Gajah and Sabah snake grass [2]. The aerial part of this plant, i.e., the leaf (Figure 1) has been shown to possess a vast spectrum of pharmacological activities, in which the anti-proliferative activities on various human cancer cell lines have been extensively studied. For instance, *C. nutans* leaf extracted with chloroform, methanol, and water has been reported to demonstrate anti-proliferative effects against various human cancer cell lines, such as the liver hepatocellular carcinoma cell line (HepG2), neuroblastoma cell line (IMR-32), lung cancer cell line (NCI-H23), gastric cancer cell line (SNU-1), colon adenocarcinoma cell line (LS-174T), erythroleukemia cell line (K-562), human cervical cancer cell line (HeLa), and Burkitt's lymphoma cell line (Raji) [3]. On the other hand, the leaf extracted with ethanol and ethyl acetate has also been reported to exhibit strong cytotoxic effects against the estrogen-dependent human breast cancer cell line, MCF-7 [4]. Meanwhile, Fong and colleagues have collected *C. nutans* leaves from different countries, such as Malaysia, Thailand and Vietnam. They extracted the collected leaves with methanol and found that the leaf extracts possess cytotoxicity and apoptotic activities against the human skin cancer cell line, D24 melanoma cells [5]. In addition to the anti-proliferative effects on various cancer cell lines, the aerial parts of the plant have also been reported to exhibit significant antitumor and apoptotic activities on the in vivo mice cancer model [6]. While most of the anticancer studies on this plant are focused on the aerial part, our group have conducted an anti-proliferative study on the underground part, i.e., the root (Figure 1) [7]. The crude methanolic and ethyl acetate extracts of *C. nutans* roots were found to exhibit anti-proliferative and apoptotic effects on breast cancer (MCF-7) and cervical cancer (HeLa) cell lines [Teoh].

Figure 1. A *Clinacanthus nutans* plant. The arrows show the leaf, stem and root parts of the plant used for extraction and volatile profiling in this study. The roots were further used for the isolation and characterization of lupeol.

Owing to its promising anti-proliferative activities, several attempts have been conducted to isolate the anti-proliferative compounds and/or active fractions from the *C. nutans* leaf. Huang and colleagues [6] have extracted the aerial parts of *C. nutans* with ethanol, followed by further fractionation on a Diaion HP-20 macroporous adsorption resin to produce a 30% ethanol fraction (CN30) and other fractions. The CN30 fraction was found to be able to induce inhibition in tumor size and weight of the HepA xenograft model mouse (hepatoma cell-injected mice). The active components in CN30 were therefore further fractionated and purified. High-performance liquid chromatography (HPLC) and mass spectrometry (LC/MS/MS) analyses revealed that there were seven active compounds present in CN30. The seven identified active components were shaftoside, apigenin 6,8-C-α-L-pyranarabinoside, orientin, isoorientin, vitexin, isovitexin, and gallic acid. After that, the same research group conducted another study to isolate a novel polysaccharide–peptide complex, which they named CNP-1-2, by a series of complex purification methods from *C. nutans* leaves [8]. This CNP-1-2 complex was found to show the strongest growth inhibition effect on the human gastric cancer cell line (SGC-7901) and was able to stimulate the activation of macrophages. On the other hand, another research group led by Mutazah and colleagues [9] has managed to fractionate two active fractions from the bark powder of *C. nutans*. These two fractions, which they named A12 and A17, were found to contain sulfur-containing compounds, including entadamide C and clinamide D, that possessed strong cytotoxic activities against MDA-MB-231 and MCF-7 breast cancer cell lines.

From the studies above, it can be postulated that the isolation of anti-proliferative compounds from *C. nutans* is not easy. This may be due to the complexities of the plant's nature, and also various fractionation/chromatographic steps which need to be optimized. Our study therefore aimed to pre-screen (profile) the volatile compounds present in every part of the plants first, followed by the selection and isolation of target compound. The details of the study are described in the subsequent sections.

2. Materials and Methods

2.1. Solvents, Chemicals and Cell Lines

All the solvents used in this study were of analytical and/or high performance liquid chromatography (HPLC) grade, and purchased from Fisher Scientific (Pittsburgh, PA, USA). The solvents used were methanol, ethyl acetate and dimethylsulfoxide (DMSO). The authentic reference standard, lupeol (L5632-25 mg, ≥94% purity) for characterization purposes was obtained from Sigma-Aldrich (St. Louis, MO, USA). Roswell Park Memorial Institute (RPMI) 1640, Phosphate-Buffered Saline (PBS), 2.5 g/L trypsin-1mmol/L EDTA, and 0.5% trypan blue stain solution were purchased from Nacalai Tesque (Kyoto, Japan), whereas the Fetal Bovine Serum (FBS) was purchased from JR Scientific (Woodland, CA, USA). The cell proliferation kit was purchased from Roche Diagnostics (Mannheim, Germany).

2.2. Plant Materials

Clinacanthus nutans (Burm. f.) Lindau plants were purchased from herbal suppliers at a local market in Kota Kinabalu, Sabah, Malaysia. The plants were verified by a botanist from the Faculty of Science and Natural Resources, Assoc. Prof. Dr. Berhaman Ahmad, and a voucher specimen (ACCN 001/2013) was deposited in the herbarium of Universiti Malaysia Sabah. The plants were first cleaned thoroughly with tap water to remove the dirt, followed by rinsing several times with distilled water and dried with a paper towel. The leaves, stems and roots were excised from the plants, freeze-dried for five days and ground into powder using a heavy-duty blender. The plant samples in powder form were stored at −80 °C until further analysis. Some of the stems of the purchased plants were re-grown in soil for a duration of 6 months for photo shooting purposes.

2.3. Extraction of *C. nutans* Leaf, Stem, and Root for Volatile Profiling

Approximately 0.5 g of each freeze-dried *C. nutans* leaf, stem, and root sample was weighed and added with 10 mL of absolute ethyl acetate, followed by maceration at 37 °C with 180 rpm agitation for three days in dark conditions. After three days of extraction, the plant extracts were filtered with Whatman filter paper, followed by drying under vacuum pressure by using a rotary evaporator at 40 °C, and then stored at −80 °C until further analysis. For each plant's part, the whole extraction process was repeated another two times to produce three replicates (n = 3) per leaf, stem and root.

2.4. Volatile Profiling of the Crude Leaf, Stem, and Root Extracts Using GC-MS

The crude *C. nutans* leaf, stem, and root ethyl acetate extracts were re-dissolved in HPLC-grade ethyl acetate to appropriate concentrations and filtered using 0.22 µm syringe filters prior to GC-MS analyses. The GC-MS analysis was carried out according to [9], with some modifications. Briefly, 1 µL aliquot of each sample was injected into the GC-MS equipment (GC model 7890 and MS model 5975C; from Agilent Technologies, Santa Clara, CA, USA). An HP-5ms capillary column (30 m × 0.250 mm inner diameter × 0.25 µm of film thickness) from Agilent Technologies was used for separation. GC-MS was performed in a splitless mode and pure helium was used as the carrier gas with a flow rate of 1 mL/min. The MS source was pre-set to 230 °C and MS quad temperature was recorded at 150 °C. The injector temperature was set to 250 °C with an injection volume of 1 µL per sample. The oven temperature was programmed from 220 °C for 10 min and continuously increased at a rate of 5 °C/min until a temperature of 300 °C was reached and then retained for 10 min. The total run time was 36 min. The post-run was fixed at 200 °C for 5 min, and the ion source temperature was constant at 200 °C. The mass analyzer was performed in a full scan mode from *m/z* 40 to 550, and mass spectra were obtained at EI = 70 eV. The identification of the compounds was performed by spectra matching using the National Institute Standard and Technology (NIST) library version 11 (NIST11). The leaf, stem, and root profiles obtained were compared to select for the most suitable plant's part to be used for the subsequent isolation of target compound.

2.5. Extraction of C. nutans Root for the Isolation of Lupeol

The selected plant part in this study was the roots, and the target compound was lupeol. First of all, 80 g of freeze-dried *C. nutans* roots were weighed and added into a conical flask containing 200 mL of absolute ethyl acetate. The extract was put inside a rotary shaker and macerated at 37 °C with 180 rpm agitation for a consecutive five days in dark conditions. After five days of extraction, the crude root extract was filtered with Whatman filter paper, transferred to a new flask, and stored at 4 °C. A repeated extraction was performed on the remaining root residue in the extraction flask with the addition of another 200 mL of ethyl acetate but only for three days. The second crude root extract was filtered and combined with the first extract, followed by drying under vacuum pressure using a rotary evaporator at 40 °C and stored at −80 °C prior to the recrystallization step.

2.6. Recrystallization of C. nutans Root Extract for the Isolation of Lupeol

The dried *C. nutans* crude root extract in Part 2.5 above was re-dissolved in 100 mL of absolute methanol and added into an Erlenmeyer flask, followed by heat-stirring on a hot plate until all the extract was fully dissolved and the methanol solvent boiled. The Erlenmeyer flask was then left to cool down inside a huge glass beaker filled with ice. While cooling down, the inner side of the Erlenmeyer flask was gently stirred by using a glass rod to induce crystal formation (nucleation). After ten minutes, the formed crystals were filtered out from the methanol solvent using a Buchner funnel under vacuum pressure. The filtered crystals were collected and left to dry inside a laminar flow for several minutes until a white amorphous powder was obtained. The collected white amorphous powder was weighed, followed by identification and characterization using various analytical instruments.

2.7. GC-MS Analyses and Comparisons of the Isolated and Authentic Reference Standard Lupeols

The isolated white amorphous compound from *C. nutans* roots and the purchased authentic standard lupeol were dissolved in HPLC-grade ethyl acetate to appropriate concentrations and were filtered using 0.22 µm syringe filters prior to GC-MS analyses. The GC-MS analysis was carried out according to Part 2.4, with some modifications on the ramping program. Briefly, the oven temperature was programmed from 220 °C for 10 min and continuously increased at a rate of 5 °C/min until a temperature of 300 °C was reached and then retained for 10 min. The total run time was 41 min. The identification of the compound was determined using the NIST11 library (Cas number for lupeol is 000545-47-1, entry number in library is #221171) and also compared with the MS spectra of the reference standard lupeol. In addition, further confirmation of the identity and purity of lupeol isolated from *C. nutans* roots was carried out by spiking the standard lupeol into the isolated lupeol sample, followed by a re-analysis with GC-MS to observe if only one overlapped peak (lupeol) was present in the chromatogram or not.

2.8. Characterization of the Isolated Lupeol from C. nutans Roots

Firstly, the melting point of the isolated lupeol was determined with SMP40 Automatic Melting Point Apparatus (Stuart, UK). Next, the infrared (IR) spectra of the isolated and standard lupeol were determined using a Cary 630 FTIR Spectrometer (Agilent Technologies, CA, USA). Finally, structural elucidation of the isolated compound was determined using 1D-FT-NMR in 600 MHz (Bruker, Germany) for both ^{1}H-NMR and ^{13}C-NMR.

2.9. Anti-Proliferative Activities of the Isolated Lupeol Compared to the Authentic Standard Lupeol

The anti-proliferative activities of isolated and standard lupeol were determined using MTT assay, which was performed using the Cell Proliferation Kit I (Roche Diagnostics, Mannheim, Germany) according to the manufacturer's protocol. Human breast cancer (MCF-7) cell line was seeded in a 96-well plate (3×10^4 cells/well) with 100 µL of RPMI 1640 medium and incubated at 37 °C with 5% CO_2 of humidified atmosphere overnight. After 24 h, cells were treated with different concentrations of the isolated compound and standard lupeol ranging from 10 to 50 µg/mL for 72 h. For positive control, cells were treated with camptothecin at the concentrations of 0.5–2.0 µg/mL for 72 h. Cells treated with DMSO were used as a control. The cell proliferation was measured at 570 nm using a microplate reader (Molecular Devices, Sunnyvale, CA, USA). The IC_{50} values (50% growth inhibitory concentration) were calculated from three independent experiments, in which there were three technical replicates (n = 3) per concentration per experiment.

2.10. Statistical Analysis

The IC_{50} values of the isolated and standard lupeol were expressed as mean ± SD of the three independent experiments (n = 3), which were calculated using a non-linear dose–response curve fitting analysis with GraphPad Prism 7 (GraphPad Software, Inc., San Diego, CA, USA). The level of significance (* $p < 0.05$, ** $p < 0.01$ and *** $p < 0.001$) was determined using one-way ANOVA followed by a post hoc Dunnett test by comparison to the untreated controls.

3. Results

3.1. Volatile Profiling of *C. nutans* Reveals Different Accumulations of Bioactive Metabolites in Different Parts of the Plant

In our volatile profiling of different parts of the *C. nutans* plant, ethyl acetate was used as the extraction solvent to extract for volatile or nonpolar compounds, and a GC-MS method, which we have optimized for the detection of volatiles or nonpolar compounds [7], was employed in this study. From the results, it was interesting to observe that the leaves, stems, and roots have different volatile profiles among each other (Figure 2). The leaves were found to contain vitamin E (α-tocopherol) as the most abundant compound, followed by squalene (the precursor of pentacyclic triterpenoids and phystosterols), and phytosterols, such as stigmasterol, β/γ-sitosterol, and campesterol. Meanwhile, the stems were shown to contain lupeol, phytosterols, squalene, and vitamin E (α-tocopherol). The most interesting part was found in the roots, where there was only one main peak present in the profile, which was designated as lupeol by the NIST11 library (with matched quality of 76). The other compounds which were observed in the leaves or stems were found to be very scarce in the root part. As lupeol has been reported to possess anti-proliferative activities (which will be discussed in the Discussion Section 4) and also the volatile profile of the root was simple, we then decided to make this compound our target in this study.

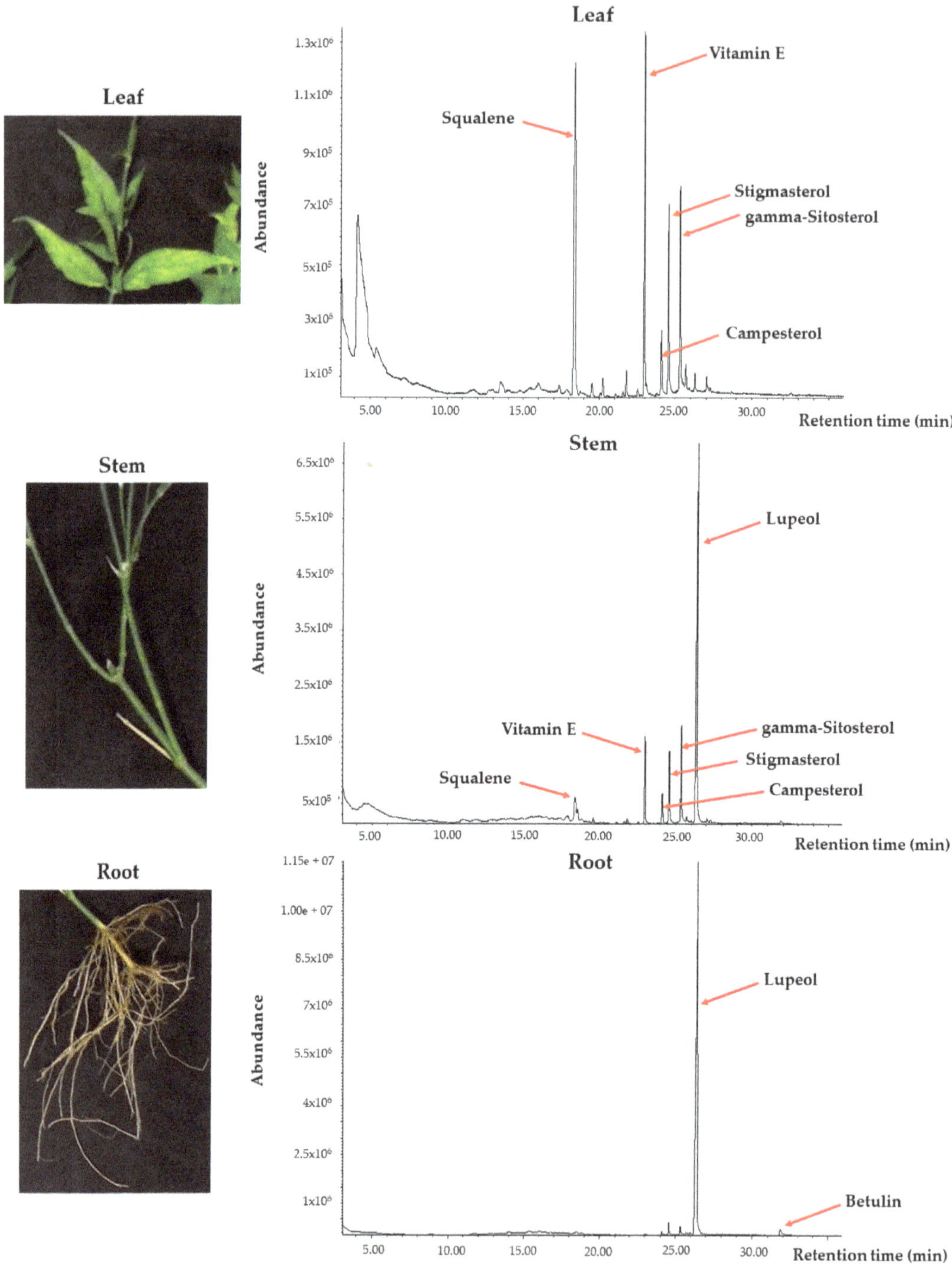

Figure 2. Volatile profiles of the leaf, stem, and root of *C. nutans* obtained using GC-MS analysis.

3.2. GC-MS Analysis Reveals the Isolation of Lupeol from C. nutans Roots

Through maceration and recrystallization methods, 1.138 g of the white amorphous compound was successfully isolated from an initial 80 g of freeze-dried *C. nutans* roots. The extraction yield was therefore 1.42% (1.138 g/80 g × 100%), or 14.2 mg/g freeze-dried root. The white amorphous compound was then re-dissolved with HPLC-grade ethyl acetate and analyzed with GC-MS. The profile was compared with standard lupeol and isolated compound spiked with standard lupeol (Figure 3). From Figure 3A, we observed that only one single main peak was shown in the profile at a retention time of 28.769 min for the isolated white amorphous compound. This isolated compound also showed almost the same retention time with the purchased authentic reference standard, lupeol, which has an Rt of 28.789 min (Figure 3B). When the electron ionization-mass spectrum (EI-MS) of the single peak from Figure 3A was compared to the EI-MS of standard lupeol, they were identical (Supplemental Figure S1A). When this isolated white amorphous compound was spiked with an equal amount of standard lupeol and loaded into GC-MS, only one single peak was detected in the profile with a retention time of 28.799 min (Figure 3C). The EI-MS also overlapped well (Supplementary Figure S1A). Lupeol has a molecular weight of 426.729 g/mol and a molecular formula of $C_{30}H_{50}O$. The EI-MS of this single peak showed the presence of the parent ion (M+) at m/z 426, which corresponds to the molecular formula of $C_{30}H_{50}O$ and most probably represented lupeol. The EI-MS also revealed the presence of fragment ions at m/z 189 and 218, which were the main characteristic of a pentacyclic triterpenoid [10]. Another key feature of pentacyclic triterpenoid is the isopropenyl fragment at m/z 411 $[M\text{-}CH_3]^+$ [11], which was also detected (Supplemental Figure S1A). Overall, the identical fragmentation patterns of EI-MS of the isolated compound with the fragmentation patterns of EI-MS of standard lupeol suggested that the isolated compound from *C. nutans* roots was lupeol.

3.3. Characterization and Structural Elucidation of the Isolated Lupeol

Characterizarion of the isolated white amorphous showed that it has a melting point of 216 °C. Meanwhile, the IR spectrum of the isolated compound was also studied (Supplemental Figure S2). The hydroxyl group (-O–H) vibration was shown as a broad intense band at 3303 cm^{-1}, and a moderately intense band was observed at 1187 cm^{-1}. The stretching and bending of the methyl group (-C–H_3) were detected at 2930 cm^{-1} with an intense band and a moderately intense band at 1450 cm^{-1}. In addition, the vibration of the (-C–H_2) methylenic part was observed at 2861 cm^{-1}, which displayed a weaker intense band. The vibration of the (-C=O) of the carboxylic acid group was also observed as an intense band at 1737 cm^{-1}. At the wavelength of 1379 cm^{-1}, there was the presence of (-C–C) vibration of the methyl group as a moderate intense band. The out of the plane (-C–H) vibration was observed at 886 cm^{-1} as a strong intense band. Those obtained IR spectra data were in good agreement with the IR spectra of lupeol, as studied and reported by [12,13].

Elucidation of the structure of the isolated compound was performed using ^{1}H-NMR (600 MHz) and ^{13}C-NMR (150 MHz). The ^{1}H-NMR spectrum of the compound showed the presence of seven methyl groups at δ0.98 (s, H_3-23), 0.79 (s, H_3-24), 0.83 (s, H_3-25), 1.03 (s, H_3-26), 1.03 (s, H_3-27), 0.81 (s, H_3-28), and 1.66 (s, H_3-30). The predominant characteristic of the lupane class of triterpenoid was found as a pair of singlets at δ4.57 and δ4.69 with ^{1}H for each singlet due to the terminal isopropenyl moiety or the vinylic proton located at carbon 29. Meanwhile, the presence of a doublet with one proton intensity found at δ3.18 (dd, J = 4.8 Hz, H-3) was caused by the proton attached to the carbon bearing the hydroxyl group or an oxymethine proton located at C-3, which is identical as reported [10]. There was also one proton with a doublet of the triplet located at δ2.37, which was assigned to 19β-H orientation, similar to the reported literature data [14]. The ^{1}H-NMR spectrum for the isolated compound can be found in Supplemental Data S1A.

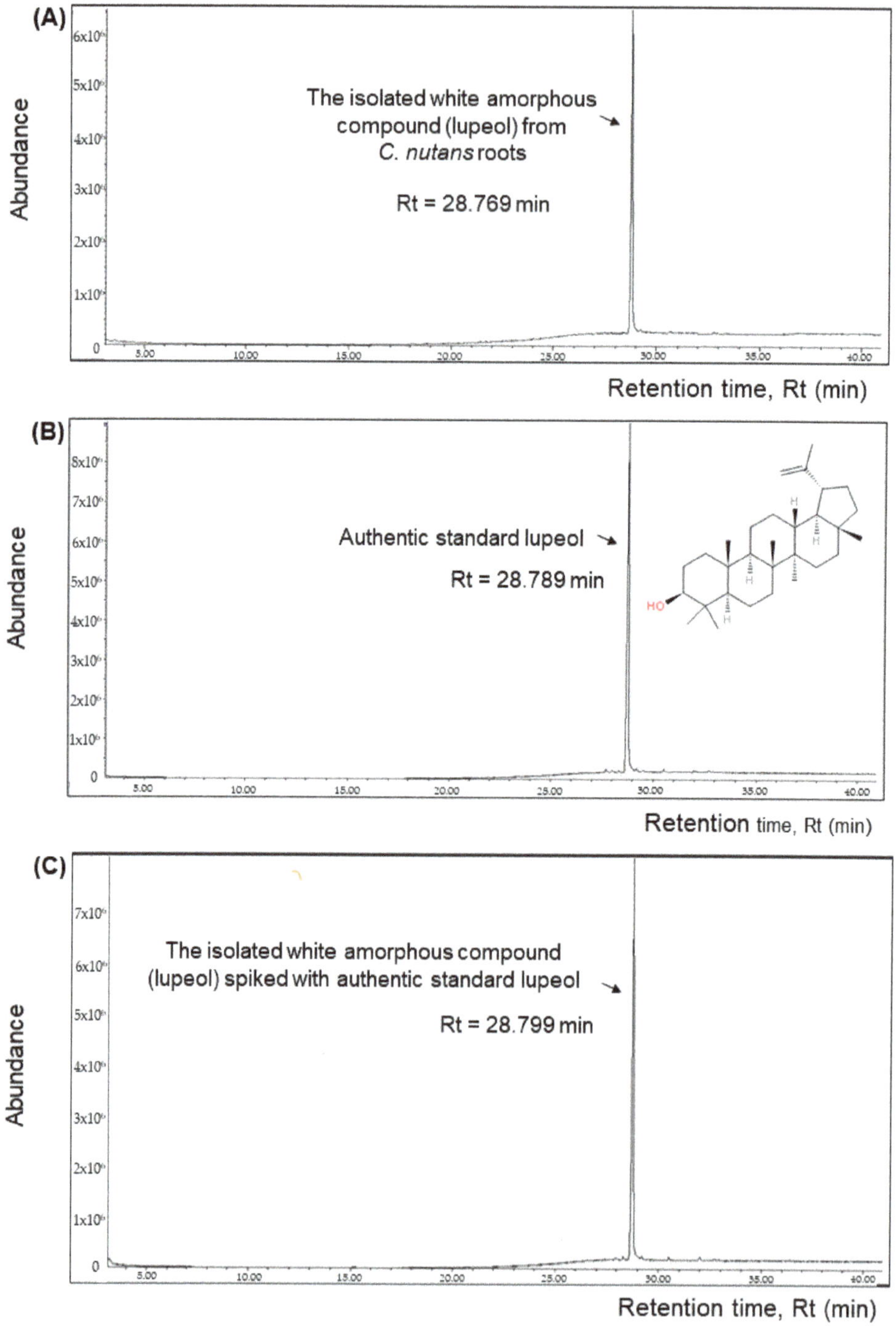

Figure 3. GC-MS profiles of the (**A**) isolated white amorphous compound from the roots of *C. nutans*, (**B**) purchased authentic lupeol, and (**C**) isolated compound spiked with authentic lupeol.

Meanwhile, the ^{13}C-NMR spectrum (150 MHz, $CDCl_3$) of the isolated compound exhibited signals at δ109.36 (C-29) and δ78.99 (C-3), which were the characteristics of protons and carbons of a methylene and oxymethine group at the positions 3 and 29 for lupeol [15]. In addition, another distinguishable characteristic of triterpenoid was found at δ150.91 for the presence of carbon C-20 [15]. The ^{13}C-NMR also revealed carbon intensity at δ 38.86 (C-1), 27.41 (C-2), 38.74 (C-4), 55.33 (C-5), 18.03 (C-6), 34.31 (C-7), 40.85 (C-8), 50.46(C-9), 37.18 (C-10), 20.95 (C-11), 25.16 (C-12), 38.07 (C-13), 42.84 (C-14), 27.47 (C-15), 35.60 (C-16), 47.99 (C-17), 48.32 (C-18), 47.99(C-19), 29.87 (C-21), 38.86 (C-22), 27.47 (C-23), 15.40 (C-24), 16.13 (C-25), 16.00 (C-26), 14.57 (C-27), 18.34 (C-28), and 19.33 (C-30), which were identical to the reported literature data by [12,15,16]. The ^{13}C-NMR spectrum for the isolated compound can be found in Supplemental Data S1B.

3.4. The Isolated Lupeol from C. nutans Roots Demonstrates Comparable Anti-proliferative Activity with Authentic Lupeol and Campthothecin

The anti-proliferative activity of the lupeol isolated from the C. nutans roots was tested against human breast cancer cell line MCF-7 using MTT assay, and the result was compared with the anti-proliferative activity of authentic standard lupeol (Figure 4). The standard lupeol was found to have an IC_{50} value of 13.624 ± 1.257 μg/mL, whereas the lupeol isolated from C. nutans exhibited a comparable IC_{50} value of 16.813 ± 1.316 μg/mL. Both authentic and isolated lupeol displayed a potent growth inhibition on MCF-7 cells in a dose-dependent manner (Figure 4). Comparing to camptothecin (IC_{50} = 0.93 μg/mL, Supplemental Figure S3), lupeol is about 15x less toxic.

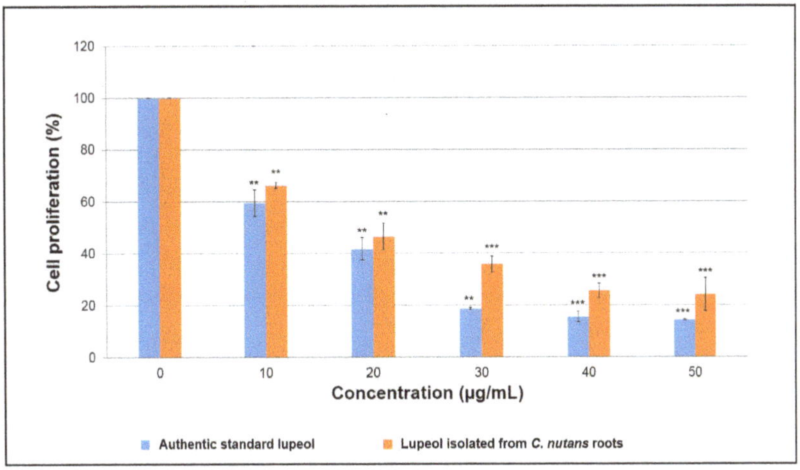

Figure 4. Anti-proliferative activities of the isolated lupeol from C. *nutans* roots compared to the authentic standard lupeol and camptothecin (Supplemental Figure S3) against MCF-7 cells after 72 h with the indicated concentrations. The IC_{50} values were expressed as mean ± SD of the three independent experiments (n = 3). Level of significance (** $p < 0.01$ and *** $p < 0.001$) for each concentration of authentic or isolated lupeol was compared to the control cells, and was determined using one-way ANOVA followed by a post hoc Dunnett test.

4. Discussion

The isolation of anti-proliferative compounds from C. *nutans* is not easy. This may be due to the complexities of the plant's nature and also various fractionation/chromatographic steps which need to be optimized. We came up with a strategy to pre-screen (profile) the volatile compounds present in every part of the plant; firstly, using a GC-MS method, we have optimized for the detection of volatile compounds [7], then only select the suitable plant part and target compound. In our present study, we found that the aerial part of the plant, i.e., the leaf, was predominated by vitamin E (α-tocopherol), squalene, and phytosterols such as stigmasterol, γ/β-sitosterol, and campesterol. Interestingly, the intensities of these compounds decreased when moving to the lower part of the plant, i.e., the stem. In the stem, lupeol was found to start to accumulate abundantly and the amount was much higher than squalene and other phytosterols. Lupeol is one of the pentacyclic triterpenes with a 30-carbon skeleton comprising four six-membered rings and one five-membered ring [17]. More intriguingly, when it came to the underground part of the plant, i.e., the root, it was found that this part contained enormous lupeol levels and some betulin. All of the detected phytosterols and pentacyclic terpenoids (lupeol and betulin) in this study have been reported to possess anti-proliferative and apoptotic activities by other researchers [18–27].

Figure 5 illustrates the biosynthesis of phytosterols and some pentacyclic triterpenoids such as lupeol and betulin. The figure was made by referring to [28–30]. The biosynthesis of the compounds begins from the reactions of two precursors called isopentenyl pyrophosphate (IPP) and dimethylallyl pyrophosphate (DMAPP), which are derived from acetyl-CoA via the mevalonic acid (MVA) pathway [28–30]. The reaction is catalyzed by the enzyme isopentenyl pyrophosphate isomerase to yield the geranylpyrophosphate (GPP) and farnesyl pyrophosphate (FPP) via the catalysis of prenyltransferase. Then, two molecules of FPPs are catalyzed by the squalene synthase to yield squalene, which is the important precursor for various phytosterols and pentacyclic triterpenoids. Via the squalene epoxidase, squalene is converted to 2,3-oxidosqualene, which is the branch point for the subsequent biosynthesis of phytosterols or pentacyclic triterpenoids [28–30]. For phytosterols, multiple enzymatic steps are involved in the plants to convert 2,3-oxidosqualene to cycloartenol and end products such as stigmasterol, sitosterol and campesterol [30]. For pentacyclic triterpenoids such as lupeol and betulin, the first step is the conversion of 2,3-oxidosqualene to lupeol by lupeol synthase, and then followed by three sequential oxidations catalyzed by cytochrome P450 monooxygenases (CyP450), leading to the generation of betulin, betulinic aldehyde, and betulinic acid [28,29]. Based on the results observed in our study, it can be postulated that squalene biosynthesis may actively take part in the aerial part (leaf). This squalene then becomes the precursor for the biosynthesis of lupeol in the stem part. Finally, the root part may either become the synthesis site or main storage site for lupeol.

To date, two studies related to the isolation of lupeol from *C. nutans* plant have been reported. One was reported by Dampawan and colleagues [31] in 1977 in which the plant part used was the stem. Another one was reported by Le and colleagues [15] in 2017 in which the plant parts used were the leaf and stem. Our study is the first report on the isolation of lupeol from the root part. Owing to its high abundance in the roots and simple root profile obtained via GC-MS analysis, we developed a simpler and straight-forward isolation method for lupeol from the plant's roots. In our study, we used an initial 80 g of freeze-dried root powder and macerated it with ethyl acetate for five days during the first extraction, followed by three days for the second extraction on the root residues with ethyl acetate, and combined the two extracts and dried them. The vacuum-dried extract was re-dissolved in methanol and underwent recrystallization to yield 1.138 g white amorphous compound, which was identified to be lupeol (yield of extraction = 1.423%). In [31], an initial two kilograms of dried and milled *C. nutans* stems were extracted by using light petroleum in Soxhlet apparatus, followed by silica gel column chromatography using ether and light petroleum for elution, then further purified by using preparative thin layer chromatography and recrystallization to yield 3.7 g of lupeol (yield = 0.185%). In [15], an initial four kilograms of dried and powdered *C. nutans* leaves were macerated with methanol and filtered, followed by solvent–solvent partitioned with hexane to yield a crude extract of 153.9 g. The crude hexane extract was later dissolved in methanol and chromatographed over Diaion HP20SS using ethanol to yield five sub-fractions (H1, H2, H3, H4, and H5). These five sub-fractions were further subjected to column chromatography using different gels to produce four pure compounds (shaftoside, stigmasterol, β-sitosterol, and lupeol). Unfortunately, the final yield for each of the pure compounds was not mentioned in the literature and thus could not compare to our method. However, compared to Dampawan, our method manages to produce a higher extraction yield of lupeol. Additionally, our method is simple and straightforward compared to [15,31] as it does not involve solvent–solvent partition, and also without the hassle to go through the column or preparative thin-layer chromatography. The successful isolation of lupeol from *C. nutans* roots in this study also suggests the importance of GC-MS- or LC-MS-based profiling of every part of the plant first prior to the selection of potential candidate compounds for the isolation process.

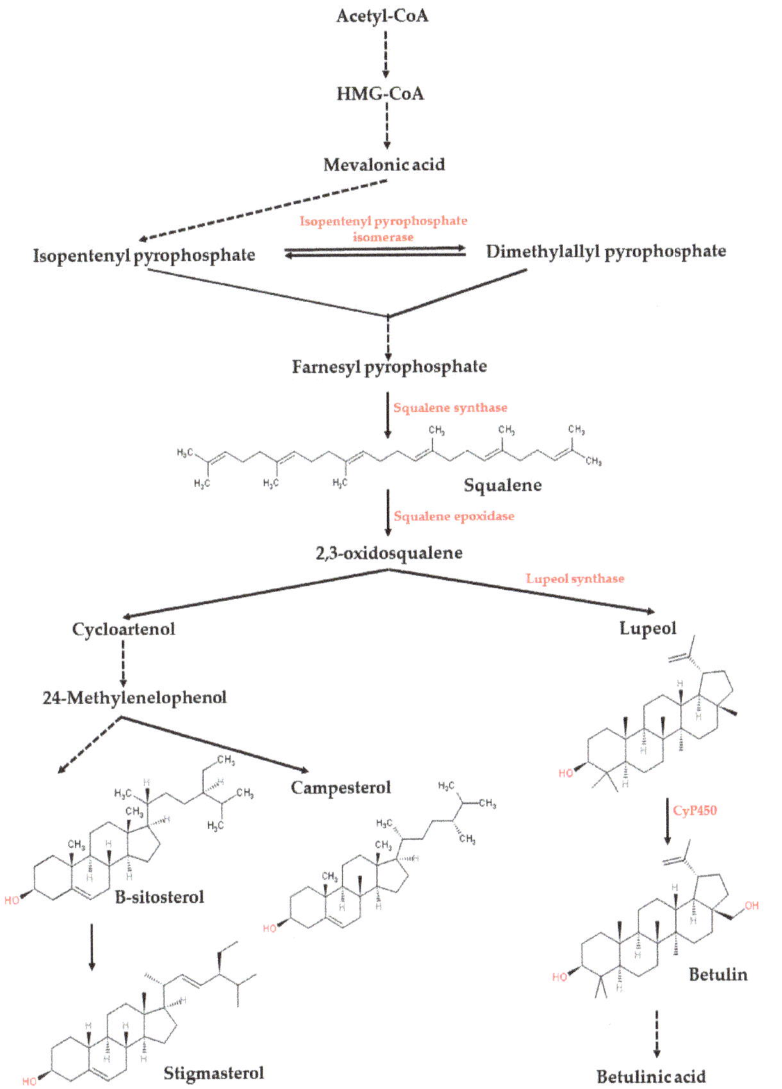

Figure 5. The biosynthesis of pentacyclic terpenoids and phystosterols. The biosynthesis structure and pathway were produced using the KingDraw v2.0 for Windows software (http://www.kingdraw.cn/en/index.html, Qingdao, China. Downloaded and installed on 1 May 2021).

In addition, the extensive characterization of the isolated compound using UV-visible spectrophotometry, Fourier transform infrared spectroscopy, GC-MS, and nuclear magnetic resonance have been able to identify the isolated compound as lupeol. Further analysis of the anti-proliferative activity of the isolated lupeol against human breast cancer cells has revealed that the isolated lupeol was able to demonstrate comparable anti-proliferative activity with the standard lupeol and positive control camptothecin. A study reported by [22] has postulated the possible mechanism of lupeol on human breast cancer cells by down-regulating the anti-apoptosis proteins, Bcl-2 and Bcl-XL, which lead to cell apoptosis. Our previous study [7] also shows that the root extract from *C. nutans* is capable of causing apoptosis of the breast cancer cells via the condensation of chromatin, down-regulation of the BCL2 gene, and the loss of mitochondrial membrane potential of the cancer cells. We therefore highly speculate that the pro-apoptotic capability of *C. nutans* roots could be highly correlated with the lupeol contained inside the roots. We therefore are interested in exploring the molecular mechanisms associated with anti-proliferative activity exerted by the isolated lupeol. Further, the profiling method presented in this study is a simple qualitative study of volatile compounds in different parts of *C. nutans* to aid in the isolation of the anti-proliferative compound. Referring to the successful GC-MS-based metabolomics study of *C. nutans* leaf extracts on α-glucosidase activity as reported by [32], we are also interested in applying this similar sophisticated GC-MS-based metabolomics in the future to explore the anti-proliferative mechanism of lupeol isolated from *C. nutans* plants harvested from different growing stages or collected from different regions.

5. Conclusions

In conclusion, comparing the plant's volatile profiles greatly narrowed down our target of study. Our strategy to profile every part of the plant first, followed by selection of most suitable plant's part and target compound are useful for further isolation and characterization of anti-proliferative compounds such as lupeol from the roots of *C. nutans*.

Supplementary Materials: The following are available online at https://www.mdpi.com/article/10.3390/pr9081383/s1: Figure S1: The electron ionization-mass spectrum (EI-MS) of the (A) isolated and (B) authentic lupeol; Figure S2: Fourier-transform infrared spectrum (FTIR) of the white amorphous compound (lupeol) isolated from *C. nutans* roots; Figure S3: Cytotoxicity assay of camptothecin against MCF-7 cell lines; Supplemental Data S1A: 1H-NMR spectrum of the isolated compound from *C. nutans* roots; Supplemental Data S1B: 13C-NMR spectrum of the isolated compound from *C. nutans* roots.

Author Contributions: Conceptualization, P.L.T. and B.E.C.; methodology, A.Y.F.C.; validation, A.Y.F.C., B.E.C., L.J. and P.L.T.; formal analysis, investigation and writing—original draft preparation, A.Y.F.C.; writing—review and editing, B.E.C., L.J. and P.L.T.; visualization, A.Y.F.C.; supervision, B.E.C. and P.L.T.; project administration, P.L.T.; funding acquisition, B.E.C. and P.L.T. All authors have read and agreed to the published version of the manuscript.

Funding: This research was funded by Universiti Malaysia Sabah (UMS) Research Grant Scheme, grant number SBK0070-SG-2013 and the Malaysia Ministry of Education, grant number FRG0360-SG-2/2013.

Acknowledgments: The authors thank Mohd. Zahid Md. Yusuff from the Universiti Kebangsaan Malaysia (UKM), Malaysia for the assistance of running the NMR experiments in this study.

Conflicts of Interest: The authors declare no conflict of interest.

References

1. Chelyn, J.L.; Omar, M.H.; Yousof, N.S.A.; Ranggasamy, R.; Wasiman, I.; Ismail, Z. Analysis of flavone C-glycosides in the leaves of *Clinacanthus nutans* (Burm. F.) Lindau by HPTLC and HPLC-UV/DAD. *Sci. World. J.* **2014**, *2014*, 1–6. [CrossRef]
2. Nik Abd Rahman, N.M.A.; Nurliyana, M.Y.; Nur Afiqah, M.N.F.N.; Osman, N.M.; Hamid, N.; Mohd Lila, M.A. Antitumor and antioxidant effects of *Clinacanthus nutans* Lindau in 4 T1 tumor-bearing mice. *BMC Complement. Altern. Med.* **2019**, *19*, 340. [CrossRef]

3. Yong, Y.K.; Tan, J.J.; Teh, S.S.; Mah, S.H.; Cheng, G.L.E.; Chiong, H.S.; Ahmad, Z. *Clinacanthus nutans* extracts are antioxidant with antiproliferative effect on cultured human cancer cell lines. *Evid. Based Complement. Altern. Med.* **2013**, *2013*, 462751. [CrossRef] [PubMed]
4. Che Sulaiman, I.S.; Basri, M.; Chan, K.W.; Ashari, A.E.; Masoumi, H.R.F.; Ismail, M. In vitro antioxidant, cytotoxic and phytochemical studies of *Clinacanthus nutans* Lindau leaf extracts. *Afr. J. Pharm. Pharmacol.* **2015**, *9*, 861–874. [CrossRef]
5. Fong, S.Y.; Piva, T.; Dekiwadia, C.; Urban, S.; Huynh, T. Comparison of cytotocity between extracts of *Clinacanthus nutans* (Burm. f.) Lindau leaves from different locations and the induction of apoptosis by the crude methanol leaf extract in D24 human melanoma cells. *BMC Complement. Altern. Med.* **2016**, *16*, 368. [CrossRef] [PubMed]
6. Huang, D.; Guo, W.; Gao, J.; Chen, J.; Olatunji, J.O. *Clinacanthus nutans* (Burm. f.) Lindau ethanol extract inhibits hepatoma in mice through upregulation of the immune response. *Molecules* **2015**, *20*, 17405–17428. [CrossRef] [PubMed]
7. Teoh, P.L.; Cheng, A.Y.F.; Liau, M.; Lem, F.F.; Kaling, G.P.; Chua, F.N.; Cheong, B.E. Chemical composition and cytotoxic properties of *Clinacanthus nutans* root extracts. *Pharm. Biol.* **2017**, *55*, 394–401. [CrossRef]
8. Huang, D.; Li, Y.; Cui, F.; Chen, J.; Sun, J. Purification and characterization of a novel polysaccharide-peptide complex from *Clinacanthus nutans* Lindau leaves. *Carbohydr. Polym.* **2016**, *137*, 701–708. [CrossRef] [PubMed]
9. Mutazah, R.; Hamid, H.A.; Mazila Ramli, A.N.; Fasihi Mohd Aluwi, M.F.; Yusoff, M.M. In vitro cytotoxicity of *Clinacanthus nutans* fractions on breast cancer cells and molecular docking study of sulphur containing compounds against caspase-3. *Food Chem. Toxicol.* **2020**, *135*, 110869. [CrossRef]
10. Agidew, E.; Reneela, P.; Tsegaye, D. Phytochemical investigation of *Sapium elliptum*. *J. Nat. Prod. Plant Res.* **2013**, *3*, 1–6.
11. Pereira, F.B.M.; Domingues, F.M.J.; Silva, A.M.S. Triterpenes from *Acacia dealbata*. *Nat. Prod. Lett.* **1996**, *8*, 97–103. [CrossRef]
12. Prachayasittikul, S.; Saraban, P.; Cherdtrakulkiat, R.; Ruchirawat, S.; Prachayasittikul, V. New bioactive triterpenoids and antimalarial activity of *Diospyros rubra* Lec. *EXCLI J.* **2010**, *9*, 1–10.
13. Saratha, V.S.; Pillai, I.; Subramaniam, S. Isolation and characterization of lupeol, a triterpenoid from *Calotropis gigantea* latex. *Int. J. Pharm. Sci. Rev. Res.* **2011**, *10*, 54–57.
14. Hague, M.A.; Hassan, M.M.; Das, A.; Begum, B.; Ali, M.Y.; Morshed, H. Phytochemical investigation of *Veronia cinerea* (Family: Asteraceae). *J. Appl. Pharm. Sci.* **2012**, *2*, 79–83. [CrossRef]
15. Le, C.-F.; Kailaivasan, T.H.; Chow, S.-C.; Abdullah, Z.; Ling, S.-K.; Fang, C.-M. Phytosterols isolated from *Clinacanthus nutans* induce immunosuppressive activity in murine cells. *Int. Immunopharm.* **2017**, *44*, 203–210. [CrossRef] [PubMed]
16. Ragasa, C.Y.; Tan, M.C.S.; Fortin, D.R.; Shen, C.-C. Chemical constituents of *Ixora philippinensis* Merr. *J. Appl. Pharm. Sci.* **2015**, *5*, 62–67. [CrossRef]
17. Saleem, M.; Kaur, S.; Kweon, M.-H.; Adhami, V.M.; Afaq, F.; Mukhtar, H. Lupeol, a fruit and vegetable based triterpene, induces apoptotic death of human pancreatic adenocarcinoma cells via inhibition of Ras signaling pathway. *Carcinogenesis* **2005**, *26*, 1956–1964. [CrossRef] [PubMed]
18. Lee, Y.S.; Tee, C.T.; Tan, S.P.; Awang, K.; Hashim, N.M.; Nafiah, M.A.; Ahmad, K. Cytotoxic, antibacterial and antioxidant activity of triterpenoids from *Kopsia singapurensis* Ridl. *J. Chem. Pharm. Res.* **2014**, *6*, 815–822.
19. Saleem, M.; Kweon, M.-H.; Yun, J.-M. A novel dietary triterpene lupeol induces fas-mediated apoptotic death of androgen-sensitive prostate cancer cells and inhibits tumor growth in a xenograft model. *Cancer Res.* **2005**, *65*, 11203–11213. [CrossRef]
20. Saleem, M. Lupeol, a novel anti-inflammatory and anti-cancer dietary triterpene. *Cancer Lett.* **2009**, *285*, 109–115. [CrossRef]
21. Saleem, M.; Murtaza, I.; Tarapore, R.S.; Suh, Y.; Adhami, V.M.; Johnson, J.J.; Siddiqui, I.A.; Khan, N.; Asim, M.; Hafeez, B.B.; et al. Lupeol inhibits proliferation of human prostate cancer cells by targeting β-catenin signaling. *Carcinogen* **2019**, *30*, 808–817. [CrossRef]
22. Pitchai, D.; Roy, A.; Ignatius, C. In vitro evaluation of anticancer potentials of lupeol isolated from *Elephantopus scaber* L. on MCF-7 cell line. *J. Adv. Pharm. Technol. Res.* **2014**, *5*, 179–184. [CrossRef]
23. Kommera, H.; Kaluderovic, G.N.; Kalbitz, J.; Paschke, R.P. Lupane triterpenoids—Betulin and betulinic acid derivatives induce apoptosis in tumor cells. *Investig. New Drugs* **2011**, *29*, 266–272. [CrossRef]
24. Awad, A.B.; Williams, H.; Fink, C.S. Effect of phytosterols on cholesterol metabolism and MAP kinase in MDA-MB-231 human breast cancer cells. *J. Nutr. Biochem.* **2003**, *14*, 111–119. [CrossRef]
25. Ghosh, T.; Maity, T.K.; Singh, J. Evaluation of antitumor activity of stigmasterol, a constituent isolated from *Bacopa monnieri* Linn aerial parts against Ehrlich Ascites carcinoma in mice. *Orient. Pharm. Exp. Med.* **2011**, *11*, 41–49. [CrossRef]
26. Chai, J.W.; Kuppusamy, U.R.; Kanthimathi, M.S. beta-Sitosterol induces apoptosis in MCF-7 cells. *Malays. J. Biochem. Mol. Biol.* **2008**, *16*, 28–30.
27. Sundarraj, S.; Thangam, R.; Sreevani, V.; Kaveri, K.; Gunasekaran, P.; Achiraman, S.; Kannan, S. γ-Sitosterol from *Acacia nilotica* L. induces G2/M cell cycle arrest and apoptosis through c-Myc suppression in MCF-7 and A549 cells. *J. Ethnopharmacol.* **2012**, *141*, 803–809. [CrossRef]
28. Jin, C.-C.; Zhang, L.-L.; Song, H.; Cao, Y.-X. Boosting the biosynthesis of betulinic acid and related triterpenoids in *Yarrowia lipolytica* via multimodular metabolic engineering. *Microb. Cell Fact.* **2019**, *18*, 77. [CrossRef]
29. Majumder, S.; Ghosh, A.; Bhattacharya, M. Natural anti-inflammatory terpenoids in *Camellia japonica* leaf and probable biosynthesis pathways of the metabolome. *Bull. Natl. Res. Centre* **2020**, *44*, 141. [CrossRef]
30. Piironen, V.; Lindsay, D.G.; Miettinen, T.A.; Toivoi, J.; Lampi, A.-M. Plant sterols: Biosynthesis, biological function and their importance to human nutrition. *J. Sci. Food Agric.* **2000**, *80*, 939–966. [CrossRef]

31. Dampawan, P.; Huntrakul, C.; Reutrakul, V. Constituents of *Clinacanthus nutans* and the crystal structure of lup-20(29)-ene-3-one. *J. Sci. Soc. Thail.* **1977**, *3*, 14–26. [CrossRef]
32. Murugesu, S.; Ibrahim, Z.; Ahmed, Q.-U.; Nik Yusoff, N.-I.; Uzir, B.-F.; Perumal, V.; Abas, F.; Saari, K.; El-Seedi, H.; Khatib, A. Characterization of α-Glucosidase Inhibitors from *Clinacanthus nutans* Lindau Leaves by Gas Chromatography-Mass Spectrometry-Based Metabolomics and Molecular Docking Simulation. *Molecules* **2018**, *23*, 2402. [CrossRef] [PubMed]

processes

MDPI

Article

Statistical Optimization of Biodiesel Production from Salmon Oil via Enzymatic Transesterification: Investigation of the Effects of Various Operational Parameters

Vegneshwaran V. Ramakrishnan [1], Deepika Dave [1,*], Yi Liu [1], Winny Routray [2] and Wade Murphy [1]

1 Centre for Aquaculture and Seafood Development, Marine Bioprocessing Facility, Fisheries and Marine Institute, Memorial University of Newfoundland, St. John's, NL A1B 4L6, Canada; Vegneshwaran.Ramakrishnan@mi.mun.ca (V.V.R.); Yi.Liu@mi.mun.ca (Y.L.); Wade.Murphy@mi.mun.ca (W.M.)

2 Department of Food Process Engineering, National Institute of Technology Rourkela, Rourkela 769008, India; routrayw@nitrkl.ac.in

* Correspondence: Deepika.Dave@mi.mun.ca

Abstract: The enzymatic transesterification of Atlantic salmon (*Salmo salar*) oil was carried out using Novozym 435 (immobilized lipase from *Candida antartica*) to produce biodiesel. A response surface modelling design was performed to investigate the relationship between biodiesel yield and several critical factors, including enzyme concentration (5, 10, or 15%), temperature (40, 45, or 50 °C), oil/alcohol molar ratio (1:3, 1:4, or 1:5) and time (8, 16, or 24 h). The results indicated that the effects of all the factors were statistically significant at *p*-values of 0.000 for biodiesel production. The optimum parameters for biodiesel production were determined as 10% enzyme concentration, 45 °C, 16 h, and 1:4 oil/alcohol molar ratio, leading to a biodiesel yield of 87.23%. The step-wise addition of methanol during the enzymatic transesterification further increased the biodiesel yield to 94.5%. This is the first study that focused on Atlantic salmon oil-derived biodiesel production, which creates a paradigm for valorization of Atlantic salmon by-products that would also reduce the consumption and demand of plant oils derived from crops and vegetables.

Keywords: biodiesel; Atlantic salmon; marine by-products; biocatalytic transesterification; response surface methodology

Citation: Ramakrishnan, V.V.; Dave, D.; Liu, Y.; Routray, W.; Murphy, W. Statistical Optimization of Biodiesel Production from Salmon Oil via Enzymatic Transesterification: Investigation of the Effects of Various Operational Parameters. *Processes* **2021**, *9*, 700. https://doi.org/10.3390/pr9040700

Academic Editor: Zongbi Bao

Received: 25 March 2021
Accepted: 14 April 2021
Published: 16 April 2021

1. Introduction

Biodiesel has been studied for decades as a promising alternative to traditional fossil fuel [1–3]. Biodiesel is a processed fuel that is produced via transesterification of triglycerides derived from naturally occurring plant oils and animal fats. During the transesterification, the triglycerides are converted into fatty acid methyl esters, with glycerol produced as a by-product. Three methods have been traditionally used to produce biodiesel, including acid catalysis, base catalysis, and non-catalytic transesterification using supercritical alcohol [4]. The most commonly used commercial process for biodiesel production is alkali-catalyzed transesterification using sodium hydroxide or potassium hydroxide due to their relatively low cost and high conversion efficiency [3]. However, chemical transesterification has multiple disadvantages [5], such as the high dependency of conversion efficiency on the content of water and free fatty acids in the raw materials and the tremendous energy consumption due to the high reaction temperature and product separation process. Moreover, the acid or base catalysts are not reusable, and extra steps of neutralization are required to dispose of them as an aqueous salt waste stream, which is less environmentally hazardous [6].

In recent years, biocatalytic transesterification, which is catalyzed by lipases, has been intensively studied for biodiesel production [5,7–9]. As lipases catalyze the reaction through interactions with the molecules at specific sites, the conversion efficiency and product purity are high, with little need for downstream processing. The biocatalytic process avoids the use of alkaline, so there is no soap formation or generation of alkaline wastewater, which reduces the workload for wastewater treatment [6]. In addition, the enzymatic transesterification can be carried out under mild operation conditions [10]. Among various lipases, the immobilized lipase from *Candida antarctica* has been widely used due to its extraordinary efficiency for biodiesel conversion [11]. After the transesterification, the immobilized lipase can be easily separated from the reaction mixture through vacuum filtration and washed for reuse. It has been reported that the enzyme can be reused for at least 10 cycles with a slight decrease in biodiesel yield [12,13].

A variety of vegetable oils extracted from corn, canola, palm, soybean, sunflower, rapeseed, coconut, and groundnut have been converted into biodiesel [14–16]. However, as most of these plant-derived oils are initially used for food, livestock feed, and oleochemical industries, biodiesel production results in increased consumption of these resources and increased crop demand [10]. Therefore, oils derived from alternative resources have been studied, such as fish oil extracted from fish by-products [17–20]. Atlantic salmon aquaculture has played an essential role in the Newfoundland aquaculture industry for decades, accounting for above 80% of the total aquaculture landing [21]. In 2017, the aquaculture landing of salmonids (mainly Atlantic salmon) in Newfoundland and Labrador was 18,822 tonnes (87%) [21]. In the Atlantic salmon processing industry, 40–50% of the fish body ends up as by-products (heads, frames, and viscera), which have been reported to be abundant in lipids (15–25%) [22]. Several studies have been reported on oil extraction from these by-products [23–27]. The extracted oil has been reported with high yield and good quality with a low extent of oxidation or hydrolysis [23]. To the authors' best knowledge, there has not been any research reported on biodiesel production from Atlantic salmon oil. This is the first study that converted enzymatically extracted Atlantic salmon oil to biodiesel. The aim of this work was to maximize the biodiesel production from the salmon oil, which can be developed as an effective way to valorize the large amounts of Atlantic salmon by-products.

The statistical design of experiments plays a vital role in determination of optimal parameters in many research areas of science and industry. In a multivariable system, only one factor can be studied in each set of experiments using the conventional protocols, and such methods require many experiments and do not represent the combined effect. Therefore, different types of design of experiment techniques, including Latin squares, full factorial, fractional factorial, response surface methodology, and Plackett–Burman and Taguchi methods, have been implemented in various studies [28,29]. The response surface methodology (RSM) is a widely used statistical modelling system to analyze a process in which the response of interest might be affected by various variables. The main objective of response surface methodology is to optimize the response as per the needs of the study. Response surface methodology reduces the number of experiments needed to determine the optimal parameters for the process [30,31]. In the present study, the transesterification process was optimized using response surface methodology to maximize the biodiesel yield, and the influence of the main reaction parameters, including enzyme concentration, oil/alcohol molar ratio, temperature, and time, was investigated.

2. Materials and Methods

2.1. Materials

Farmed Atlantic salmon (*Salmo salar*) by-products, including heads, frames, and viscera, were collected in Styrofoam boxes on ice from a salmon aquaculture processing plant, Atlantic Canada. The by-products were shipped overnight to the Marine Bioprocessing Facility, Marine Institute of Memorial University of Newfoundland, St. John's, NL, Canada, and immediately processed once received. The salmon raw materials were minced three times using a Hobart grinder (Model 4146 The Hobart MFG. Co. Ltd., Peterborough, UK) using different plate sizes (17, 15, and 13 mm). The raw materials were vacuum-packed in plastic bags with each weighing approximately 500 g, and stored at −28 °C until use.

Alcalase 2.4 L, boron trifluoride, tricosanoic acid methyl ester, Novozym 435 (immobilized lipase from *Candida antarctica*), *tert*-butanol, tetrahydrofuran, N,O-bis(trimethylsilyl) trifluoroacetamide (BSTFA), and PUFA No. 3 (from menhaden oil) analytical standard were purchased from Sigma-Aldrich, Oakville, ON, Canada. Hexane and methanol were purchased from Fisher Scientific, Nepean, ON, Canada.

2.2. Salmon Oil Extraction

The frozen ground salmon materials were thawed in cold running water. An amount of 500 g salmon by-products was weighed in a 1 L Mason jar. To the weighed raw material, 500 mL water was added (1:1 *w*/*w*). The mixture was gently stirred using a magnetic stirrer, and the pH was adjusted to the optimal value of 8 for Alcalase by adding a 10% sodium hydroxide solution. The mixture was heated in a water bath to 80 °C for 5 min to deactivate the endogenous enzymes in the fish materials. Afterwards, the mixture was cooled down to the optimal temperature of 55 °C for Alcalase. The enzymatic hydrolysis was initiated by adding 1% (*w*/*w* of the raw fish materials) of Alcalase, and the mixture was digested in an incubator shaker (Thermo-Scientific Max Q 6000, Marietta, OH, USA) at 140 rpm for 2 h. The hydrolysis was stopped by heating the mixture at 90 °C for 10 min to deactivate the enzyme. The mixture was cooled down to room temperature and centrifuged at 10,000 rpm for 20 min. After centrifugation, four layers were obtained: the oil layer on top, an emulsion layer, an aqueous layer of protein hydrolysates, and the sludge on the bottom. The oil was collected and stored at −80 °C for subsequent analyses.

2.3. Enzymatic Transesterification of Salmon Oil

The concentration of Novozym 435 used for the enzymatic transesterification was investigated at three different levels (5, 10, or 15% *w*/*w*). Other factors studied included reaction temperature (40, 45, or 50 °C), time (8, 16, or 24 h), and oil/alcohol molar ratio (1:3, 1:4, or 1:5) for the optimization of biodiesel production (Figure 1). The homogenized salmon oil (2 g) was placed into a 50 mL conical flask, and an appropriate amount of methanol was added based on the stoichiometric oil/alcohol molar ratio (1:3, 1:4, or 1:5), followed by the addition of *tert*-butanol (75 wt% based on oil weight). The mixture was heated at an appropriate temperature (40, 45, or 50 °C) in a reciprocal shaking water bath at 200 rpm for 5–10 min. The reaction was initiated by adding the appropriate amount of immobilized enzyme (5, 10, or 15% *w*/*w*). After the desired reaction time (8, 16, or 24 h), the enzyme was filtered by vacuum filtration as per the method reported by Nelson et al. [32]. Then, 100 μL of the filtrate was taken and analyzed using a gas chromatograph (GC) (Section 2.5).

2.4. Fatty Acid Composition Analysis of Salmon Oil

The extracted salmon oil was analyzed using the AOAC Official Method 991.39 [33]. First, 25 ± 0.1 mg of tricosanoic acid methyl ester (the internal standard) was weighed into a 25 mL volumetric flask and diluted to the line with hexane. Next, 1.0 mL portions of this stock solution were pipetted into screw cap glass test tubes, and the solvent was evaporated in a gentle stream of nitrogen (N_2). The tubes were stored in the freezer (−80 °C) if not used immediately. A total of 25 ± 0.1 mg of the extracted salmon oil was weighed into a glass test tube containing the internal standard, followed by the addition of 1.5 mL of 0.5 M methanolic NaOH. The mixture was blanketed with N_2, capped, and heated at 100 °C for 5 min. After the test tube was cooled, 2 mL of boron trifluoride (in 14% methanol) was added. The solution was blanketed with N_2, capped, and heated at 100 °C for 30 min. Afterwards, the mixture was cooled to 30–40 °C, and 1 mL of hexane was added. The solution was blanketed with N_2, capped, and mixed vigorously for 30 s while still warm. Next, 5 mL of saturated NaCl solution was immediately added to the test tube, and the mixture was blanketed with N_2, capped, and mixed thoroughly. The sample was cooled to room temperature and left still for the formation of two layers. The hexane layer was transferred to another test tube, blanketed with N_2, and capped tightly. The aqueous layer was extracted again with 1 mL of hexane. Afterwards, the hexane extracts were collected together and concentrated to approximately 1 mL under N_2. Finally, 1 μL of the sample was injected into the GC for analysis. The gas chromatogram of the salmon oil is illustrated in Figure 2.

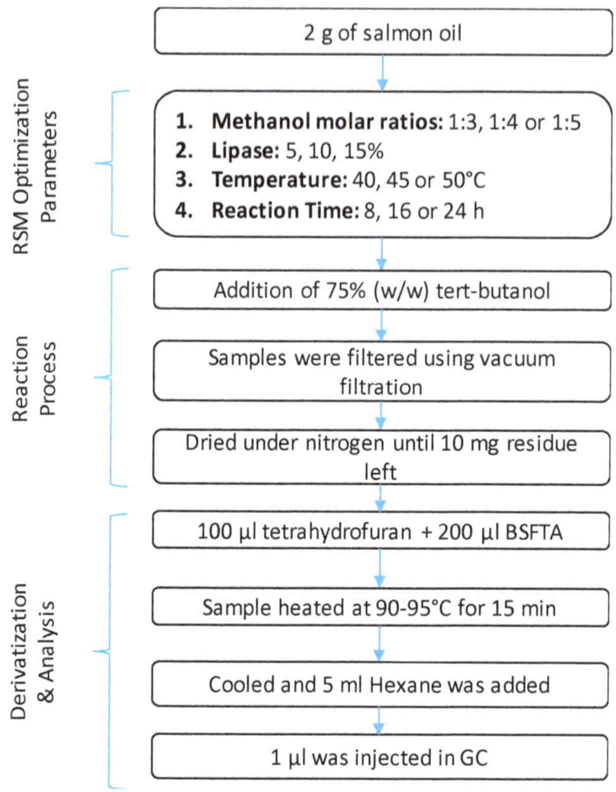

Figure 1. Experiment design of enzymatic transesterification of salmon oil for production of biodiesel.

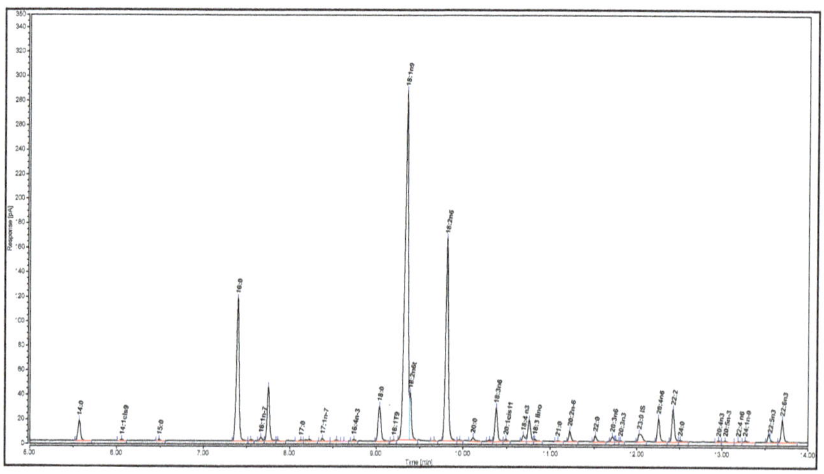

Figure 2. Gas chromatogram of Atlantic salmon oil.

The area percentage of the fatty acid was calculated as follows:

$$\text{Area percent of the fatty acid (\%)} = A_X/(A_T - A_{IS})*100 \quad (1)$$

where

A_X = area counts of individual methyl ester;
A_T = total area counts for chromatogram;
A_{IS} = area counts of the internal standard;
The fatty acid composition of the salmon oil.

2.5. *Fatty Acid Methyl Ester (FAME) Composition Analysis of Biodiesel*

The sample for analysis was prepared as described by Nelson et al. [32]. First, 100 μL of the filtrate after reaction and filtration (Section 2.3) was concentrated under N_2 until approximately 10 mg was left. Then, the sample was dissolved in 100 μL tetrahydrofuran, followed by the addition of 200 μL BSTFA. The mixture was heated in a water bath at 90–95 °C for 15 min. After the mixture was cooled to room temperature, 5 mL hexane was added. An aliquot of 1 μL of the mixture was injected into the GC for analysis. The gas chromatogram of the biodiesel is illustrated in Figure 3.

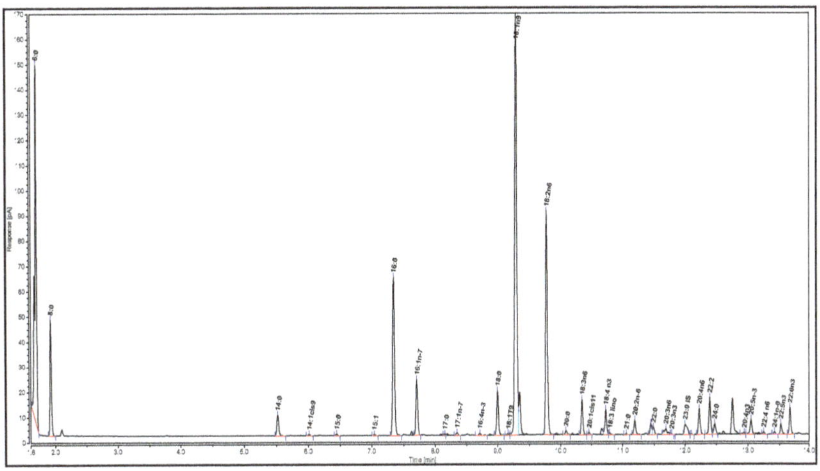

Figure 3. Gas chromatogram of biodiesel.

The conversion yield of FAME was calculated as follows:

$$\text{FAME Conversion(\%)} = \text{Total methyl ester peak area after enzymatic reaction/Total methyl ester peak area of fish oil} \quad (2)$$

2.6. *GC Settings for Analysis*

The composition analysis of fish oil and biodiesel was carried out using Trace 1300 gas chromatograph (GC) with flame ionization detector (FID) (Thermo Fisher, Canada). The column used for the analysis was the TR-FAME Trace GC capillary column with an internal diameter of 0.22 mm, film thickness of 0.25 μm, and a length of 30 m. The inlet temperature was maintained at 255 °C under split mode. The split flow of the carrier gas was maintained at 100 mL/min, and the purge flow was 2 mL/min. The inlet pressure was maintained at 81.7 kPa. The temperature of the flame ionization detector was maintained at 270 °C. The ignition threshold was 1 pA, and the data collection rate was 10 Hz. The gas settings included an airflow at 350 mL/min, makeup gas flow at 35 mL/min, and hydrogen flow at 35 mL/min. The GC oven temperature was initially maintained at 130 °C for 3 min, ramped to 250 °C at the rate of 10 °C/min, and held for 15 min. The total run time for one sample was 30 min.

2.7. Statistical Analysis

The response surface methodology was used for the experimental modelling design and analysis to describe the relationship between the response (yield) and critical factors (enzyme concentration, oil/alcohol molar ratio, time, and temperature). The experimental design consisted of four factors, generating $2^4 = 16$ runs. The reactions were carried out in duplicates, and the design contained six center points, therefore yielding 38 runs. The initial design was modelled to fit linear modelling. However, curvature was observed in data analysis, which indicated that the average response points detected at the center points were higher or lower than the average response at the corner cube points. Therefore, the linear model was inadequate to describe the relationship between the response and the factors completely. A total of 11 curvature points were added to modify the model, and the experiments were carried out. All the data were analyzed with a one-way analysis of variance (ANOVA) at a 95% confidence level. The effect of each factor was determined by estimating the main products and two-way interactions. All experimental design and data analyses were performed using Minitab 17.1.

3. Results

3.1. Statistical Modeling of Biodiesel Production

The experiment conditions and corresponding biodiesel yield of the forty-nine transesterification reactions are shown in Table 1. The highest yield of 91.86% was obtained from the 16 h reaction at 45 °C with 15% enzyme concentration and 1:4 oil/alcohol molar ratio (Run order 41). A quadratic model fits the results with an R^2 value of 99.31%, suggesting that this model can explain 99.31% of the yield variation. The regression equation that was developed for fitting the quadratic model for the biodiesel conversion was as follows:

$$\begin{aligned}\text{Yield} = &-722.4 + 1.629\ \text{EC} + 28.31\ \text{Temp} - 0.582\ \text{Time} + 83.02\ \text{Ratio} + 0.0018\ \text{EC*EC} - 0.3296\ \text{Temp}\times\text{Temp} \\ &- 0.1065\ \text{Time}\times\text{Time} - 9.516\ \text{Ratio}\times\text{Ratio} + 0.00417\ \text{EC}\times\text{Temp} - 0.05352\ \text{EC}\times\text{Time} \\ &- 0.0354\ \text{EC}\times\text{Ratio} + 0.08611\ \text{Temp}\times\text{Time} - 0.1679\ \text{Temp}\times\text{Ratio} + 0.3099\ \text{Time}\times\text{Ratio}\end{aligned} \quad (3)$$

Table 1. Biodiesel conversion yield from enzymatic transesterification of Atlantic salmon oil.

StdOrder	RunOrder	CenterPt	Blocks	EC	Temp	Time	Ratio	Yield (%)
9	1	1	1	5	40	8	5	61.07
10	2	1	1	15	40	8	5	72.77
19	3	0	1	10	45	16	4	84.26
17	4	0	1	10	45	16	4	86.27
13	5	1	1	5	40	24	5	71.07
3	6	1	1	5	50	8	3	42.61
18	7	0	1	10	45	16	4	87.77
5	8	1	1	5	40	24	3	56.55
16	9	1	1	15	50	24	5	76.00
14	10	1	1	15	40	24	5	78.54
15	11	1	1	5	50	24	5	73.00
1	12	1	1	5	40	8	3	56.98
11	13	1	1	5	50	8	5	46.16
7	14	1	1	5	50	24	3	61.94
12	15	1	1	15	50	8	5	59.39
4	16	1	1	15	50	8	3	59.35
2	17	1	1	15	40	8	3	68.38
6	18	1	1	15	40	24	3	64.39
8	19	1	1	15	50	24	3	63.45
22	20	1	2	5	50	8	3	42.09
27	21	1	2	15	50	24	3	62.79

Table 1. *Cont.*

StdOrder	RunOrder	CenterPt	Blocks	EC	Temp	Time	Ratio	Yield (%)
24	22	1	2	5	40	24	3	55.81
25	23	1	2	15	40	24	3	61.28
21	24	1	2	15	40	8	3	67.2
35	25	1	2	15	50	24	5	74.03
38	26	0	2	10	45	16	4	87.24
20	27	1	2	5	40	8	3	56.88
26	28	1	2	5	50	24	3	61.28
33	29	1	2	15	40	24	5	75.28
28	30	1	2	5	40	8	5	61.18
34	31	1	2	5	50	24	5	69.32
37	32	0	2	10	45	16	4	86.66
31	33	1	2	15	50	8	5	58.18
29	34	1	2	15	40	8	5	71.07
23	35	1	2	15	50	8	3	58.35
36	36	0	2	10	45	16	4	88.18
30	37	1	2	5	50	8	5	45.25
32	38	1	2	5	40	24	5	72.82
42	39	−1	3	10	50	16	4	76.32
39	40	−1	3	5	45	16	4	83.12
40	41	−1	3	15	45	16	4	91.86
45	42	−1	3	10	45	16	3	73.75
49	43	0	3	10	45	16	4	87.64
44	44	−1	3	10	45	24	4	85.48
46	45	−1	3	10	45	16	5	82.11
48	46	0	3	10	45	16	4	85.45
41	47	−1	3	10	40	16	4	82.09
43	48	−1	3	10	45	8	4	75.78
47	49	0	3	10	45	16	4	88.24

The model indicated a relationship between biodiesel yield and the factors at a 0.05 level of significance. As indicated in the Pareto chart (Figure 4), the most influential factor was time (C), followed by enzyme concentration (A), oil/alcohol molar ratio (D), the interaction between temperature and time (BC), temperature (B), and other interactions between factors. The analysis of variance of the model summarized the linear terms, square terms, and interaction of factors (Table 2). The effects of enzyme concentration (*p*-value: 0.000), temperature (*p*-value: 0.000), time (*p*-value: 0.000), and molar ratio (*p*-value: 0.000) on biodiesel conversion yield were statistically significant (*p*-value < 0.05).

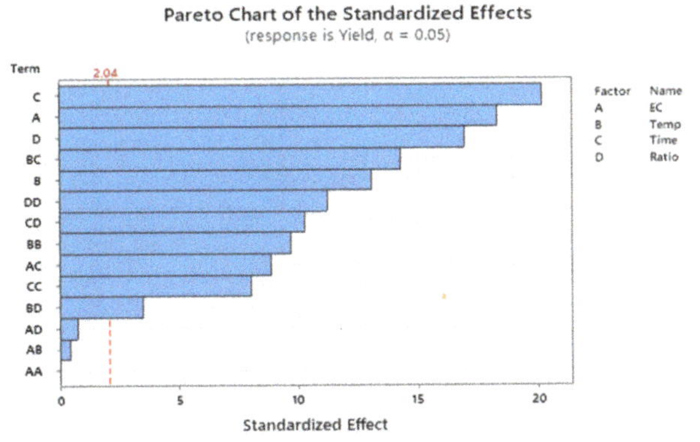

Figure 4. Pareto chart for the effect of factors on biodiesel conversion yield (EC: enzyme concentration; Temp: temperature; Ratio: oil/alcohol molar ratio).

Table 2. Analysis of variance on biodiesel conversion yield.

Source	DF	*p*-Value	VIF
Model	16	0.000	
Blocks	2	0.115	
1		0.181	1.40
2		0.087	1.40
Linear	4	0.000	
EC	1	0.000	1.00
Temp	1	0.000	1.00
Time	1	0.000	1.00
Ratio	1	0.000	1.00
Square	4	0.000	
EC*EC	1	0.959	4.03
Temp*Temp	1	0.000	4.03
Time*Time	1	0.000	4.03
Ratio*Ratio	1	0.000	4.03
2-Way Interaction	6	0.000	
EC*Temp	1	0.668	1.00
EC*Time	1	0.000	1.00
EC*Ratio	1	0.468	1.00
Temp*Time	1	0.000	1.00
Temp*Ratio	1	0.001	1.00
Time*Ratio	1	0.000	1.00
Error	32		
Lack-of-Fit	26	0.230	
Pure Error	6		
Total	48		

EC: enzyme concentration; Temp: temperature; Ratio: oil/alcohol molar ratio.

As indicated in Table 2, interaction terms including EC*Time, Temp*Time, Temp*Ratio, and Time*Ratio had a significant influence ($p < 0.05$) on the biodiesel conversion yield. However, EC*Temp and EC*Ratio did not significantly affect the biodiesel yield ($p > 0.05$). The square terms Temp*Temp, Time*Time, and Ratio*Ratio had p-values of 0.000, suggesting a significant quadratic effect and the relationship between temperature, time, and ratio of biodiesel yield followed a curved line rather than a straight line. The square term EC*EC had a p-value of 0.959, so it did not significantly contribute to the biodiesel conversion yield.

The p-values of 0.181 and 0.087 for blocks 1 and 2 (Table 2), respectively, indicated no significant block effect. Therefore, the data collected in those two blocks did not significantly affect the yield. The small p-values for linear terms (p-value: 0.000), square terms (p-value: 0.000), and interactions (p-value: 0.000) indicated there was curvature in the response surface. The ANOVA results showed a p-value of 0.230 for the lack of fit, suggesting a significant model fit with the data. The s-value (standard error of the regression) in a model indicates the standard distance that data values fall from the regression line or the standard deviation of the residuals. The s-value of the current model was 1.362, which was quite low and therefore suggested a good fit.

As presented in Table 2, the variance inflation factors (VIFs) for most of the predictors were close to 1, which suggested that the predictors were not correlated. The VIFs of four interactions (EC$\times$EC, Temp$\times$Temp, Time$\times$Time and Ratio$\times$Ratio) were 4.03, suggesting that they were moderately correlated. None of the results showed VIF values higher than 5, which would result in severe multicollinearity, and the model would become unstable and difficult to interpret.

3.2. Effect of Operating Parameters

3.2.1. Temperature

Figure 5 shows counterplots of biodiesel yield vs. enzyme concentration and temperature, with time and oil/alcohol molar ratio held constant at different conditions. In Figure 5a, which had a hold value of 8 h (time) and 1:3 (oil/alcohol molar ratio), the highest biodiesel yield (>70%) was obtained at 43–45 °C and 13–15% enzyme concentration. When the hold values were changed to 16 h (time) and 1:4 (oil/alcohol molar ratio) (Figure 5b), the highest yield was >90%, with the optimal setting of around 45 °C and 13% enzyme concentration. In Figure 5c, which had a hold value of 24 h (time) and 1:5 (oil/alcohol molar ratio), the highest biodiesel yield was >84% at around 45 °C and 14% enzyme concentration. In all the three plots, it is evident that when the temperature was increased to above 45 °C, the yield decreased significantly. The highest biodiesel yield was always obtained at around 45 °C regardless of the enzyme concentration. This effect was further supported in Table 2, which indicated no significant interaction between the two factors (enzyme concentration and temperature) for biodiesel yield (p-value of 0.668).

As observed in the present study, the effect of temperature on biodiesel conversion agrees with previously reported research. Chen et al. studied the enzymatic conversion of waste cooking oils into biodiesel and reported the highest biodiesel yield of 87% at 40 °C using immobilized *Rhizopus oryzae* lipase [34]. It was observed that the further increase of temperature to above 40 °C decreased the biodiesel yield. Dizge and Keskinler produced biodiesel from canola oil using immobilized *Thermomyces lanuginosus* lipase [35]. The effect of temperature on enzymatic transesterification for biodiesel production was studied from 30 to 70 °C, and 40 °C was the optimal temperature, resulting in the highest biodiesel yield of 85.8%. With a further increase of temperature to above 50 °C, the biodiesel yield was decreased drastically since the enzyme lost its activity significantly above 50 °C. Nie et al. performed lipase-catalyzed methanolysis of salad oil using immobilized *Candida sp.* 99–125 lipase [36]. The highest yield was observed at 40 °C and decreased drastically above that. As claimed by the authors, high temperatures can accelerate the reaction process; however, the enzyme will be denatured if the temperatures are too high. This was supported by Pinyaphong et al. in their study of methanolysis of fish oil catalyzed by *Carcia papaya* lipase [37]. The reaction was performed at 30–60 °C, and a maximum biodiesel conversion yield of 83% was obtained at 40 °C. The results showed that reaction temperature significantly influenced enzyme activity and stability, and a high temperature can lead to deactivation of the lipase.

3.2.2. Reaction Time

Figure 6 indicates the relationship between biodiesel yield and enzyme concentration and time, with temperature and oil/alcohol molar ratio held constant at different conditions. In Figure 6a, the biodiesel yield was increased when enzyme concentration was increased from 5 to 15% and time was increased from 8 to 15 h, and the highest yield (>72%) was achieved at 14.5% enzyme concentration within 13–15 h of reaction time. However, Figure 6b indicates that biodiesel yield >90% was obtained when the reaction time was 18 h and the enzyme concentration was 14.5%. Figure 6c shows that further increase in time did not increase biodiesel yield, which was >75% after 22 h of reaction with 13% enzyme concentration. Therefore, the optimal setting for the maximum biodiesel yield (>90%) could be achieved after around 18 h of reaction with 14.5% enzyme concentration. The biodiesel yield was significantly affected by the interaction between enzyme concentration and time, supported by the p-value of 0.000 (Table 2).

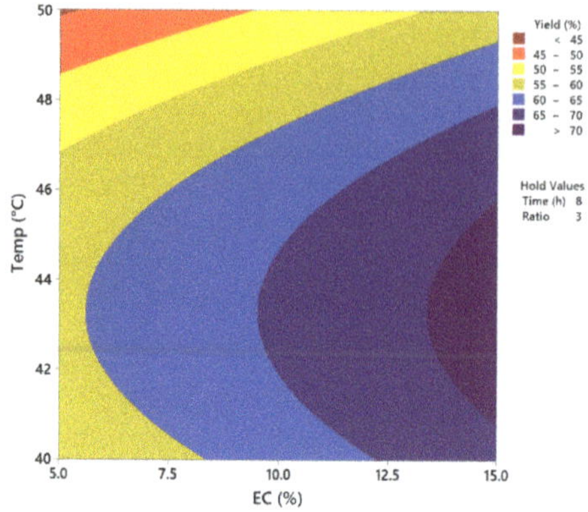

(a) EC*Temp at Hold values of Time-8 h and Ratio-1:3

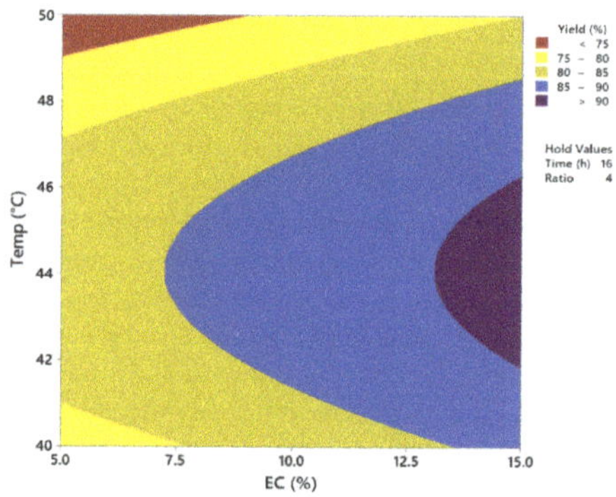

(b) EC*Temp at Hold values of Time-16 h and Ratio-1:4

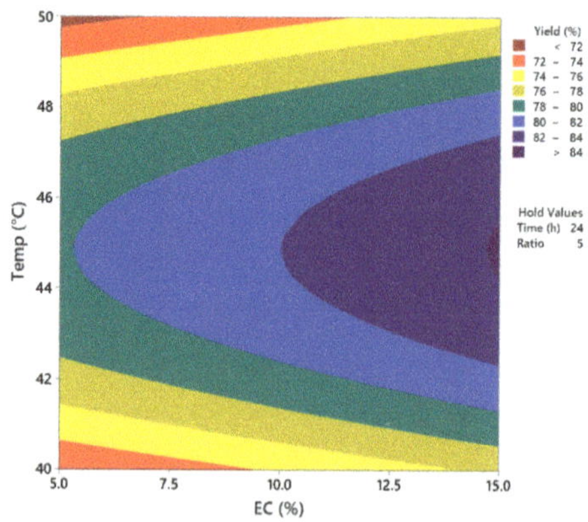

(c) EC*Temp at Hold values of Time-24 h and Ratio-1:5

Figure 5. Counterplots of biodiesel yield vs. enzyme concentration and temperature at (**a**) 8 h and 1:3; (**b**) 16 h and 1:4; (**c**) 24 h and 1:5.

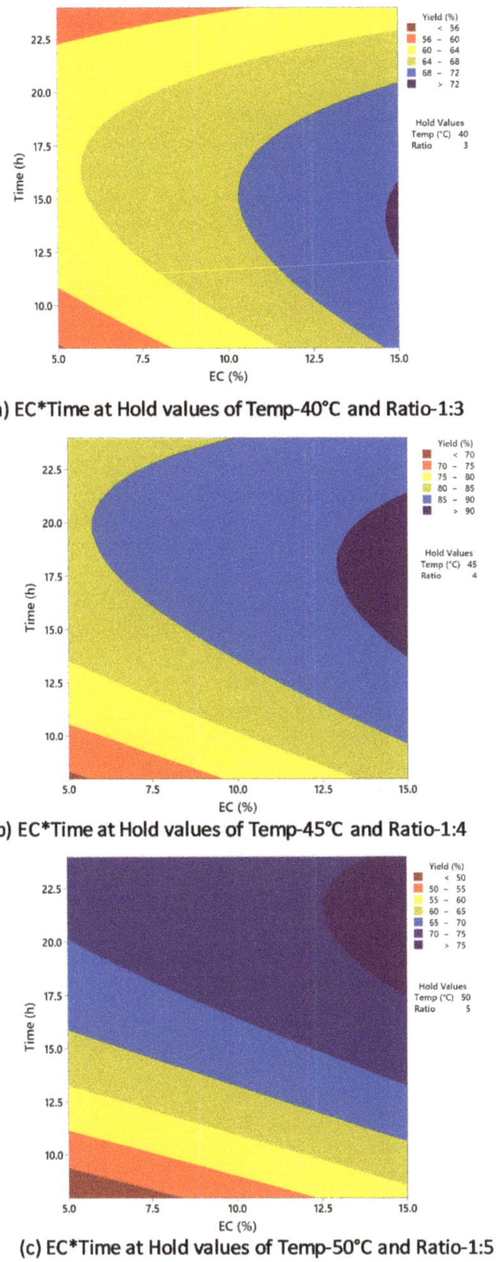

Figure 6. Counterplots of biodiesel yield vs. enzyme concentration and time at (**a**) 40 °C and 1:3; (**b**) 45 °C and 1:4; (**c**) 50 °C and 1:5.

In the present study, the increase of time resulted in the increase of biodiesel production yield in most of the parameters. However, when the time was increased from 16 to 24 h, there was no significant increase in biodiesel yield, and the optimum time for biodiesel production was limited to 16 h. Pinyaphong et al. reported a maximum biodiesel conversion yield of 83% from fish oil from a reaction of 18 h [37]. Nelson et al. carried out biodiesel production from tallow oil and reported the highest yield of 83.8% after 16 h of reaction [32]. Li et al. performed transesterification of rapeseed oil for biodiesel production using Novozyme 435, and a 12 h reaction resulted in the highest biodiesel yield of 95% [38]. Azcar et al. studied the biodiesel production from waste frying oil using Novozyme 435 and reported the highest biodiesel yield of 100% after 4 h of reaction [39]. Compared to the optimized reaction time of 16 h in the present study, their reaction time was four times shorter. The biodiesel yield was higher, mainly due to the lower content of unsaturated fatty acids in frying oil than that of fish oil.

3.2.3. Oil/Alcohol Molar Ratio

As indicated in Table 1, the increase in the oil/alcohol molar ratio positively affected biodiesel production from 1:3 to 1:4. However, there was no significant increase when the molar ratio was increased to 1:5, and in some experiments, the biodiesel yield was reduced. Figure 7 indicates the influence of enzyme concentration and oil/alcohol molar ratio on biodiesel yield with temperature and time held constant at different conditions. The results from Figure 7a–c indicate that whether the enzyme concentration was increased or decreased, the highest biodiesel yield of >90% was achieved with around 1:4 oil/alcohol molar ratio. Further increase in the molar ratio did not increase the biodiesel yield. The interaction between enzyme concentration and oil/alcohol molar ratio had no significant effect on biodiesel yield, as supported by the p-value of 0.468 (Table 2).

Li et al. reported in their study of lipase-catalyzed transesterification of rapeseed oil using Novozyme 435 that the highest biodiesel conversion yield of 95% was obtained with 1:4 oil/alcohol molar ratio with *tert*-butanol as a solvent [38]. Nelson et al. reported a maximum biodiesel conversion of 83.8% with a 1:3 oil/alcohol molar ratio using tallow and 25% *Candida antarctica* (SP 435) with hexane as the solvent [32]. Ognjanovic et al. reported biodiesel conversion of above 99% with 1:3 oil/methanol molar ratio after 50 h of reaction, but the enzyme activity rapidly decreased [40]. To stabilize the enzyme, methyl acetate was used as an alternative to methanol in the ratio of 1:12, and 93.6% biodiesel yield was achieved from sunflower oil using Novozyme 435 enzyme. Chen et al. reported a maximum biodiesel conversion yield of 89% at 1:4 oil/alcohol molar ratio from waste cooking oil [34]. As suggested by Hernández-Martín and Otero, the amount of alcohol used for biodiesel conversion should be slightly higher than the stoichiometric amount equal to the number of fatty acids in the oil to compensate for the thermodynamic or kinetic constraints [41]. Furthermore, the use of excess alcohol in biodiesel production has been claimed to maintain a uniform suspension of catalysts in the reaction medium, increase reaction rates, minimize diffusion limitations, and retain glycerol formation without deactivating the catalytic pores [42,43].

3.2.4. Enzyme Concentration

The effect of enzyme concentration (5, 10, and 15% *w*/*w*) of Novozyme 435 on biodiesel conversion yield was studied. The highest biodiesel yield was obtained using 15% enzyme. The results from Figures 5–7 suggest that while visualizing the biodiesel yield with respect to time, temperature, and molar ratio, the highest yield was always obtained with above 10% enzyme concentration.

Marín-Suárez et al. studied the transesterification of waste fish oil for biodiesel production using three commercial immobilized enzymes: Lipozyme RM IM, Lipozyme TL IM, and Novozym 435 [44]. They reported the highest biodiesel yield of 82.91% with 50% Novozyme 435 using excess ethanol per mol of oil. Pinyaphong et al. reported a maximum biodiesel conversion yield of 83% from fish oil using 20% *Carica papaya* lipase [37]. Chen et al. reported a maximum biodiesel conversion yield of 89% from waste cooking oil using 30% *Rhizopus oryzae* lipase [34]. Nelson et al. carried out biodiesel production from tallow oil using 25% *Candida antartica* (SP 435) and reported the highest yield of 83.8% [32].

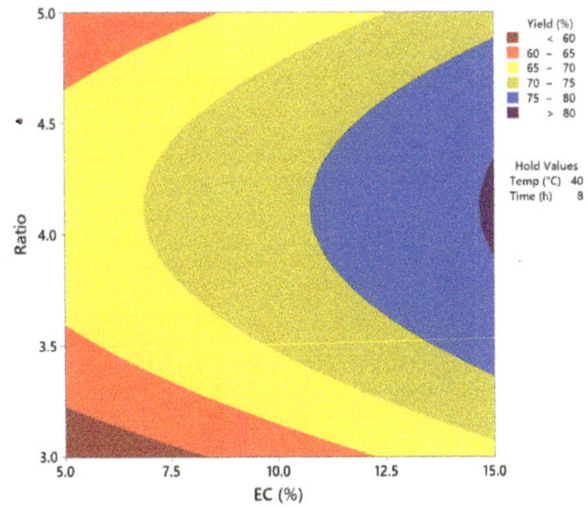

(a) EC*Ratio at Hold values of Temp-40°C and Time-8 h

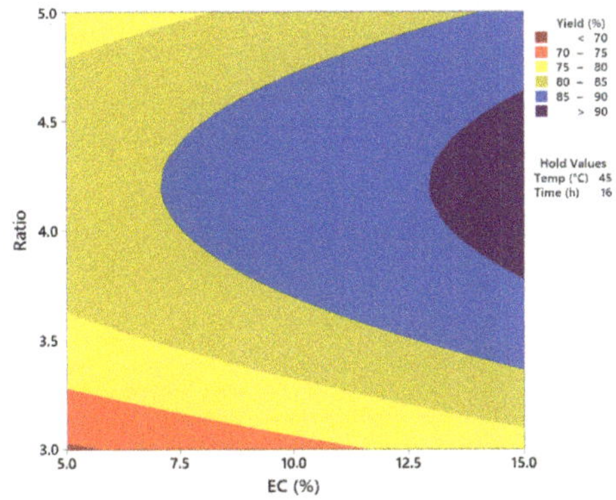

(b) EC*Ratio at Hold values of Temp-45°C and Time-16 h

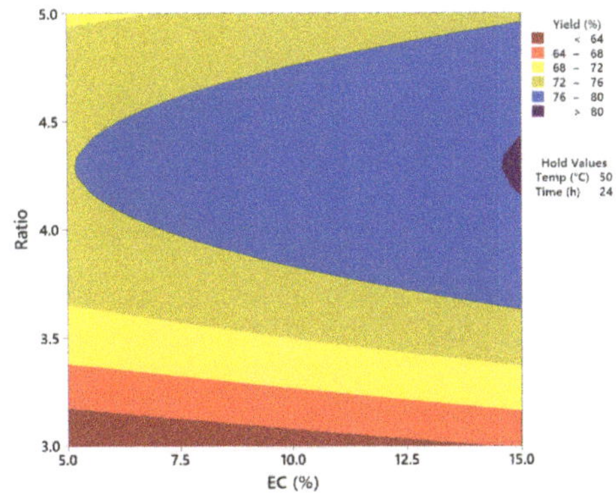

(c) EC*Ratio at Hold values of Temp-50°C and Time-24 h

Figure 7. Counterplots of yield vs. enzyme concentration and oil/alcohol ratio at (**a**) 40 °C and 8 h; (**b**) 45 °C and 16 h; (**c**) 50 °C and 24 h.

3.3. Optimization of Biodiesel Production

The response surface methodology was used to model the experiments statistically and optimize the parameters for biodiesel production. As indicated in Figure 8, the optimized reaction conditions were 17.53 h of transesterification at 44.24 °C with 15% enzyme concentration and oil/alcohol molar ratio of 1:4.23, resulting in a predicted maximum biodiesel yield of 92.46%. The highest experimental yield was 91.86% from 16 h of reaction at 45 °C with 15% enzyme concentration and oil/alcohol molar ratio of 1:4 (Table 1, Run order 41), which was in agreement with the prediction. In addition to the maximum biodiesel yield conditions, the modelling also provided five alternative solutions for optimizing the biodiesel production process (Figure 8). With 10% enzyme concentration, a biodiesel yield of 87.24% can be obtained. The increase of 10% to 15% enzyme concentration (50% increase) to achieve only a 5% higher yield (87.24% to 92.46%) may not be economically feasible. Therefore, the optimum parameters for biodiesel production were determined as 10% enzyme concentration, 45 °C, 16 h and 1:4 oil/alcohol molar ratio for a predicted biodiesel yield of 87.24%. The experiment performed under these conditions resulted in a biodiesel yield of 88.24% (Table 1, Run order 49), which was consistent with the predicted value.

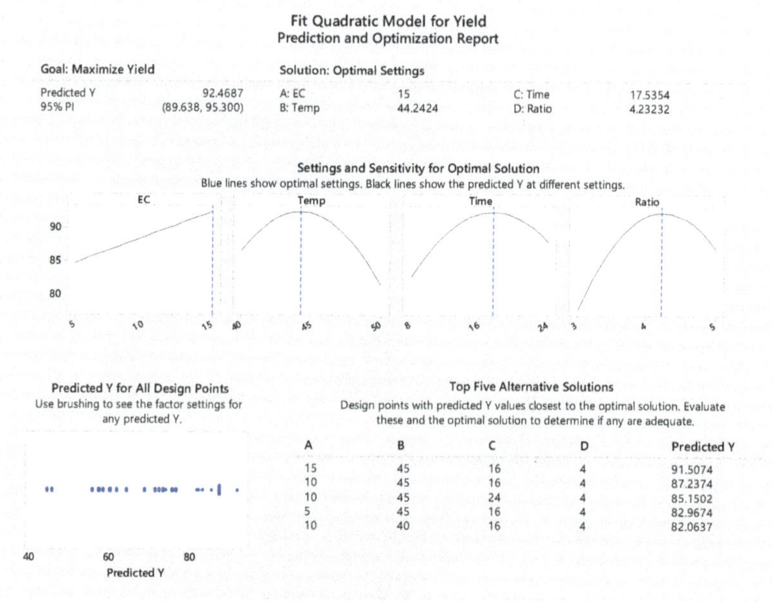

Figure 8. Prediction and optimization report for biodiesel conversion yield.

Step-Wise Addition of Methanol

As indicated in several previously reported studies about biodiesel production, methanol would lead to lipase inactivation during the transesterification [45–47]. Therefore, step-wise addition of methanol has been adopted to reduce its concentration in the system, thereby minimizing its influence on lipase activity and improving the oils' conversion to biodiesel [48–50]. In the present study, based on the optimized conditions for biodiesel production, the efficiency of step-wise addition of methanol was evaluated to further improve biodiesel yield. Instead of a one-time addition of methanol as 1:4 oil/alcohol molar ratio, the experiment was started with 1:2 oil/methanol molar ratio for the first 8 h, and then another portion of methanol was added following the 1:2 oil/methanol molar ratio for the subsequent 8 h of reaction. The biodiesel yield obtained from this experiment was 94.5%, which was approximately 6% higher than the results with a one-time addition of methanol (88.24%, Table 1, Run order 49). Shimada et al. produced biodiesel from waste edible oil with *Candida antartica* lipase by step-wise addition of methanol [51] and reported more than 90% biodiesel yield. They also reported the lipase catalyst retained activity for more than 100 cycles of production. The same group also performed biodiesel production using vegetable oil with *Candida antartica* lipase and reported a 98.4% biodiesel yield after 72 h by step-wise addition of methanol [52]. Watanable et al. successfully produced biodiesel from degummed soybean oil using a three-step methanolysis with a yield of 93.8% in the presence of *Candida antartica* lipase [53]. Hajar et al. applied a three-step methanolysis of canola oil using *Candida antartica* lipase in a solvent-free system and produced biodiesel with a yield of 99.57% [54].

3.4. Investigation of Glycerol as the By-Product

In the present study, free glycerol was not detected from reactions at all the oil/alcohol molar ratios, reaction temperatures, and times studied. This could be due to the low alcohol concentration present in the reaction system and the fact that the whole reaction was carried out using a small amount of starting material. Theoretically, 3 moles of alcohol reacts with 1 mole of triglycerides to give 3 moles of FAME and 1 mole of glycerol (by-product). According to the optimization, 2 g of oil was reacted with methanol with a 1:4 molar ratio and 15% enzyme concentration at 45 °C for 17.53 h to result in a maximum biodiesel conversion yield of 92.46%. The remaining balance of 7.54% observed in this study was made of intermediates and bound glycerols, such as monoacylglycerol (monoglycerides), diacylglycerol (diglycerides), and triacylglycerol (triglycerides). Glycerol is immiscible with oil and biodiesel and has a higher density than that of any other component in the liquid phase of the reaction system. However, it was difficult to identify and separate the glycerol phase on a laboratory scale using immobilized enzymes because the glycerol phase was relatively thin and colorless, as reported in many studies in the literature [55,56]. It has also been reported that the formation of glycerol during transesterification can inhibit the reaction. The free glycerol can clog the lipase's active sites, and the bound glycerol can restrict the mass transfer between the substrate and enzyme. Therefore, it is essential to remove glycerol from the reaction system to improve transesterification efficiency. In the present study, *tert*-butanol dissolved both methanol and glycerol, which eliminated the adverse effects of glycerol accumulation during the transesterification process. As reported in previous studies, *tert*-butanol is not a substrate for lipases, and it does not act on tertiary alcohols [57–59].

4. Conclusions

In the present study, biodiesel production from salmon oil extracted from Atlantic salmon by-products was optimized using response surface methodology. The four parameters, reaction time, temperature, enzyme concentration and oil/alcohol molar ratio, significantly influenced biodiesel production. The maximum biodiesel yield was 92.46% from transesterification for 17.53 h at 44.24 °C with 15% enzyme concentration, oil/alcohol molar ratio of 1:4.23, and the use of 75% (*w*/*w*) *tert*-butanol as the solvent. In comparison to the five alternative optimization solutions provided by the modelling, an increase of 10% to 15% enzyme concentration (50% increase) resulted in only a 5% higher yield (87.24% to 92.46%), which may not be economically feasible. Therefore, the optimum parameters of biodiesel production were determined as 10% enzyme concentration, 45 °C, 16 h, and 1:4 oil/alcohol molar ratio, resulting in a biodiesel yield of 87.24%. By applying step-wise addition of methanol, the biodiesel yield was further increased to 94.5%.

The present study was the first to produce biodiesel from Atlantic salmon oil, which can be a potential replacement of plant-derived oils in conventional biodiesel production. As the salmon oil is extracted from Atlantic salmon heads and frames, this will also be a promising way to valorize the salmon by-products. The results of the present study are promising for the biocatalytic production of biodiesel on an industrial scale. However, there are still some disadvantages in the optimized process, such as the use of *tert*-butanol as the solvent. Future studies will be focused on improving the process by reducing the amount of the solvent, recovery of the solvent for reuse, or completely removing the solvent from the system. The physicochemical properties of FAMEs in the biodiesel obtained were not analyzed in the present study, since the whole process was performed on a small scale and the amount of biodiesel obtained was small. Future studies should be carried out on a larger scale to produce biodiesel using the optimized parameters, and its properties should be assessed and compared against traditional fuel. In addition, the reusability of the lipase and reaction kinetics should be studied as they are essential if the process will be implemented on a larger scale.

Author Contributions: Conceptualization, D.D. and W.M.; methodology, V.V.R.; validation, V.V.R.; formal analysis, V.V.R. and W.R.; investigation, V.V.R.; resources, W.M.; data curation, V.V.R. and Y.L.; writing—original draft preparation, V.V.R.; writing—review and editing, Y.L., V.V.R., D.D., W.R. and W.M.; visualization, V.V.R. and Y.L.; supervision, D.D.; project administration, D.D.; funding acquisition, D.D. All authors have read and agreed to the published version of the manuscript.

Funding: This work was funded by the Ocean Frontier Institute, through an award from the Canada First Research Excellence Fund.

Conflicts of Interest: The authors declare no conflict of interest. The funders had no role in the design of the study; in the collection, analyses, or interpretation of data; in the writing of the manuscript, or in the decision to publish the results.

References

1. Ambat, I.; Srivastava, V.; Sillanpää, M. Recent advancement in biodiesel production methodologies using various feedstock: A review. *Renew. Sustain. Energy Rev.* **2018**, *90*, 356–369. [CrossRef]
2. Manaf, I.S.A.; Embong, N.H.; Khazaai, S.N.M.; Rahim, M.H.A.; Yusoff, M.M.; Lee, K.T.; Maniam, G.P. A review for key challenges of the development of biodiesel industry. *Energy Convers. Manag.* **2019**, *185*, 508–517. [CrossRef]
3. Thangaraj, B.; Solomon, P.R.; Muniyandi, B.; Ranganathan, S.; Lin, L. Catalysis in biodiesel production-A review. *Clean Energy* **2019**, *3*, 2–23. [CrossRef]
4. Singh, D.; Sharma, D.; Soni, S.L.; Sharma, S.; Sharma, P.K.; Jhalani, A. A review on feedstocks, production processes, and yield for different generations of biodiesel. *Fuel* **2020**, *262*, 116553. [CrossRef]
5. Lv, L.; Dai, L.; Du, W.; Liu, D. Progress in enzymatic biodiesel production and commercialization. *Processes* **2021**, *9*, 355. [CrossRef]
6. de Lima, A.L.; Mota, C.J.A. Biodiesel: A Survey on Production Methods and Catalysts. In *Jatropha, Challenges for a New Energy Crop*; Mulpuri, S., Carels, N., Bahadur, B., Eds.; Springer: Singapore, 2019; pp. 475–491.
7. Sendzikiene, E.; Santaraite, M.; Makareviciene, V. Lipase-Catalysed In Situ Transesterification of Waste Rapeseed Oil to Produce Diesel-Biodiesel Blends. *Processes* **2020**, *8*, 1118. [CrossRef]
8. Zhong, L.; Feng, Y.; Wang, G.; Wang, Z.; Bilal, M.; Lv, H.; Jia, S.; Cui, J. Production and use of immobilized lipases in/on nanomaterials: A review from the waste to biodiesel production. *Int. J. Biol. Macromol.* **2020**, *152*, 207–222. [CrossRef]
9. Ching-Velasquez, J.; Fernandez-Lafuente, R.; Rodrigues, R.C.; Plata, V.; Rosales-Quintero, A.; Torrestiana-Sanchez, B.; Tacias-Pascacio, V.G. Production and characterization of biodiesel from oil of fi sh waste by enzymatic catalysis. *Renew. Energy* **2020**, *153*, 1346–1354. [CrossRef]
10. Moazeni, F.; Chen, Y.C.; Zhang, G. Enzymatic transesterification for biodiesel production from used cooking oil, a review. *J. Clean. Prod.* **2019**, *216*, 117–128. [CrossRef]
11. Ortiz, C.; Ferreira, M.L.; Barbosa, O.; Santos, J.C.S.; Rodrigues, R.C.; Berenguer-murcia, Á.; Briand, L.E.; Fernandez-lafuente, R. Novozym 435: The "perfect" lipase immobilized biocatalyst? *Catal. Sci. Technol.* **2019**, *9*, 2380–2420. [CrossRef]
12. Vijayan, G. *Enzymatic Transesterification of Fish Oil for the Biodiesel Production*; Dalhousie University: Halifax, NS, Canada, 2013.
13. Kumar, S. *Enzymatic Transesterification of Waste Animal Fats for Production of Biodiesel*; Dalhousie University: Halifax, NS, Canada, 2013.
14. Kianimanesh, H.R.; Abbaspour-aghdam, F.; Derakhshan, M.V. Biodiesel production from vegetable oil: Process design, evaluation and optimization. *Pol. J. Chem. Technol.* **2017**, *19*, 49–55. [CrossRef]
15. Santaraite, M.; Sendzikiene, E.; Makareviciene, V.; Kazancev, K. Biodiesel Production by Lipase-Catalyzed In Situ Transesterification of Rapeseed Oil Containing a High Free Fatty Acid Content with Ethanol in Diesel Fuel Media. *Energies* **2020**, *13*, 2588. [CrossRef]
16. Veljkovi, V.B.; Biberd, M.O.; Bankovi, I.B.; Djalovi, I.G.; Tasi, M.B.; Nje, Z.B.; Stamenkovi, O.S. Biodiesel production from corn oil: A review. *Renew. Sustain. Energy Rev.* **2018**, *91*, 531–548. [CrossRef]
17. Kara, K.; Ouanji, F.; Lotfi, E.M.; El Mahi, M.; Kacimi, M.; Ziyad, M. Biodiesel production from waste fish oil with high free fatty acid content from Moroccan fish-processing industries. *Egypt. J. Pet.* **2018**, *27*, 249–255. [CrossRef]
18. Yari, N.; Mostafaei, M.; Naderloo, L.; Ardebili, S.M.S. Energy indicators for microwave-assisted biodiesel production from waste fish oil. *Energy Sources Part A Recover. Util. Environ. Eff.* **2019**, 1–12. [CrossRef]
19. Makoure, D.; Arhaliass, A.; Echchelh, A.; Legrand, J. Valorization of Fish By-Products Using Reactive Extrusion for Biodiesel Production and Optimization. *Waste Biomass Valoriz.* **2020**, *11*, 6285–6293. [CrossRef]
20. Samat, A.F.; Muhamad, N.A.S.; Rasib, N.A.A.; Hassan, S.A.M.; Sohaimi, K.S.A.; Iberahim, N.I. The Potential of Biodiesel Production derived from Fish Waste. *IOP Conf. Ser. Mater. Sci. Eng.* **2018**, *318*, 012017. [CrossRef]

21. Fisheries and Land Resources Seafood Industry Year in Review. 2019. Available online: https://www.gov.nl.ca/ffa/files/2019-SIYIR-WEB.pdf (accessed on 22 January 2021).
22. Liu, Y.; Ramakrishnan, V.V.; Dave, D. Lipid class and fatty acid composition of oil extracted from Atlantic salmon by-products under different optimization parameters of enzymatic hydrolysis. *Biocatal. Agric. Biotechnol.* **2020**, *30*, 101866. [CrossRef]
23. Liu, Y.; Ramakrishnan, V.V.; Dave, D. Enzymatic hydrolysis of farmed Atlantic salmon by-products: Investigation of operational parameters on extracted oil yield and quality. *Process. Biochem.* **2021**, *100*, 10–19. [CrossRef]
24. Dave, D.; Ramakrishnan, V.R.; Pohling, J.; Cheema, S.K.; Trenholm, S.; Manuel, H.; Murphy, W. Investigation on oil extraction methods and its influence on omega-3 content from cultured salmon. *J. Food Process. Technol.* **2014**, *5*, 1000401. [CrossRef]
25. Routray, W.; Dave, D.; Ramakrishnan, V.V.; Murphy, W. Production of high quality fish oil by enzymatic protein hydrolysis from cultured Atlantic salmon by-products: Investigation on effect of various extraction parameters using central composite rotatable design. *Waste Biomass Valoriz.* **2018**, *9*, 2003–2014. [CrossRef]
26. Rubio-Rodríguez, N.; Sara, M.; Beltrán, S.; Jaime, I.; Sanz, M.T.; Rovira, J. Supercritical fluid extraction of fish oil from fish by-products: A comparison with other extraction methods. *J. Food Eng.* **2012**, *109*, 238–248. [CrossRef]
27. Głowacz-Różyńska, A.; Tynek, M.; Malinowska-Pańczyk, E.; Martysiak-Żurowska, D.; Pawłowicz, R.; Kołodziejska, I. Comparison of oil yield and quality obtained by different extraction procedures from salmon (Salmo salar) processing byproducts. *Eur. J. Lipid Sci. Technol.* **2016**, *118*, 1759–1767. [CrossRef]
28. Zieniuk, B.; Wołoszynowska, M.; Białecka-Florjańczyk, E. Enzymatic Synthesis of Biodiesel by Direct Transesterification of Rapeseed Cake. *Int. J. Food Eng.* **2020**, *16*, 1–10. [CrossRef]
29. Behera, S.K.; Meena, H.; Chakraborty, S.; Meikap, B.C. Application of response surface methodology (RSM) for optimization of leaching parameters for ash reduction from low-grade coal. *Int. J. Min. Sci. Technol.* **2018**, *28*, 621–629. [CrossRef]
30. Kumari, M.; Gupta, S.K. Response surface methodological (RSM) approach for optimizing the removal of trihalomethanes (THMs) and its precursor's by surfactant modified magnetic nanoadsorbents (sMNP)-An endeavor to diminish probable cancer risk. *Sci. Rep.* **2019**, *9*, 18339. [CrossRef] [PubMed]
31. Chumuang, N.; Punsuvon, V. Response Surface Methodology for Biodiesel Production Using Calcium Methoxide Catalyst Assisted with Tetrahydrofuran as Cosolvent. *J. Chem.* **2017**, *2017*, 4190818. [CrossRef]
32. Nelson, L.A.; Foglia, T.A.; Marmer, W.N. Lipase-catalyzed production of biodiesel. *JAOCS J. Am. Oil Chem. Soc.* **1996**, *73*, 1191–1195. [CrossRef]
33. Association of Official Analytical Chemists. *Official Methods of Analysis of AOAC International*, 17th ed.; Association of Official Analytical Chemists: Gaithersburg, MD, USA, 2000.
34. Chen, G.; Ying, M.; Li, W. Enzymatic Conversion of Waste Cooking Oils Into Alternative Fuel—Biodiesel. *Appl. Biochem. Biotechnol.* **2006**, *132*, 911–921. [CrossRef]
35. Dizge, N.; Keskinler, B. Enzymatic production of biodiesel from canola oil using immobilized lipase. *Biomass Bioenergy* **2008**, *32*, 1274–1278. [CrossRef]
36. Nie, K.; Xie, F.; Wang, F.; Tan, T. Lipase catalyzed methanolysis to produce biodiesel: Optimization of the biodiesel production. *J. Mol. Catal. B Enzym.* **2006**, *43*, 142–147. [CrossRef]
37. Pinyaphong, P.; Sriburi, P.; Phutrakul, S. Biodiesel fuel production by methanolysis of fish oil derived from the discarded parts of fish catalyzed by Carica papaya lipase. *World Acad. Sci. Eng. Technol.* **2011**, *76*, 466–472.
38. Li, L.; Du, W.; Liu, D.; Wang, L.; Li, Z. Lipase-catalyzed transesterification of rapeseed oils for biodiesel production with a novel organic solvent as the reaction medium. *J. Mol. Catal. B Enzym.* **2006**, *43*, 58–62. [CrossRef]
39. Azcar, L.; Ciudad, G.; Muoz, R.; Jeison, D.; Toro, C.; Navi, R. Feasible Novozym 435-Catalyzed Process to Fatty Acid Methyl Ester Production from Waste Frying Oil: Role of Lipase Inhibition. In *Enzyme Inhibition and Bioapplications*; InTech: Philadelphia, PA, USA, 2012.
40. Ognjanovic, N.; Bezbradica, D.; Knezevic-Jugovic, Z. Enzymatic conversion of sunflower oil to biodiesel in a solvent-free system: Process optimization and the immobilized system stability. *Bioresour. Technol.* **2009**, *100*, 5146–5154. [CrossRef] [PubMed]
41. Hernández-Martín, E.; Otero, C. Different enzyme requirements for the synthesis of biodiesel: Novozym® 435 and Lipozyme® TL IM. *Bioresour. Technol.* **2008**, *99*, 277–286. [CrossRef]
42. Rodrigues, R.C.; Volpato, G.; Wada, K.; Ayub, M.A.Z. Enzymatic synthesis of biodiesel from transesterification reactions of vegetable oils and short chain alcohols. *J. Am. Oil Chem. Soc.* **2008**, *85*, 925–930. [CrossRef]
43. Noureddini, H.; Gao, X.; Philkana, R.S. Immobilized Pseudomonas cepacia lipase for biodiesel fuel production from soybean oil. *Bioresour. Technol.* **2005**, *96*, 769–777. [CrossRef] [PubMed]
44. Marín-Suárez, M.; Méndez-Mateos, D.; Guadix, A.; Guadix, E.M. Reuse of immobilized lipases in the transesterification of waste fish oil for the production of biodiesel. *Renew. Energy* **2019**, *140*, 1–8. [CrossRef]
45. Lotti, M.; Pleiss, J.; Valero, F.; Ferrer, P. Effects of methanol on lipases: Molecular, kinetic and process issues in the production of biodiesel. *Biotechnol. J.* **2015**, *10*, 22–30. [CrossRef]
46. Lotti, M.; Pleiss, J.; Valero, F.; Ferrer, P. Enzymatic Production of Biodiesel: Strategies to Overcome Methanol Inactivation. *Biotechnol. J.* **2018**, *13*, 1700155. [CrossRef] [PubMed]
47. Hama, S.; Noda, H.; Kondo, A. How lipase technology contributes to evolution of biodiesel production using multiple feedstocks. *Curr. Opin. Biotechnol.* **2018**, *50*, 57–64. [CrossRef] [PubMed]

48. Andrade, T.A.; Errico, M.; Christensen, K.V. Influence of the reaction conditions on the enzyme catalyzed transesterification of castor oil: A possible step in biodiesel production. *Bioresour. Technol.* **2017**, *243*, 366–374. [CrossRef]
49. Al Basir, F.; Roy, P.K. Study on enzyme inhibition in biodiesel synthesis: Effect of stepwise addition of methanol and removal of glycerol. *Energy Ecol. Environ.* **2019**, *4*, 75–84. [CrossRef]
50. Norjannah, B.; Ong, H.C.; Masjuki, H.H. Effects of methanol and enzyme pretreatment to Ceiba pentandra biodiesel production. *Energy Sources Part A Recover. Util. Environ. Eff.* **2017**, *39*, 1548–1555. [CrossRef]
51. Shimada, Y.; Watanabe, Y.; Sugihara, A.; Tominaga, Y. Enzymatic alcoholysis for biodiesel fuel production and application of the reaction to oil processing. *J. Mol. Catal. B Enzym.* **2002**, *17*, 133–142. [CrossRef]
52. Shimada, Y.; Watanabe, Y.; Samukawa, T.; Sugihara, A.; Noda, H.; Fukuda, H.; Tominaga, Y. Conversion of vegetable oil to biodiesel using immobilized Candida antarctica lipase. *JAOCS J. Am. Oil Chem. Soc.* **1999**, *76*, 789–793. [CrossRef]
53. Watanabe, Y.; Shimada, Y.; Sugihara, A.; Tominaga, Y. Conversion of degummed soybean oil to biodiesel fuel with immobilized Candida antarctica lipase. *J. Mol. Catal. B Enzym.* **2002**, *17*, 151–155. [CrossRef]
54. Hajar, M.; Shokrollahzadeh, S.; Vahabzadeh, F.; Monazzami, A. Solvent-free methanolysis of canola oil in a packed-bed reactor with use of Novozym 435 plus loofa. *Enzym. Microb. Technol.* **2009**, *45*, 188–194. [CrossRef]
55. Xu, Y.; Nordblad, M.; Nielsen, P.M.; Brask, J.; Woodley, J.M. In situ visualization and effect of glycerol in lipase-catalyzed ethanolysis of rapeseed oil. *J. Mol. Catal. B Enzym.* **2011**, *72*, 213–219. [CrossRef]
56. Watanabe, Y.; Shimada, Y.; Sugihara, A.; Tominaga, Y. Stepwise ethanolysis of tuna oil using immobilized Candida antarctica lipase. *J. Biosci. Bioeng.* **1999**, *88*, 622–626. [CrossRef]
57. Tiosso, P.C.; Carvalho, A.K.F.; De Castro, H.F.; De Moraes, F.F.; Zanin, G.M. Utilization of immobilized lipases as catalysts in the transesterification of non-edible vegetable oils with ethanol. *Braz. J. Chem. Eng.* **2014**, *31*, 839–847. [CrossRef]
58. Taher, H.; Al-Zuhair, S.; Al-Marzouqi, A.H.; Haik, Y.; Farid, M.M. A review of enzymatic transesterification of microalgal oil-based biodiesel using supercritical technology. *Enzym. Res.* **2011**, *2011*. [CrossRef] [PubMed]
59. Royon, D.; Daz, M.; Ellenrieder, G.; Locatelli, S. Enzymatic production of biodiesel from cotton seed oil using t-butanol as a solvent. *Bioresour. Technol.* **2007**, *98*, 648–653. [CrossRef] [PubMed]

processes

MDPI

Article

Solubility of Biocompounds 2,5-Furandicarboxylic Acid and 5-Formylfuran-2-Carboxylic Acid in Binary Solvent Mixtures of Water and 1,4-Dioxane

Linli Xu [1,†], Jianhui Fu [1,†], Cunbin Du [1], Qianqian Xu [1,2,*], Baojian Liu [3] and Zongbi Bao [2,3,*]

1 School of Pharmaceutical and Materials Engineering, Taizhou University, Taizhou 318000, China
2 Key Laboratory of Biomass Chemical Engineering of Ministry of Education, Zhejiang University, Hangzhou 310027, China
3 Institute of Zhejiang University-Quzhou, 99 Zheda Road, Quzhou 324000, China
* Correspondence: xuqq@tzc.edu.cn (Q.X.); baozb@zju.edu.cn (Z.B.)
† These authors contributed equally to this work.

Abstract: The solubility of 2,5-furandicarboxylic acid (FDCA) and its synthetic intermediates (e.g., 5-formylfuran-2-carboxylic acid, FFCA) provides fundamental information for the preparation and purification of the value-added biocompound FDCA. We measured the solubility of FDCA and FFCA in binary water + 1,4-dioxane mixtures with different mixing ratios at 303.15 K–342.15 K. The obtained solubility values were correlated with the Jouyban-Acree-van't Hoff model, and the preferential solvation theory was used to study the microscopic dissolution mechanism. The solubility of FDCA/FFCA increases with increasing temperature, and pure 1,4-dioxane dissolves more solutes than pure water. FFCA shows higher solubility than FDCA. In the binary solvent mixtures, the phenomenon of co-solvency exists for both FDCA and FFCA, i.e., at a 1,4-dioxane mole fraction of about 0.60, FDCA and FFCA dissolve the most. Acceptable mean percentage deviations (MPD) (5.5% and 6.9%) are obtained for FDCA and FFCA (Jouyban-Acree-van't Hoff model). The calculated preferential solvation parameters show different dissolution mechanisms at different solvent compositions. When the 1,4-dioxane mole fraction is 0.17~0.62/0.63, FDCA/FFCA are preferentially solvated by 1,4-dioxane. Otherwise, they are preferentially solvated by water. A trend similar to the "co-solvency phenomenon" is observed in the differences in solubility of FFCA and FDCA. This study gives important guidance for the use of binary water and 1,4-dioxane solvents in practical FDCA purification.

Keywords: solubility; binary solvent mixtures; water + 1,4-dioxane; 2,5-furandicarboxylic acid; 5-formylfuran-2-carboxylic acid; co-solvency phenomenon

Citation: Xu, L.; Fu, J.; Du, C.; Xu, Q.; Liu, B.; Bao, Z. Solubility of Biocompounds 2,5-Furandicarboxylic Acid and 5-Formylfuran-2-Carboxylic Acid in Binary Solvent Mixtures of Water and 1,4-Dioxane. *Processes* **2022**, *10*, 2480. https://doi.org/10.3390/pr10122480

Academic Editor: Xiaodong Wu

Received: 23 October 2022
Accepted: 18 November 2022
Published: 23 November 2022

1. Introduction

The compound 2,5-furandicarboxylic acid (FDCA), with two carboxylic acid groups symmetrically attached to the furan ring, is a normal human urinary and microbial metabolite [1]. It is included in the top 15 biobased platform compounds by the US Department of Energy (DOE) as a promising substitute for petrochemical monomers such as terephthalic acid (TPA) [2,3]. Biomass feedstocks for the production of FDCA include fructose, glucose, sucrose, high fructose corn syrup, and starch. The oligosaccharides or polysaccharides in these sugars are generally hydrolyzed first to C6 sugars, then the C6 sugars are cyclodehydrated to 5-hydroxymethylfurfural (HMF), and finally, FDCA is synthesized by catalytic oxidation of 5-HMF [4]. Companies currently producing FDCA include Avantium, Corbion, Toronto Chemicals, Alfa Aesar, Synvina, Novamont, AVA and TCI However, due to the low synthetic yield and great purification difficulties, FDCA production is still not reaching its full potential [3].

The pathway products of FDCA often include oxidation intermediates such as 5-formylfuran-2-carboxylic acid (FFCA). Due to the similarity in structure between FDCA and FFCA (Figure 1), separating the two presents a significant challenge. Despite the inherent difficulties in removing FFCA from FDCA, purification of FDCA is urgently needed, because the presence of FFCA seriously affects the application of FDCA. For example, FFCA containing only a mono-carboxylic acid group would terminate chain growth during the polymerization of FDCA, thus reducing the quality of the desired polymer [5,6]. In addition, pale yellow FFCA is a known color body that affects the color of FDCA-based downstream polymer products [7]. Most conventional separation processes for FDCA and FFCA use catalytic hydrogenation to reduce FFCA to structurally dissimilar compounds such as 5-hydroxymethylfuran-2-carboxylic acid, 5-methylfuran-2-carboxylic acid and furan-2-carboxylic acid, and then separate these compounds from FDCA. Such a process not only requires high temperature and pressure, but it is also difficult to avoid the reduction of FDCA. Moreover, the recovery of the catalyst makes the operation very complicated. Other separation methods, such as the FDCA salt method, may produce a significant amount of wastewater, leading to serious environmental problems [8].

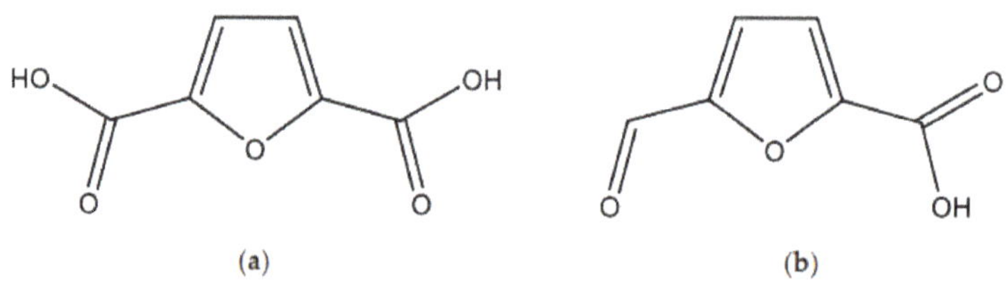

Figure 1. The chemical structures of FDCA and FFCA. (**a**) FDCA; (**b**) FFCA.

Extraction or crystallization are mild separation/purification methods. To evaluate the possibility of using these methods, solubility data of FDCA and FFCA in appropriate solvents (pure solvents or solvent mixtures) provide fundamental information. Furthermore, solubility data are also necessary for the selection of suitable reaction solvents for the preparation of FDCA [9]. However, based on the available documents, there is insufficient research on the solubility for FDCA. Zhang et al. reported the solubility of FDCA in eight pure solvents (methanol, acetonitrile, water, acetic acid, methyl isobutyl ketone, ethyl acetate, 1-butanol, and isobutanol) and two binary solvent mixtures (water + acetic acid and water + acetonitrile) [9]. The solubility data of FDCA in three binary solvent mixtures, namely methanol +water, acetic acid + water, and ethanol + water, were reported by Ban et al. [10]. Moreover, to our knowledge, the solubility of FFCA in solvents has rarely been reported. It is clear that the solubility of FFCA and FDCA in different solvent systems needs to be further investigated to serve as a basic database for promoting the production and purification of FDCA.

Since water is formed during the synthesis of FDCA by the oxidation of 5-HMF [11], it is particularly important to examine the solubility in aqueous solvents for the production and purification of FDCA. The solvent 1,4-dioxane readily forms hydrogen bonds with water and is completely miscible with water. Binary mixtures formed by water and 1,4-dioxane in different mixing ratios cover a wide range of polarities and can react with FDCA and other intermediates through both hydrophobic and hydrophilic interactions [12–15]. Therefore, in this study, water and 1,4-dioxane with different mixing proportions are selected as solvents, and the solubility values of FDCA and FFCA in these binary mixtures are measured at 303.15 K–342.15 K. The obtained solubility values are correlated with the Jouyban-Acree-van't Hoff model, which is the preferred and recommended co-solvency model for binary solvents at a variety of different temperatures with acceptable model accuracy [16–20]. Moreover, we use the preferential solvation theory to understand the microscopic dissolution of FDCA/FFCA in binary water and 1,4-dioxane mixtures better. In addition, the differences in solubility of FDCA and FFCA are compared.

2. Materials and Methods

2.1. Chemicals

FDCA (>98.0%), FFCA (>98.0%) and 1,4-dioxane were obtained from Aladdin Chemistry Co., Ltd. (Shanghai, China) and used without further purification. Deionized water was used throughout all of the experiments. HPLC-grade methanol and acetic acid were purchased from Tedia Co. (Shanghai, China) and used as the mobile phase for HPLC analysis.

2.2. Apparatus and Method

Solubility data were measured using a static equilibrium method [21]. An excess amount of solute (FDCA/FFCA) and the solvent (water + 1,4-dioxane of specified composition) were added to a jacketed, single-mouthed glass vessel (20 mL). The magnetically stirred vessel was sealed with a stopper to prevent the evaporation of solvent. The temperature of the solution in the vessel was controlled by circulating water through the vessel jacket. The actual temperature of the solution inside the vessel was further checked with a mercury glass thermometer. Dissolution equilibrium was achieved by continuous stirring for 6 h followed by being at rest for 5 h (tested in preliminary experiments). The supernatant of about 1 mL was transferred to a volumetric flask with a syringe, weighed and diluted to get a solution with the desired concentration. All experiments were repeated 3 times to check the reproducibility, and the average value was determined.

2.3. Analysis Method

The FDCA/FFCA concentration in the samples was determined by HPLC. Measurements were performed using an Agilent chromatograph equipped with an Ultimate LP-C18 column (250 mm $\times$ 4.6 mm, 5 μm) and a mobile phase of methanol:water:acetic acid = 40:60:1 at a flow rate of 1 mL$\cdot$min^{-1}. The wavelength of the detector was set at 265 nm. Each sample was analyzed 3 times, and the relative error was within $\pm$1%.

To confirm the accuracy of this experimental procedure, the solubility data of FDCA in methanol were measured and compared with published data [9]. It was found that the deviations of the solubility were 2.1%, 1.9%, 1.8%, 1.7%, 1.9%, 2.9% and 2.2% at (313.15, 318.15, 323.15, 328.15, 333.15, 338.15, and 343.15) K, respectively. The good agreement of the data proves the reliability of our experimental method.

2.4. Powder X-ray Diffraction and Differential Scanning Calorimetry Identification

The raw FDCA/FFCA and their excess solids after solubility equilibration were characterized by powder X-ray diffraction (PXRD) to check if there are hydrates or 1,4-dioxane solvates of FDCA/FFCA before and after the solubility measurements. The X-ray diffraction characterization was performed with a D8 Advance X-ray diffractometer using Cu Kα radiation (λ = 1.54184 nm). The XRD scanning speed was 5$^\circ\cdot$min$^{-1,}$ and data were obtained at an angle (2-Theta) of 5–80°. In addition, excess solids were collected after solubility measurements in each binary solvent and mixed together for both FDCA and FFCA. DSC analysis for FDCA/FFCA raw and excess solids mixture was performed using a Pyris-Diamond (PerkinElmer) differential scanning calorimeter calibrated with an indium standard. Approximately 5 mg of sample was added, and the heating rate was 10° K$\cdot$min^{-1} (atmosphere, nitrogen).

3. Results and Discussion

3.1. XRD and DSC Identification Results

The XRD and DSC curves with characteristic peaks for FDCA/FFCA raw and excess solids after solubility equilibration are shown in Figure 2. For XRD experiments, excess solids of six different 1,4-dioxane mole fractions (x_1 = 0, 0.20, 0.40, 0.60, 0.80, 1.00, x_1 is the mole fraction of 1,4-dioxane (1) in the mixture of 1,4-dioxane (1) + water (2) in the absence of FDCA/FFCA) were analyzed separately. Peak positions and shapes of PXRD plots for solid phases after equilibration with solvents 1,4-dioxane (1) + water (2), in various proportions, are not significantly different from those of the original raw powder for both FDCA and FFCA. The identical diagrams for FDCA or FFCA obtained before and after the experiments indicate that the binary solvents with varying mixing ratios do not induce solute form changes during solubility equilibration. The DSC curves of the samples do not show endothermic peaks before they reach their melting temperature. Comparing the DSC curves for raw FDCA and FFCA in Figure 2 with the literature [5], the peak temperatures agree well. The DSC peak shape and position of the excess FDCA/FFCA mixture for each binary solvent are not different from that of the raw FDCA/FFCA. Therefore, FDCA and FFCA do not form hydrates or 1,4-dioxane solvates in the binary solvents studied, and the changes in solubility are mainly due to different intermolecular interactions between solute and solvents.

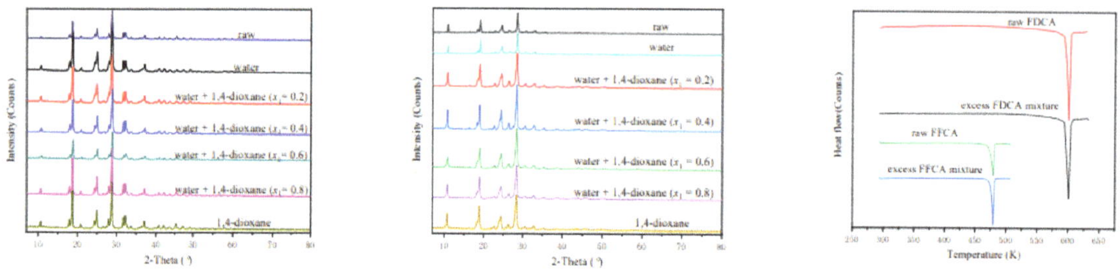

(a) (b) (c)

Figure 2. PXRD and DSC results of FDCA and FFCA raw materials and excess solids after solubility equilibration. (**a**) PXRD plots for raw FDCA and excess solids of FDCA after equilibration with binary water + 1,4-dioxane solvents; (**b**) PXRD plots for raw FFCA and excess solids of FFCA after equilibration with binary water + 1,4-dioxane solvents; (**c**) DSC curves for raw FDCA/FFCA and their excess solids mixture after solubility measurements in each binary solvent; x_1 is the mole fraction of 1,4-dioxane (1) In binary water (2) + 1,4-dioxane (1) Mixtures in the absence of FDCA/FFCA.

3.2. *Solubility Results*

The mole fraction solubility values (x_m) of FDCA and FFCA in water + 1,4-dioxane with different 1,4-dioxane mole fractions (x_1 = 0, 0.20, 0.40, 0.60, 0.80, 1.00) at 303.15 K to 343.15 K are listed and plotted in Table 1 and Figure 3.

As shown in Table 1 and Figure 3, the solubility of both FDCA and FFCA increases with increasing temperature for a given solvent mixture composition. All solubility values of FFCA in the binary water + 1,4-dioxane mixtures are higher than those of FDCA under the same conditions, indicating that the aldehyde group in FFCA has a higher affinity for water and 1,4-dioxane. The solubility of both FDCA and FFCA in pure 1,4-dioxane (x_1 = 1) is higher than in pure water (x_1 = 0) but not the maximum solubility. At a given temperature, the solubility reaches a maximum when the mole fraction of 1,4-dioxane is about 0.60 for both FDCA and FFCA. According to the Hildebrand solubility parameter theory, this phenomenon of co-solvency occurs when the solubility parameter (a measure of polarity and defined as the square root of the cohesive energy density) of the solute (FDCA or FFCA) is in the range of solubility parameters for the solvent components (water and 1,4-dioxane) that make up the binary solvent [22–25]. Indeed, the solubility parameter of FDCA is 26.9 MPa$^{1/2}$ [26], and that of FFCA is 27.3 MPa$^{1/2}$ (calculated by Fedors' method, Table S1), both of which are between the values of 1,4-dioxane (20.47 MPa$^{1/2}$) and water (47.86 MPa$^{1/2}$).

Table 1. Experimental mole fraction solubility x_m of FDCA and FFCA in binary 1,4-dioxane + water mixture solvents.

				FDCA				
T (K)	x_1	$x_m \times 10^3$	T (K)	x_1	$x_m \times 10^3$	T (K)	x_1	$x_m \times 10^3$
303.15	0	0.132	313.15	0	0.191	323.15	0	0.252
	0.20	3.525		0.20	5.191		0.20	6.528
	0.40	10.934		0.40	12.392		0.40	14.923
	0.60	14.244		0.60	15.131		0.60	16.931
	0.80	11.552		0.80	12.527		0.80	13.575
	1.00	2.386		1.00	3.1860		1.00	3.644
333.15	0	0.383	343.15	0	0.494			
	0.20	8.335		0.20	10.276			
	0.40	16.747		0.40	18.840			
	0.60	17.733		0.60	19.308			
	0.80	14.857		0.80	16.178			
	1.00	4.373		1.00	6.054			
				FFCA				
T(K)	x_1	$x_m \times 10^3$	T(K)	x_1	$x_m \times 10^3$	T(K)	x_1	$x_m \times 10^3$
303.15	0	1.271	313.15	0	2.168	323.15	0	2.531
	0.20	29.421		0.20	37.628		0.20	49.005
	0.40	69.041		0.40	82.790		0.40	93.303
	0.60	83.903		0.60	99.376		0.60	121.330
	0.80	72.231		0.80	80.155		0.80	102.283
	1.00	43.636		1.00	56.325		1.00	71.574
333.15	0	3.771	343.15	0	5.941			
	0.20	59.817		0.20	65.567			
	0.40	103.458		0.40	105.974			
	0.60	128.289		0.60	138.055			
	0.80	113.543		0.80	119.001			
	1.00	82.575		1.00	85.083			

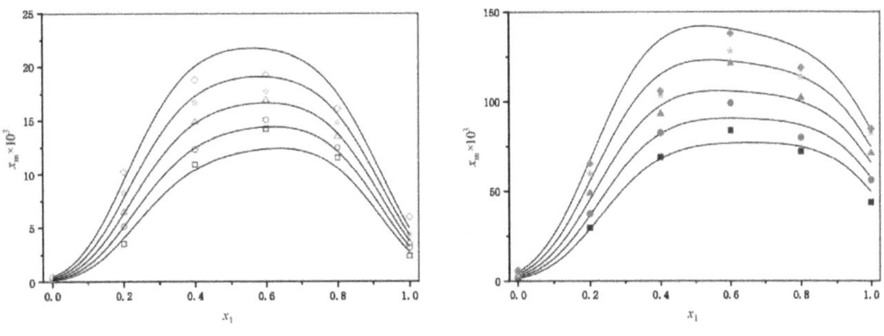

(a) (b)

Figure 3. Mole fraction solubility x_m of FDCA and FFCA in water + 1,4-dioxane binary solvent mixtures with varying 1,4-dioxane mole fractions at 303.15 − 343.15 K; (**a**) FDCA solubility: □, T = 303.15 K; ○, T = 313.15 K; △, T = 323.15 K; ☆, T = 333.15 K; ◇, T = 343.15 K; (**b**) FFCA solubility: ■, T = 303.15 K; •, T = 313.15 K; ▲, T = 323.15 K; ★, T = 333.15 K; ◆, T = 343.15 K; lines were calculated with the Jouyban-Acree-van't Hoff model.

3.3. Jouyban-Acree-van't Hoff Model Correlation

The solubility data of FDCA and FFCA in water + 1,4-dioxane mixtures at various temperatures were correlated using the Jouyban-Acree-van't Hoff model [12]:

$$\ln x_{m,T} = x_1\left(\alpha_1 + \frac{\beta_1}{T}\right) + x_2\left(\alpha_2 + \frac{\beta_2}{T}\right) + x_1 x_2 \sum_{i=0}^{2} \frac{J_i (x_1 - x_2)^i}{T} \quad (1)$$

where $x_{m,T}$ is the mole fraction solubility of FDCA/FFCA in binary water and 1,4-dioxane mixtures at temperature T (K), x_1 and x_2 are the mole fractions of solvents 1 (1,4-dioxane) and 2 (water) in the binary mixtures in the absence of solute (FDCA/FFCA), J_i, α_1, β_1, α_2, β_2 stand for the model constants.

In order to check the correlation accuracy, the mean percentage deviations (MPD) were calculated according to Equation (2):

$$\text{MPD} = \frac{100}{N}\sum\left(\frac{|\text{Calculated value} - \text{Observed value}|}{\text{Observed value}}\right) \quad (2)$$

where N refers to the number of solubility data points.

The experimental data points of FDCA and FFCA dissolved in binary water + 1,4-dioxane mixtures (30 data points for each system) were fitted to Equation (1). The model constants obtained were used to back-calculate the solubility data, and the overall MPD could be attained. The model constants are listed in Table 2, along with the MPD values. The MPD values for FDCA and FFCA systems are 5.5% and 6.9%, respectively. The low MPD values confirm the correlation ability of the Jouyban-Acree-van't Hoff model. In other words, the Jouyban-Acree-van't Hoff model is sufficiently accurate to correlate the solubility values of FDCA/FFCA in binary water + 1,4-dioxane solvent mixtures and could be used as a practical strategy to predict further solubility values at unmeasured solvent mixing ratios and temperatures in these two systems.

Table 2. The Jouyban-Acree-van't Hoff model constants and the mean percentage deviations (MPD) for solubility values of FDCA and FFCA in water + 1,4-dioxane mixtures.

Solute	α_1	β_1	α_2	β_2	J_0	J_1	J_2	MPD%
FDCA	−0.857	−1527.5	2.316	−3408.7	3604.69	−1372.052	2556.36	5.5%
FFCA	1.544	−1378.7	4.226	−3264.0	2645.19	−1852.031	1623.40	6.9%

3.4. Preferential Solvation of FDCA/FFCA in Binary Mixtures of 1,4-Dioxane and Water

Preferential solvation refers to a phenomenon that the local distribution of solvent molecules around the solute molecules differs from the bulk distribution of solvent molecules due to size and interaction differences between the solute and solvent molecules [27–29]. In this study, the preferential solvation theory involving the inverse Kirkwood buff integrals (IKBI) method [30,31] is used to characterize the dissolution of FDCA/FFCA in binary water and 1,4-dioxane mixtures. The preferential solvation parameter, which is the difference between the local mole fraction of one solvent component near to the solute and the bulk mole fraction of the solvent component in the binary solvent mixtures, was defined as follows:

$$\delta_{1,3} = x_{1,3}^L - x_1 = -\delta_{2,3} \quad (3)$$

where $x_{1,3}^L$ is the local mole fraction of 1,4-dioxane (1) in the vicinity of solute FDCA/FFCA (3), x_1 is the mole fraction of 1,4-dioxane (1) in binary 1,4-dioxane (1) and water (2) mixtures in the absence of solute FDCA/FFCA, and $\delta_{1,3}$ and $\delta_{2,3}$ are the preferential solvation parameters for solute FDCA/FFCA (3) solvated by 1,4-dioxane (1), and solvated by water (2), respectively.

If the value of the preferential solvation parameter is $\delta_{1,3} > 0$, solute FDCA or FFCA (3) is preferentially solvated by 1,4-dioxane (1). On the other hand, if $\delta_{1,3} < 0$ (i.e., $\delta_{2,3} > 0$), FDCA or FFCA (3) is preferentially solvated by water (2).

The calculation of the preferential solvation parameters for FDCA/FFCA in binary mixtures of 1,4-dioxane and water is described in detail in the Supplementary Materials (pages 3–8) [32–34]. The calculated values of the preferential solvation parameters ($\delta_{1,3}$) at 303.15 K, plotted as a function of solvent mixing ratio, are shown in Figure 4.

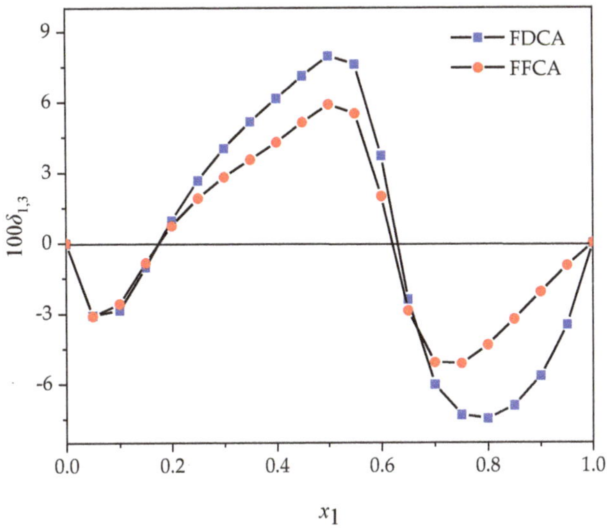

Figure 4. Preferential solvation parameter ($\delta_{1,3}$) values of FDCA and FFCA in binary water + 1,4-dioxane with different solvent mixing ratios at 303.15 K.

In Figure 4, we can see that in the binary water and 1,4-dioxane solvent mixtures with compositions of $0 < x_1 < 0.17$, the $\delta_{1,3}$ values for both FDCA and FFCA are negative, suggesting that both FDCA and FFCA are preferentially solvated by water. When the mixing ratio of 1,4-dioxane increases to $0.17 < x_1 < 0.63$ for FDCA and $0.17 < x_1 < 0.62$ for FFCA, the $\delta_{1,3}$ values become positive, implying that FDCA and FFCA are preferentially solvated by 1,4-dioxane in this region of the mixture composition where the co-solvency phenomenon also occurs. A higher local 1,4-dioxane concentration than the bulk 1,4-dioxane concentration could be related to the breaking of the slightly ordered structure of water molecules (connected by hydrogen bonds of their hydroxyl groups) [32]. At a higher 1,4-dioxane: water ratio to pure 1,4-dioxane ($0.62/0.63 < x_1 < 1$ as in 1,4-dioxane-rich mixtures), the $\delta_{1,3}$ values for FDCA and FFCA become negative again, implying that they are again preferentially solvated by water. It can be found in Figure 4 that both FDCA and FFCA are preferentially solvated by water in water-rich and 1,4-dioxane-rich mixtures. This may be due to the fact that water can act as both a Lewis acid and a Lewis base, but 1,4-dioxane can only act as a Lewis base (According to the Kamlet-Taft hydrogen bond donor parameter and hydrogen bond acceptor parameter, $\alpha = 1.17$, $\beta = 0.18$ for water and $\alpha = 0.00$, $\beta = 0.37$ for 1,4-dioxane) [35–38].

3.5. Comparison of the Solubility of FDCA and FFCA in Water and 1,4-Dioxane Mixtures

Table 1 and Figure 3 clearly show that the solubility of FFCA in binary mixtures of water and 1,4-dioxane is higher than FDCA under the same conditions. To get a clearer picture of the exact difference in solubility of FFCA and FDCA in this binary solvent system, we constructed a plot of the solubility differences between FFCA and FDCA as a function of mixing ratio of water and 1,4-dioxane at different temperatures.

Figure 5 shows that in the binary water and 1,4-dioxane solvent mixture with a certain composition (e.g., $x_1 = 0.80$), the solubility difference between FFCA and FDCA increases with increasing temperature. At a certain temperature (e.g., $T = 323.15$ K, green color), the solubility difference between FFCA and FDCA first increases with the increase of 1,4-dioxane mole fraction, reaches a maximum value and then decreases. The trend of the change is similar to that of the solubility values, i.e., it is a trend similar to the "co-solvency phenomenon". In pure 1,4-dioxane ($x_1 = 1$), the solubility differences between FFCA and FDCA are larger than the small differences in pure water ($x_1 = 0$). Since the solubility differences in pure water ($x_1 = 0$) are all close to zero at the measured temperatures, and the value increases slightly with increasing temperature, it can be concluded that pure water below 343.15 K is not suitable as a crystallization or extraction reagent for the separation of FDCA and FFCA. However, the solubility differences between FFCA and FDCA in pure 1,4-dioxane and binary water + 1,4-dioxane mixtures are substantial and greater at higher temperatures. It is expected that an increase in the solubility difference between FFCA and FDCA will lead to an increase in the separation of these two biocompounds by crystallization or extraction. Therefore, the most efficient crystallization conditions for the separation of FDCA and FFCA using binary water + 1,4-dioxane solvents are expected to involve a 1,4-dioxane mole fraction of around 0.60 and high temperatures.

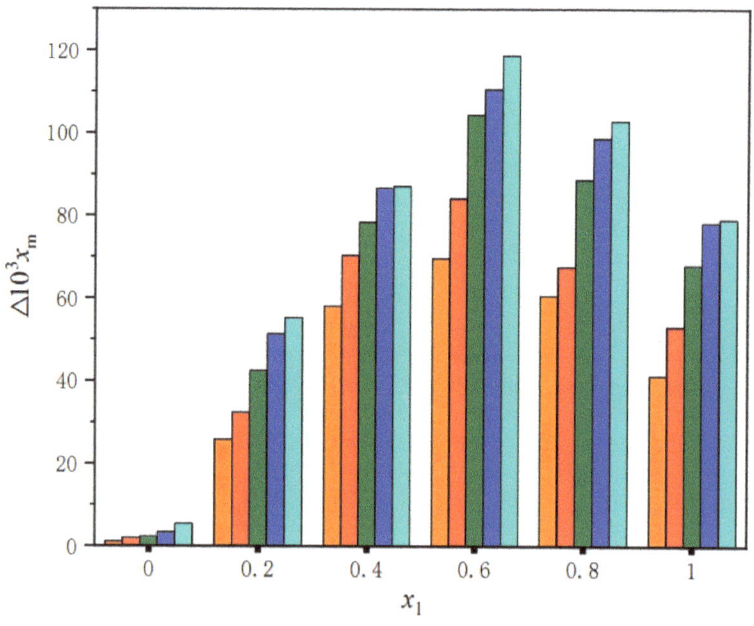

Figure 5. The mole fraction solubility difference $\triangle 10^3 x_m$ between FFCA and FDCA in water + 1,4-dioxane solvent mixtures (FFCA solubility–FDCA solubility) at different solvent mixing ratios and different temperatures. Orange = 303.15 K; Red = 313.15 K; Green = 323.15 K; Blue = 333.15 K; Cyan = 343.15 K.

4. Conclusions

In this study, the solubility of two biocompounds, FDCA and FFCA, was determined in binary water + 1,4-dioxane solvents with different mixing ratios at different temperatures. The solubility increases with increasing temperature at constant solvent composition. FFCA exhibits higher solubility in these binary solvents. At constant temperature, the solubility reaches a maximum at a certain 1,4-dioxane mole fraction, and the co-solvency phenomenon can be attributed to the fact that the solubility parameter of FDCA/FFCA is in the range for water and 1,4-dioxane. The Jouyban-Acree-van't Hoff model was applied to correlate the solubility data, and satisfactory fitting results were obtained. The calculated preferential solvation parameters clearly show different dissolution mechanisms at different mixing ratios of water and 1,4-dioxane. The differences in solubility of FDCA and FFCA were compared at different solvent mixing ratios and temperatures. It is hoped to separate FDCA and FFCA at a 1,4-dioxane mole fraction of about 0.60 and high temperatures. This study will help in the selection of solvents for the separation of the structurally similar biocompounds FDCA and FFCA.

Supplementary Materials: The following supporting information can be downloaded at: https://www.mdpi.com/article/10.3390/pr10122480/s1, Table S1: Calculation of solubility parameters δ for FFCA by Fedors' method along with molar volume; Table S2: Correlation volume V_{cor} and the preferential solvation parameters $\delta_{1,3}$ of FDCA and FFCA in binary water + 1,4-dioxane mixtures with different solvent mixing ratio at 303.15 K; Table S3: Gibbs energy of transfer (kJ·mol^{-1}) of FDCA and FFCA in 1,4-dioxane (1) + water (2) mixtures at 303.15 K; Table S4: D values (kJ· mol^{-1}) of FDCA and FFCA in 1,4-dioxane (1) + water (2) mixtures at 303.15 K; Table S5: $G_{1,3}$ and $G_{2,3}$ values (cm^3·mol^{-1}) of solute (3) in 1,4-dioxane (1) + water (2) mixtures at 303.15 K.

Author Contributions: Conceptualization, methodology, formal analysis, writing—original draft preparation, writing—review and editing, resources, visualization, project administration, funding acquisition, Q.X.; investigation, data curation, L.X. and J.F.; software, C.D.; validation, B.L.; supervision, Z.B. All authors have read and agreed to the published version of the manuscript.

Funding: This research was funded by "the Fundamental Research Funds for the Central Universities" (No. 226-2022-00055).

Acknowledgments: We are grateful for the financial support by "the Fundamental Research Funds for the Central Universities" (No. 226-2022-00055) and the support from Domestic Visiting Scholar's "Professional Development Project for Teachers" (FX2021080).

Conflicts of Interest: The authors declare no conflict of interest.

References

1. Teong, S.P.; Yi, G.S.; Zhang, Y.G. Hydroxymethylfurfural production from bioresources: Past, present and future. *Green Chem.* **2014**, *16*, 2015–2026. [CrossRef]
2. Bozell, J.J.; Petersen, G.R. Technology development for the production of biobased products from biorefinery carbohydrates—The US Department of Energy's "Top 10" revisited. *Green Chem.* **2010**, *12*, 539–554. [CrossRef]
3. Pandey, S.; Dumont, M.J.; Orsat, V.; Rodrigue, D. Biobased 2,5-furandicarboxylic acid (FDCA) and its emerging copolyesters' properties for packaging applications. *Eur. Polym. J.* **2021**, *160*, 110778. [CrossRef]
4. van Putten, R.J.; van der Waal, J.C.; de Jong, E.; Rasrendra, C.B.; Heeres, H.J.; de Vries, J.G. Hydroxymethylfurfural, a versatile platform chemical made from renewable resources. *Chem. Rev.* **2013**, *113*, 1499–1597. [CrossRef]
5. Motagamwala, A.H.; Won, W.Y.; Sener, C.; Alonso, D.M.; Maravelias, C.T.; Dumesic, J.A. Toward biomass-derived renewable plastics: Production of 2,5-furandicarboxylic acid from fructose. *Sci. Adv.* **2018**, *4*, eaap9722. [CrossRef]
6. Zheng, X.; Li, C.Q.; Mao, H.F.; Liu, X.H.; Guo, Y.; Wang, Y.Q. Boosting 2,5-Furandicarboxylic acid production via coating carbon over CeO2 in a Pt catalyst. *Ind. Crops Prod.* **2022**, *186*, 115168. [CrossRef]
7. Den Ouden, H.J.C.; Sololovskii, V.; Boussie, T.R.; Diamond, G.M.; Dias, E.L.; Zhu, G.; Torssell, S. Purified 2,5-furandicarboxylic acid pathway products. International Application No. PCT/US2018/041694, 17 January 2019.
8. Huang, Y.T.; Wong, J.J.; Chen, J.H. Method for purifying crude of 2,5-furandicarboxylic acid by crystallization. U.S. Patent No. 10,344,010, 9 July 2019.
9. Zhang, Y.Z.; Guo, X.; Tang, P.; Xu, J. Solubility of 2,5-furandicarboxylic acid in eight pure solvents and two binary solvent systems at 313.15−363.15 K. *J. Chem. Eng. Data* **2018**, *63*, 1316–1324. [CrossRef]
10. Ban, H.; Pan, T.; Cheng, Y.W.; Wang, L.J.; Li, X. Solubilities of 2,5-furandicarboxylic acid in binary acetic acid + water, methanol + water, and ethanol + water solvent mixtures. *J. Chem. Eng. Data* **2018**, *63*, 1987–1993. [CrossRef]
11. Cong, H.Y.; Yuan, H.B.; Tao, Z.K.; Bao, H.L.; Zhang, Z.M.; Jiang, Y.; Huang, D.; Liu, H.L.; Wang, T.F. Recent advances in catalytic conversion of biomass to 2,5-furandicarboxylic acid. *Catalysts* **2021**, *11*, 1113. [CrossRef]
12. Spiegel, A.J.; Noseworthy, M.M. Use of non-aqueous solvents in parenteral products. *J Pharm. Sci.* **1963**, *52*, 917–927. [CrossRef]
13. Paruta, A.N.; Irani, S.A. Dielectric solubility profiles in dioxane–water mixtures for several antipyretic drugs. Effect of substituents. *J. Pharm. Sci.* **1965**, *54*, 1334–1338. [CrossRef]
14. Rathi, P.B.; Mourya, V.K. Extended Hildebrand solubility approach: Satranidazole in mixtures of dioxane and water. *Indian J. Pharm. Sci.* **2011**, *73*, 315–319. [PubMed]

15. Chen, Y.H.; Luo, Z.Y.; Ren, Z.X.; Shen, L.B.; Li, R.R.; Chen, L.; Du, C.B. The interactions and thermodynamic parameters of lenvatinib mesylate in pure and mixed solvents at several temperatures. *J. Chem. Thermodyn.* **2022**, *176*, 106922. [CrossRef]
16. Jouyban, A. Review of the cosolvency models for predicting drug solubility in solvent mixtures: An update. *J. Pharm. Pharm. Sci.* **2019**, *22*, 466–485. [CrossRef] [PubMed]
17. Dadmand, S.; Kamari, F.; Acree, W.E., Jr.; Jouyban, A. Solubility prediction of drugs in binary solvent mixtures at various temperatures using a minimum number of experimental data points. *AAPS PharmSciTech* **2019**, *20*, 10. [CrossRef]
18. Mirheydari, S.N.; Barzegar-Jalali, M.; Acree, W.E., Jr.; Shekaari, H.; Shayanfar, A.; Jouyban, A. Comparison of the solubility models for correlation of drug solubility in ethanol + water binary mixtures. *J. Solut. Chem.* **2019**, *48*, 1079–1104. [CrossRef]
19. Barzegar-Jalali, M.; Rahimpour, E.; Martinez, F.; Shayanfar, A.; Jouyban, A. Generally trained models to predict drug solubility in methanol + water mixtures. *J. Mol. Liq.* **2018**, *264*, 631–644. [CrossRef]
20. Jouyban, A.; Fakhree, M.A.A.; Hamzeh-Mivehroud, M.; Acree, W.E., Jr. Modelling the deviations of solubilities in water-dioxane mixtures from predicted solubilities by the Jouyban-Acree model. *J. Drug Del. Sci. Technol.* **2007**, *17*, 359–363. [CrossRef]
21. Liang, R.S.; Bao, Z.B.; Su, B.G.; Xing, H.B.; Ren, Q.L. Solubility of Vitamin D_3 in six organic solvents at temperatures from (248.2 to 273.2) K. *J. Chem. Eng. Data* **2012**, *57*, 2328–2331. [CrossRef]
22. Pena, M.A.; Reillo, A.; Escalera, B.; Bustamante, P. Solubility parameter of drugs for predicting the solubility profile type within a wide polarity range in solvent mixtures. *Int. J. Pharm.* **2006**, *321*, 155–161. [CrossRef]
23. Hildebrand, J.H.; Prausnitz, J.M.; Scott, R.L. *Regular and related solutions*, 1st ed.; Van Nostrand Reinhold: New York, NY, USA, 1970; pp. 26–27.
24. Martin, A.; Newburger, J.; Adjei, A. New solubility equation. *J. Pharm. Sci.* **1979**, *68*, 4–5.
25. Barton, A.F.M. Solubility parameters. *Chemical Reviews.* **1975**, *75*, 731–753. [CrossRef]
26. Joshi, A.S.; Alipourasiabi, N.; Vinnakota, K.; Coleman, M.R.; Lawrence, J.G. Improved polymerization and depolymerization kinetics of poly(ethylene terephthalate) by co-polymerization with 2,5-furandicarboxylic acid. *RSC Adv.* **2021**, *11*, 23506–23518. [CrossRef] [PubMed]
27. Ben-Naim, A. Theory of preferential solvation of nonelectrolytes. *Cell Biophys.* **1988**, *12*, 255–269. [CrossRef]
28. Banerjee, D.; Laha, A.K.; Bagchi, S. Preferential solvation in mixed binary solvent. *J. Chem. Soc., Faraday Trans.* **1995**, *91*, 631–636. [CrossRef]
29. Pallewela, G.N.; Smith, P.E. Preferential solvation in binary and Ternary mixtures. *J. Phys. Chem. B* **2015**, *119*, 15706–15717. [CrossRef]
30. Martínez, F.; Jouyban, A.; Acree, W.E., Jr. Solubility of phenobarbital in aqueous cosolvent mixtures revisited: IKBI preferential solvation analysis. *Phys. Chem. Liq.* **2017**, *55*, 432–443. [CrossRef]
31. Du, C.B.; Li, R.R.; Chen, L. Dissolution thermodynamic properties calculation and intermolecular interaction analysis of diacerein in different pure and mixed solvents. *J. Chem. Thermodyn.* **2022**, *173*, 106850. [CrossRef]
32. Du, C.B.; Luo, Y.L.; Huang, C.Y.; Li, R.R. The solubility measurement and thermodynamic models correlation of baclofen in twelve pure organic solvents. *J. Chem. Eng. Data* **2022**, *67*, 2655–2661. [CrossRef]
33. Rodrigueza, G.A.; Delgadoa, D.R.; Fleming, M. Preferential solvation of indomethacin and naproxen in ethyl acetate + ethanol mixtures according to the IKBI method. *Phys. Chem. Liq.* **2014**, *52*, 533–545. [CrossRef]
34. Jimenez, D.M.; Cardenas, Z.J.; Delgado, D.R.; Pena, M.A.; Martinez, F. Solubility temperature dependence and preferential solvation of sulfadiazine in 1, 4-dioxane+ water co-solvent mixtures. *Fluid Phase Equilibr.* **2015**, *397*, 26–36. [CrossRef]
35. Li, X.B.; Cong, Y.; Li, W.T.; Yan, P.Y.; Zhao, H.K. Thermodynamic modelling of solubility and preferential solvation for ribavirin (II) in co-solvent mixtures of (methanol, n-propanol, acetonitrile or 1,4-dioxane) + water. *J. Chem. Thermodyn.* **2017**, *115*, 74–83. [CrossRef] [PubMed]
36. Bani-Yaseen, A.D.; Al-Balawi, M. Solvatochromic, spectral, and geometrical properties of nifenazone: A DFT/TD-DFT and experimental study. *Phys. Chem. Chem. Phys.* **2014**, *16*, 15519–15526. [CrossRef] [PubMed]
37. Delgado, D.R.; Pena, M.A.; Martinez, F. Preferential solvation of some sulfonamides in 1,4-dioxane + water co-solvent mixtures at 298.15 K according to the inverse Kirkwood-Buff integrals method. *Rev. Acad. Colomb. Cienc.* **2014**, *38*, 104–114.
38. Xu, Q.Q.; Su, B.G.; Bao, Z.B.; Yang, Y.W.; Yang, Q.W.; Ren, Q.L. Microgeometry-independent equation for measuring infinite dilution activity coefficients using gas-liquid chromatography with static-wall-coated open-tubular columns. *J. Chromatogr. A* **2020**, *1624*, 461264. [CrossRef]

MDPI
St. Alban-Anlage 66
4052 Basel
Switzerland
www.mdpi.com

Processes Editorial Office
E-mail: processes@mdpi.com
www.mdpi.com/journal/processes

www.ingramcontent.com/pod-product-compliance
Lightning Source LLC
Chambersburg PA
CBHW040314270926
51850CB00007BC/1332
* 9 7 8 3 0 3 6 5 8 9 9 8 5 *